2012 IEEE Silicon Nanoelectronics Workshop

(SNW 2012)

Honolulu, Hawaii, USA
10 – 11 June 2012

IEEE Catalog Number: CFP12SNW-PRT
ISBN: 978-1-4673-0996-7

Copyright © 2012 by the Institute of Electrical and Electronic Engineers, Inc
All Rights Reserved

Copyright and Reprint Permissions: Abstracting is permitted with credit to the source. Libraries are permitted to photocopy beyond the limit of U.S. copyright law for private use of patrons those articles in this volume that carry a code at the bottom of the first page, provided the per-copy fee indicated in the code is paid through Copyright Clearance Center, 222 Rosewood Drive, Danvers, MA 01923.

For other copying, reprint or republication permission, write to IEEE Copyrights Manager, IEEE Service Center, 445 Hoes Lane, Piscataway, NJ 08854. All rights reserved.

This publication is a representation of what appears in the IEEE Digital Libraries. Some format issues inherent in the e-media version may also appear in this print version.

IEEE Catalog Number: CFP12SNW-PRT
ISBN 13:   978-1-4673-0996-7
ISSN: 2161-4636

Additional Copies of This Publication Are Available From:

Curran Associates, Inc
57 Morehouse Lane
Red Hook, NY 12571 USA
Phone: (845) 758-0400
Fax: (845) 758-2633
E-mail: curran@proceedings.com
Web: www.proceedings.com

2012 IEEE Silicon Nanoelectronics Workshop (SNW 2012)

Honolulu, Hawaii, USA
10-11 June 2012

IEEE Catalog Number: CFP12SNW-POD
ISBN: 978-1-46730-996-7

Welcome Message

The 2012 IEEE Silicon Nanoelectronics Workshop is a satellite workshop of the 2012 VLSI Symposia sponsored by the IEEE Electron Device Society. It is the seventeenth workshop in the annual series, which showcases original work on nanometer-scale devices and technologies that utilize silicon or which are based on silicon substrates. The program this year includes 8 invited talks, 27 oral presentations, and 47 poster papers contributed by researchers from around the world. The Program Chair, Thomas Skotnicki, and I would like to thank the members of the Technical Program Committee and all of the attendees for their contributions and participation to make this workshop a success. We hope that you enjoy the workshop and the wonderful venue in Honolulu, Hawaii!

Tsu-Jae King Liu

General Chair, 2012 Silicon Nanoelectronics Workshop

Committee Members for the 2012 IEEE Silicon Nanoelectronics Workshop

General Chair
Tsu-Jae King Liu, *University of California at Berkeley*

Program Chair
Thomas Skotnicki, *STMicroelectronics*

Program Committee
Kristin De Meyer, *IMEC*
Simon Deleonibus, *LETI*
Kazuhiko Endo, *AIST*
Stephen Goodnick, *Arizona State University*
Toshiro Hiramoto, *University of Tokyo*
Ru Huang, *Peking University*
Adrian Ionescu, *EPFL*
Raj Jammy, *SEMATECH*
Malgorzata Jurczak, *IMEC*
Dong-Won Kim, *Samsung Electronics*
Atsuhiro Kinoshita, *Toshiba*
Bunji Mizuno, *Panasonic*
Yoshio Nishi, *Stanford University*
Yukinori Ono, *University of Toyama*
Mikael Östling, *KTH*
Byung-Gook Park, *Seoul National University*
Wolfgang Porod, *University of Notre Dame*
Heike Riel, *IBM Zurich*
Ken Rim, *IBM*
Shintaro Sato, *AIST*
Michiharu Tabe, *Shizuoka University*
Yasuo Takahashi, *Hokkaido University*
Ken Uchida, *Tokyo Institute of Technology*
Yee-Chia Yeo (National University of Singapore)

2012 IEEE Silicon Nanoelectronics Workshop

Hilton Hawaiian Village, Honolulu, HI USA
June 10-11, 2012

Technical Program

Opening Remarks

Sunday, June 10, 8:30
Tsu-Jae King Liu, *University of California at Berkeley*, General Chair

Session 1: Plenary & Towards Zero Power Electronics

Sunday, June 10, 8:40
Co-chairs: Toshiro Hiramoto, *University of Tokyo* and Thomas Skotnicki, *STMicroelectronics*

8:40	**1-1**	(Plenary Invited) **Innovative thermal energy harvesting for zero power electronics**, S. Monfray[1], O. Puscasu[1,2], G. Savelli[2], U. Soupremanien[2], E. Ollier[2], C. Guerin[2], L.G. Fréchette[3], E. Léveille[3], G. Mirshekari[3], C. Maitre[1], P. Coronel[2], K. Domanski[4], P. Grabiec[4], P. Ancey[1], D. Guyomar[5], V. Bottarel[6], G. Ricotti[6], F. Boeuf[1], F. Gaillard[2], and T. Skotnicki[1], *[1]STMicroelectronics, France [2]CEA Liten, France [3]Université de Sherbrooke, Canada [4]ITE, Poland, [5]INSA Lyon, France, [6]STMicroelectronics, Italy*	1
9:10	**1-2**	(Plenary Invited) **New type steep-S device using the bipolar action**, D. Hisamoto, S. Saito, A. Shima, H. Yoshimoto, and K. Torii, *Hitachi, Ltd., Japan*	5
9:40	**1-3**	**Experimental Demonstration of Temperature Stability of Si-Tunnel FET over Si-MOSFET**, S. Migita, K. Fukuda, Y. Morita, and H. Ota, *AIST, Japan*	7
09:55	**1-4**	**Scale laws for enhanced power for MEMS based heat energy harvesting**, O. Puscasu[1,2], S. Monfray[1], F. Boeuf[1], G. Savelli[3], F. Gaillard[3], D. Guyomar[2], T. Skotnicki[1], *[1]STMicroelectronics, [2]INSA Lyon, [3]CEA Liten, France*	9

Session 2: Thermal Management & Nanoscale Memory

Sunday, June 10, 10:30
Co-chairs: Byung-Gook Park, *Seoul National University* and Tsu-Jae King Liu, *UC Berkeley*

10:30	**2-1**	(Invited) **Energy-Efficiency and Thermal Management in Nanoscale Devices**, *A.D. Liao, Z.-Y. Ong, A.Y. Serov, F. Xiong, and Eric Pop, University of Illinois at Urbana-Champaign, USA*	11
11:00	**2-2**	**Comparative Study of Tri-Gate- and Double-Gate-Type Poly-Si Fin-Channel Split-Gate Flash Memories**, Y.X. Liu[1], T. Kamei[2], T. Matsukawa[1], K. Endo[1], S. O'uchi[1], J. Tsukada[1], H. Yamauchi[1], Y. Ishikawa[1], T. Hayashida[2], K. Sakamoto[1], A. Ogura[2], and M. Masahara[1,2], *[1]AIST [2]Meiji University, Japan*	15
11:15	**2-3**	**Variation-Aware Study of BJT-based Capacitorless DRAM Cell Scaling Limit**, M.H. Cho, W. Kwon, N. Xu, and T.-J.K. Liu, *University of California at Berkeley, USA*	17

| 11:30 | 2-4 | Investigation into the Effect of the Variation of Gate Dimensions on Program Characteristics in 3D NAND Flash Array, J.Y. Seo, Y. Kim, S.H. Park, W. Kim, D.-B. Kim, J.-H. Lee, H. Shin, and B.-G. Park, *Seoul National University, Korea* | 19 |

| 11:45 | 2-5 | A novel Gate-All-Around Ultra-Thin p-channel Poly-Si TFT Functioning as Transistor and Flash Memory with Silicon Nanocrystals, H.-B. Chen[1], S.-H. Lin[2], J.-J. Wu[1], Y.-C. Wu[2], and C.-Y. Chang[1], *[1]National Chiao Tung University [2]National Tsing Hua University, Taiwan ROC* | 21 |

Session 3: Advanced Channel and Gate Stack Materials

Sunday, June 10, 13:30
Co-chairs: Kristin De Meyer, *IMEC* and Dong-Won Kim, *Samsung Electronics*

| 13:30 | 3-1 | (Invited) **Graphene for More Moore and More Than Moore Applications,** M.C. Lemme, S. Vaziri, A.D. Smith, J. Li, S. Rodriguez, A. Rusu, M. Ostling, *KTH Royal Institute of Technology, Sweden* | 23 |

| 14:00 | 3-2 | **High Performance Ω-Gate Ge FinFET Featuring Low Temperature Si_2H_6 Passivation and Implantless Schottky-Barrier NiGe Metallic Source/Drain,** B. Liu[1], X. Gong[1], G. Han[1], P.S.Y. Lim[1], Y. Tong[1], Y. Yang[1], N. Daval[2], M. Pulido[2], D. Delprat[2], B.-Y. Nguyen[2], and Y.-C. Yeo[1], *[1]National University of Singapore, Singapore [2]Soitec, France* | 26 |

| 14:15 | 3-3 | **High-performance pMOSFETs with High-k Gate Dielectric and Dislocation-free Epitaxial Si/Ge Super-lattice Channel,** L.-J. Liu[1], K.-S. Chang-Liao[1], C.-H. Fu[1], H.-C. Hsieh[1], C.-C. Lu[1], T.-K. Wang[1], P. Y. Gu[2], and M.J. Tsai[2], *[1]National Tsing Hua University, [2]Industrial Technology Research Institute, Taiwan ROC* | 28 |

| 14:30 | 3-4 | **Counter Dipole Layer Formation in SiO_2/High-k/SiO_2/Si Gate Stacks,** S. Hibino, T. Nishimura, K. Nagashio, K. Kita, and A. Toriumi, *University of Tokyo, Japan* | 30 |

| 14:45 | 3-5 | **Simultaneous Carrier Transport Enhancement and Variability Reduction in Si MOSFETs by Insertion of Partial Monolayers of Oxygen,** R.J. Mears[1], N. Xu[2], N. Damrongplasit[2], H. Takeuchi[1], R.J. Stephenson[1], N.W. Cody[1], A. Yiptong[1], X. Huang[1], M. Hytha[1], and T.-J.K. Liu[2], *[1]Mears Technologies [2]University of California at Berkeley, USA* | 32 |

| 15:00 | 3-6 | **Transport in Graphene on Boron Nitride,** D.K. Ferry, *Arizona State University, USA* | 34 |

Session 4: Spintronic Devices

Sunday, June 10, 15:35
Chair: Stephen Goodnick, *Arizona State University*

15:35   **4-1**   (Invited) **Magnetic Tunnel Junction for Magnetoresistive Random Access**   **36** **Memory and Beyond,** H. Ohno, *Tohoku Univeristy, Japan*

16:05   **4-2**   **Systolic Architectures and Applications for Nanomagnet Logic,** M. Niemier[1],   **38** X. Ju[2], M. Becherer[2], G. Csaba[1], X.S. Hu[1], D. Schmitt-Landsiedel[2], P. Lugli[2], and W. Porod[1], [1]*University of Notre Dame, USA* [2]*Technical University of Munich, Germany*

16:20   **4-3**   **Analysis of static noise margin and power-gating efficiency of a new**   **40** **nonvolatile SRAM cell using pseudo-spin-MOSFETs,** Y. Shuto, S. Yamamoto, and S. Sugahara, *Tokyo Institute of Technology, Japan*

Poster Session 1: Advanced Memory and Channel Materials

Sunday, June 10, 16:40 – 19:00
Chair: Yee-Chia Yeo, *National University of Singapore*

16:40   **Poster introductions** (1 minute each)

Session 5: Emerging Memory Devices

Monday, June 11, 8:30
Co-chairs: Simon Deleonibus, *LETI* and Malgorzata Jurczak, *IMEC*

8:30   **5-1**   (Invited) **Recent Progress of Resistive Switching Random Access Memory**   **42** **(RRAM),** Y. Wu, S. Yu, X. Guan, and H.-S.P. Wong, *Stanford University, USA*

9:00   **5-2**   **Bidirectional Selection Device Characteristics of Ultra-Thin (<3nm) TiO_2**   **46** **layer for 3D Vertically Stackable ReRAM Application,** J. Woo[1], J. Park[2], J. Shin[2], G. Choi[1], S. Kim[2], W. Lee[2], S. Park[2], D. Lee[1], E. Cha[2], and H. Hwang[1], Pohang [1]*University of Science and Technology* [2]*Gwangju Institute of Science and Technology, Republic of Korea*

9:15   **5-3**   **Co-existed Unipolar and Bipolar Resistive Switching Effect of HfOx-Based**   **48** **RRAM,** B. Chen, B. Gao, Y.H. Fu, R. Liu, L. Ma, P. Huang, F.F. Zhang, L.F. Liu, X.Y. Liu, J.F. Kang, and G.J. Lian, *Peking University, China*

9:30   **5-4**   **4kb nonvolatile nanogap memory (NGpM) with 1 ns programming**   **50** **capability,** T. Takahashi[1], S. Furuta[1], Y. Masuda[1], S. Kumaragurubaran[1], T. Sumiya[1], M. Ono[1], Y. Hayashi[2], T. Shimizu[3], H. Suga[3], M. Horikawa[3], and Y. Naitoh[3], [1]*Funai Electric Advanced Applied Technology Research Institute* [2]*Tsukuba Device Solution Center* [3]*AIST, Japan*

9:45   **5-5**   **Characteristics of Metal/Ferroelectric (PVDF-TrFE)/Graphene (MFG)**   **52** **Device,** H.J. Hwang, E.J. Paek, J.H. Yang, C.G. Kang, and B.H. Lee, *Gwangju Institute of Science and Technology, Korea*

Session 6: Single Electron Devices & Quantum Transport

Monday, June 11, 10:20

Co-chairs: Michiharu Tabe, *Shizuoka University* and Wolfgang Porod, *Notre Dame University*

10:20	6-1	(Invited) **Silicon Single-Electron Transfer Devices: Ultimate Control of Electric Charge**, A. Fujiwara, G. Yamahata, K. Nishiguchi, G.P. Lansbergen, and Y. Ono, *NTT Corporation, Japan*	54
10:50	6-2	**Reinvestigation of Dot Formation Mechanisms in Silicon Nanowire Channel Single-Electron/Hole Transistors Operating at Room Temperature**, R. Suzuki, M. Nozue, T. Saraya, and T. Hiramoto, *University of Tokyo, Japan*	56
11:05	6-3	**Quantum Transport Property in FETs with Deterministically Implanted Single-Arsenic Ions Using Single-ion Implantation**, M. Hori[1], T. Shinada[1], F. Guagliardo[2], G. Ferrari[2], and E. Prati[3], [1]*Waseda University, Japan* [2]*Politecnico di Milano,* [3]*CNR-IMM, Italy*	58
11:20	6-4	**High-frequency properties of Si single-electron transistor**, H. Takenaka[1], M. Shinohara[1], T. Uchida[1], M. Arita[1], A. Fujiwara[2], Y. Ono[2], K. Nishiguchi[2], H. Inokawa[3], and Y. Takahashi[1], [1]*Hokkaido University* [2]*NTT Corporation* [3]*Shizuoka University, Japan*	60
11:35	6-5	**Negative Differential Resistance Devices with Ultra-High Peak-to-Valley Current Ratio Based on Silicon Nanowire Structure**, S. Shin, M.W. Ryu, and K.R. Kim, *Ulsan National Institute of Science and Technology, Korea*	62
11:50	6-6	**Mapping of single donors in nano-scale MOSFETs at low temperature**, J. Verduijn[1,2], G.C. Tettamanzi[1], R. Wacquez[3], B. Roche[3], B. Voisin[3], X. Jehl[3], M. Sanquer[3], S. Rogge[1,2], [1]*University of New South Wales, Australia* [2]*Delft University of Technology, The Netherlands* [3]*CEA-LETI, France*	64

Session 7: Nanoscale Phenomena

Monday, June 11, 13:30

Co-chairs: Kazuhiko Endo, *AIST* and Yukinori Ono, *University of Toyama*

13:30	7-1	(Invited) **A Single Atom Transistor**, M.Y. Simmons, *University of New South Wales, Australia*	66
14:00	7-2	**Statistical Variability Study of a 10nm Gate Length SOI FinFET Device**, B. Cheng[1], A.R. Brown[2], X. Wang[1], and A. Asenov[1,2], [1]*University of Glasgow* [2]*Gold Standard Simulations, United Kingdom*	67
14:15	7-3	**Reduced Drain Current Variability in Fully Depleted Silicon-on-Thin-BOX (SOTB) MOSFETs**, T. Mizutani[1], Y. Yamamoto[2], H. Makiyama[2], T. Tsunomura[2], T. Iwamatsu[2], H. Oda[2], N. Sugii[2], and T. Hiramoto[1], [1]*University of Tokyo* [2]*Low-power Electronics Association & Project, Japan*	69
14:30	7-4	**The Impact of the Carrier Transport on the Random Dopant Induced Drain Current Variation in the Saturation Regime of Advanced Strained-Silicon CMOS Devices**, E.R. Hsieh[1], S.S. Chung[1], C.H. Tsai[2], R.M. Huang[2], C.T. Tsai[2], and C.W. Liang[2], [1]*National Chiao Tung University* [2]*United Microelectronics Corporation, Taiwan ROC*	71

14:45   **7-5**   **On the Statistical Trap-Response (STR) Method for Characterizing Random**   73
Trap Occupancy and NBTI Fluctuation, J. Zou[1], C. Liu[1], R. Wang[1], X. Xu[1], J. Liu[2], H. Wu[2], Y. Wang[1], R. Huang[1], [1]*Peking University* [2]*Semiconductor Manufacturing International Corporation, China*

15:00   **7-6**   **Statistical distribution of RTS amplitudes in 20nm SOI FinFETs**, X. Wang[1],   75
A.R. Brown[2], B. Cheng[1], and A. Asenov[1,2], [1]*University of Glasgow* [2]*Gold Standard Simulations, United Kingdom*

Poster Session 2: Nanoscale/Quantum Devices and Phenomena

Monday, June 11, 15:20 – 17:30
Chair: Thomas Skotnicki, *STMicroelectronics*

15:20   **Poster introductions** (1 minute each)

Poster Session 1: Advanced Memory and Channel Materials

Sunday, June 10, 16:40 – 19:00

P1-1   **Self-Improvement of Cell Stability in SRAM by Post Fabrication Technique**, A.   77
Kumar, T. Saraya, S. Miyano, and T. Hiramoto, *University of Tokyo, Japan*

P1-2   **Improving the Endurance of Floating Gate NAND Flash Cells with Junction-Free**   N/A
Structure, I. Joo[1,2], S. Hur[1], C. Lee[1], S. Lee[1], H. Park[1], J. Song[1], H. Lee[1], Y. Jun[1], and I. Chung[2], [1]*Samsung Electronics* [2]*SungKyunKwan University, Korea*

P1-3   **Low Standby Power Charge Trap Flash Memory with Tunneling Field Effect**   79
Transistor, M.S. Han[1], J.H. Lee[2], D. Seo[1], C.-D. Park[1], Y. Oh[1], and I.H. Cho[1], [1]*Myongji University,* [2]*Seoul National University, Korea*

P1-4   **Charge-trap flash memory devices fabricated with nano-scale patterns on the Si_3H_4**   81
trapping layer, H.-M. An[1], K.H. Kim[1], H.-D. Kim[1], W.-J. Cho[2], and T.G. Kim[1], [1]*Korea University,* [2]*Kwangwoon University, Korea*

P1-5   **Simulation of Charge Trapping Memory with Silicon Nanocrystals Embedded in**   83
Silicon Nitride Layer, Y. Peng, X. Liu, G. Du, Y. Yang, and J. Kang, *Peking University, China*

P1-6   **Nanodot-type Floating Gate Memory with High-density Nanodot Array Formed**   85
Utilizing *Listeria* Dps, H. Kamitake[1], K. Ohara[1], M. Uenuma[1], B. Zheng[1], Y. Ishikawa[1], I. Yamashita[1,2], and Y. Uraoka[1], [1]*Nara Institute of Science and Technology* [2]*Panasonic Corporation, Japan*

P1-7   **Impacts of Silicon Nanocrystal Incorporation on the Transfer Characteristics of**   87
Poly-Silicon nanowire SONOS Devices, K.-H. Lee, H.-C. Lin, and T.-Y. Huang, *National Chiao Tung University, Taiwan ROC*

P1-8   **3-D Stacked NAND Flash Memory Having Lateral Bit-Line Layers and Vertical Gate**,   89
J.-W. Lee, M.-K. Jeong, B.-G. Park, H. Shin and J.-H. Lee, *Seoul National University, Korea*

P1-9 Effect of Cu Insertion Layer between Top Electrode and Switching Layer on Resistive Switching Characteristics, S. Jung, J.-H. Oh, K.-C. Ryoo, S. Kim, J.-H. Lee, H. Shin, and B.-G. Park, *Seoul National University, Korea* ... 91

P1-10 Self-compliance Unipolar Resistive Switching and Mechanism of $Cu/SiO_2/TiN$ RRAM Devices, D. Yu, L.F. Liu, P. Huang, F.F. Zhang, B. Chen, B. Gao, Y. Hou, D.D. Han, Y. Wang, J.F. Kang, and X. Zhang, *Peking University, China* ... 93

P1-11 Stable Resistive Switching Characteristics Observed in SiN-based Resistive Switching Memory Devices by using RF-sputtering methods, H.-D. Kim, S.M. Hong, H.-M. An, K.H. Kim, Y. Seo, M. Song, D. Li, and T.G. Kim, *Korea University, Korea* ... N/A

P1-12 Rectifying Characteristics and Implementation of n-Si/HfO_2 based Devices for 1D1R-based Cross-Bar Memory Array, F. F. Zhang, P. Huang, B. Chen, D. Yu, Y.H. Fu, L. Ma, B. Gao, L.F. Liu, X.Y. Liu, and J.F. Kang, *Peking University, China* ... 95

P1-13 Oxygen-induced High-*k* Degradation in TiN/HfSiO Gate Stacks, T. Hosoi, Y. Odake, K. Chikaraishi, H. Arimura, N. Kitano, T. Shimura, and H. Watanabe , *Osaka University, Japan* ... 97

P1-14 Metal/Ge Schottky Barrier Modulation With C-Containing Layer by Chemical Bath, W. Wang, J. Wang, M. Zhao, R. Liang, and J. Xu, *Tsinghua University, China* ... 99

P1-15 Orientation and Size Effects on Ballistic Electron Transport Properties in Gate-All-Around Rectangular Germanium Nanowire FETs, S. Mori, N. Morioka, J. Suda, and T. Kimoto, *Kyoto University, Japan* ... 101

P1-16 Quantum Transport Simulation of III-V MOSFETs based on Wigner Monte Carlo Approach, Y. Maegawa, S. Koba, H. Tsuchiya, and M. Ogawa, *Kobe University, Japan* ... 103

P1-17 Mechanisms of Ambient Dependent Mobility Degradation in the Graphene MOSFETs on SiO_2 Substrate, Y.G. Lee, C.G. Kang, C. Cho, Y.H. Kim, H.J. Hwang, J.J. Kim, U.J. Jung, E. J. Park, M.W. Kim, and B.H. Lee, *Gwangju Institute of Science and Technology, Korea* ... 105

P1-18 Electronic Band Structures of Graphene Nanomeshes, R. Sako, N. Hasegawa, H. Tsuchiya, and M. Ogawa, *Kobe University, Japan* ... 107

P1-19 Band Structure and Electron Transport in Multi-Junction Graphene Nanoribbons, N. Hasegawa, R. Sako, H. Tsuchiya, and M. Ogawa, *Kobe University, Japan* ... 109

P1-20 Graphene-Diamond-Silicon Devices with Increased Current-Carrying Capacity: sp^2-Carbon-sp^3-Carbon-on-Silicon Technology, J. Yu[1], G. Liu[1], A.V. Sumant[2], and A. A. Balandin[1], *[1]University of California at Riverside, [1]Argonne National Laboratory, USA* ... 111

P1-21 *Selective* Gas Sensing with a *Single* Graphene-on-Silicon Transistor, A.A. Balandin[1], S. Rumyantsev[2], G. Liu[1], M.S. Shur[2], and R.A. Potyrailo[3], *[1]University of California at Riverside [2]Rensselaer Polytechnic Institute, [3]GE Global Research, USA* ... 113

P1-22 Graphene Fillers for Ultra-Efficient Thermal Interface Materials, K.M.F. Shahil, V. Goyal, R. Gulotty, and A.A. Balandin, *University of California at Riverside, USA* ... 115

P1-23 Silicon Microfabrication Technologies for THz applications, C. Jung-Kubiak, J. Gill, T. Reck, C. Lee , J. Siles, G. Chattopadhyay, R. Lin, K. Cooper and I. Mehdi, *Jet Propulsion Laboratory, California of Technology* ... 117

Poster Session 2: Nanoscale/Quantum Devices and Phenomena

Monday, June 11, 15:20 – 17:30

P2-1   **Simulation Study on Process Conditions for High-Speed Silicon Photodetector and Quantum-Well Structuring for Increased Number of Wavelength Discriminations**, S. Cho[1], H. Kim[2], M.-C. Sun[2], T.I. Kamins[1], B.-G. Park[2], and J.S. Harris, Jr.[1], *Stanford University, USA [2]Seoul National University, Korea*   119

P2-2   **Nano-Transfer Printing of Functioning MIM Tunnel Diodes**, Mario Bareiß[1], B. Weiler[1], D. Kälblein[2], U. Zschieschang[2], H. Klauk[2], G. Scarpa[1], B. Fabel[1], P. Lugli[1], and W. Porod[3], *[1]Technische Universität München, Germany [2]Max Planck Institute for Solid State Research, Germany [3]University of Notre Dame, USA*   *121*

P2-3   **Fabrication and evaluation of heavily P-doped Si quantum dot and back-gate induced Si quantum dot**, J. Kamioka[1], T. Kodera[1,2], K., Horibe[1], Y. Kawano[1], and S. Oda[1], *[1]Tokyo Institute of Technology [3]University of Tokyo Japan*   123

P2-4   **Microwave manipulation of electrons in silicon quantum dots**, T. Ferrus[1], A. Rossi[1], T. Kodera[2,3], T. Kambara[2], W. Lin[2], S. Oda[2], and D.A. Williams[1], *[1]Hitachi Cambridge Laboratory, United Kingdom [2]Tokyo Institute of Technology [3]University of Tokyo Japan*   125

P2-5   **Charge sensing of a Si triple quantum dot system using single electron transistors**, R. Mizokuchi, T. Kodera, K. Horibe, Y. Kawano, and S. Oda, *Tokyo Institute of Technology, Japan*   127

P2-6   **Fabrication and characterization of Si/SiGe quantum dots with capping gate**, T. Kodera[1,2], Y. Fukuoka[1], K. Takeda[2], T. Obata[2], K. Yoshida[2], K. Sawano[3], K. Uchida[1], Y. Shiraki[3], S. Tarucha[2], and S. Oda[1], *Tokyo Institute of Technology [2]University of Tokyo [3]Tokyo City University, Japan*   129

P2-7   **Single Ge quantum dot placement along with self-aligned electrodes for effective management of single electron tunneling**, I. H. Chen, K. H. Chen, and P. W. Li, *National Central University, Taiwan ROC*   131

P2-8   **Single-electron transport through a single donor at elevated temperatures**, E. Hamid, D. Moraru, T. Mizuno and M. Tabe, *Shizuoka University, Japan*   133

P2-9   **The Interplay of Self-Heating Effects and Static RTF in Nanowire Transistors**, D. Vasileska, A. Hossain, and S.M. Goodnick, *Arizona State University, USA*   135

P2-10   **Effect of Interfacial States on the technological variability of Trigate MOSFETs**, E. González-Marín, F.G. Ruiz, A. Godoy, I.M. Tienda-Luna, F. Gámiz, *Universidad de Granada, Spain*   137

P2-11   **Evolution of Channel Trap Distribution under Bias Stress in Polysilicon Thin Film Transistors evaluated by Charge Pumping Method**, C.N. Manh[1], J.S. Chang[1], T.-Y. Jang[1], M. Hasan[1], H. Yang[1], J.K. Jeong[1], B. Kim[2], J. Ahn[2], K. Hwang[2], and R. Choi[1], *[1]Inha University [2]Samsung Electronics Co., Ltd., Korea*   N/A

P2-12   **Physical Model for Random Telegraph Noise Amplitudes and Implications**, R.G. Southwick III[1], K.P. Cheung[1], J.P. Campbell[1], S.A. Drozdov[2], J.T. Ryan[1], J.S. Suehle[1], and A.S. Oates[3], *[1]National Institute of Standards and Technology, [2]University of Maryland USA [3]Taiwan Semiconductor Manufacturing Company Ltd., Taiwan ROC*   139

P2-13 Optoelectrical Lifetime Evaluation of Single Holes in SOI MOSFET, W. Du[1], D.S.   141
Putranto[1,2], H. Satoh[1], A. Ono[1], P.S. Priambodo[2], D. Hartanto[2], and H. Inokawa[1]
[1]*Shizuoka University, Japan* [2]*University of Indonesia, Indonesia*

P2-14 *Ab initio* analysis of donor state deepening in Si nano-channels, D. Moraru[1], Y.   143
Kuzuya[1], E. Hamid[1], T. Mizuno[1], M. Tabe[1], and H. Mizuta[2], [1]*Shizuoka University,*
Japan [2]*University of Southampton, United Kingdom*

P2-15 Channel Length-Dependent Series Resistance?, J.P. Campbell[1], K.P. Cheung[1], S.A.   145
Drozdov[2], R.G. Southwick[1], J.T. Ryan[1], A.S. Oates[3], J.S. Suehle[1], [1]*National Institute of*
Standards and Technology, [2]*University of Maryland USA* [3]*Taiwan Semiconductor*
Manufacturing Company Ltd., Taiwan ROC

P2-16 Effects of Amorphous Silicon Atomic Density Variation on Series and Contact   147
Resistances in Nanoscale Thin-Film Structures, M.W. Ryu, S.-H. Kim, and K.R. Kim,
Ulsan National Institute of Science and Technology, Korea

P2-17 Evaluation of Scattering in Asymmetric Quasi-Ballistic DG-MOSFET, G. Liu, G. Du, T.   149
Lu, X. Liu, P. Zhang, and X. Zhang, *Peking University, China*

Fabrication and Characterization of a Pi-Gate Ultrathin Body Junctionless poly-Si TFTs   151
J. Wu, H. Chen, M. Han, Y. Wu, C. Chang

P2-18 Orientational and Si-SiO$_2$ roughness topology dependence of electron mobilities in   N/A
silicon gate-all-around nanowire FETs, M. Bescond and E. Dib, *Technologies*
Château-Gombert, France

P2-19 Junctionless poly-Si TFTs, J.-J. Wu[1], H.-B. Chen[1], M.-H. Han[1], Y.-C. Wu[2], and C.-.   N/A
Chang[1], [1]*National Chiao Tung University* [2]*National Tsing Hua Unviersity, Taiwan ROC*

P2-20 Quantum Drift-Diffusion and Quantum Energy Balance Simulation of Nanowire   153
Junctionless Transistors, O. Badami, N. Kumar, D. Saha, and S. Ganguly, *Indian*
Institute of Technology Bombay, India

P2-21 Characteristics and Sensitivity of p-Type Junctionless Gate-All-Around Nanowire   155
Transistor, M.-H. Han[1], Y.-R. Jhan[2], J.-J. Wu[1], H.-B. Chen[1], Y.-C. Wu[2], and C.-Y. Chang[1],
[1]*National Chiao Tung University* [2]*National Tsing Hua University, Taiwan ROC*

P2-22 Analysis of Hysteresis Characteristics of Fabricated SiNW Biosensor in Aqueous   157
Environment with Reference Electrode, J.H. Lee[1], J. Lee[2], M.-C. Sun[1], W.H. Lee[2], M.
Uhm[2], S. Hwang[2], I.-Y. Chung[3], D.M. Kim[2], D.H. Kim[2], and B.-G. Park[1], [1]*Seoul National*
University [3]*Kookmin University* [3]*Kwangwoon University, Korea*

P2-23 Investigation on Hump Effects of L-shaped Tunneling Field-Effect Transistors, S.W.   159
Kim[1], W.Y. Choi[2], H. Kim[1], M.-C. Sun[1,3], H. W. Kim[1], and B.-G. Park[1], [1]*Seoul National*
University [2]*Sogang University* [3]*Samsung Electronics Co., Ltd., Korea*

P2-24 Device Structure for the Characterization of Nanowire Thermocouples, G.P.   161
Szakmany, P.M. Krenz, A.O. Orlov, G.H. Bernstein, and W. Porod, *University of Notre*
Dame, USA

Innovative thermal energy harvesting for zero power electronics

S.Monfray[1], O.Puscasu[1,2], G. Savelli[2], U.Soupremanien[2], E.Ollier[2], C.Guerin[2], L.G. Fréchette[3], E.Léveille[3], G. Mirshekari[3], C.Maitre[1],P.Coronel[2], K.Domanski[4], P.Grabiec[4], P.Ancey[1], Daniel Guyomar[5], V.Bottarel[6], G.Ricotti[6], F. Boeuf[1], F. Gaillard[2], Thomas Skotnicki[1]

[1]STMicroelectronics (Crolles 2) SAS, 850 Jean Monnet st., 38926, Crolles Cedex, France
[2]CEA Liten,17 rue des martyrs 38054 Grenoble Cedex 9, France
[3]Université de Sherbrooke, 2500 boul. Université, Sherbrooke, QC, Canada, J1K 2R1
[4]ITE, 32/46 LotnikowAvenue 02-668 Warsaw, Poland
[5]LGEF, INSA Lyon, 8, Rue de la Physique 69621 Villeurbanne Cedex, France
[6]STMicroelectronics, 20010 Cornaredo, Milan, Italy

Abstract

Thermal gradients, commonly present in our environment (fluid lines, warm fronts, electronics) are sources of energy rarely used today. This paper aims to present innovative approaches of thin and/or flexible thermal energy harvesters for smart and autonomous sensor network applications. The harvester system will be based on the collaborative work of interrelated energy nodes/units, which will be either piezo-thermofluidic converters (use of rapid thermal cycles of a working fluid) or piezo-thermomechanic converters (use of the mechanical energy developed by rapid snapping of micro-switches). The two kinds of energy nodes convert a heat flux into storable electrical energy through a piezoelectric transducer. Miniaturization of the energy nodes will lead to increased thermal transfer rates and consequently increased harvested power. To effectively use thermal energy sources in varying environments, the nodes will be adaptive versus different thermal gradients (in a predefined temperature range) and will possibly influence each other. The concept is unique in the sense that it is based on a matrix structure of micro or mini energy nodes which will work together in a collective approach to optimize the harvested energy, and which do not require the use of radiators as classical Seebeck approach, thanks to the controlled thermal resistance. This opens the door to new properties and features of the object, with better performances. It could therefore be declined on flexible substrates, allowing conformability around the sources of potential heat for low power applications.

I. Introduction

Heat is one of the most abundant energy sources that can be converted into electricity in order to power circuits. Harvesting systems that use wasted heat open new ways to power autonomous systems when the energy consumption is low, or to create systems of power generators when the conversion efficiency is high. The solutions currently implemented are limited in their potential use for several reasons: 1/ energy recovery can be done by maintaining a large thermal gradient between the faces of the recovery system, which is difficult to achieve without the integration of heat sinks, 2/ rare, costly and difficult to integrate materials need to be used for applications at room temperatures, 3/ heat sources in our environment are generally non-planar forms (i.e. pipes) and the creation of flexible recovery systems would be a great value. Significant efforts have been made to harvest heat using thermoelectric generators (TEGs), based on the Seebeck effect [1]. They require rare materials based on bismuth telluride to work at ambient temperatures or complex nanostructures. These materials show low thermal insulating properties, which makes keeping temperature gradients difficult and the use of a heat sink necessary.

An innovative way of harvesting heat, that avoids the difficulties mentioned above is presented in this paper. It enables the fabrication on thin modules that work without a heat sink at temperatures close to ambient (see fig-1). The key point of the integration of this technology is the ability to keep an important gradient on a device body by intelligent control over the thermal flow.

Fig1: Innovative product features planned: Flexibility, low cost, large area

II. Working principle: thermomecanic & thermofluidic conversions

To provide an innovative system of conversion of heat into electricity, the concept is based on two key principles: 1-the conversion of a continuous stream of heat into mechanical impulses, 2-the conversion of these mechanical pulses into stored electrical energy. The conversion of the continuous flow of heat into mechanical impulses will be addressed according to two concepts. The first concept uses a bistable bimetal that swings periodically between hot and cold points (*thermomechanical* system), while the second uses the periodic explosion and condensation of micro-droplets of liquid (*thermofluidic* system). The mechanical energy supplied either by the bimetal or the explosion of the droplet is then used to distort a piezoelectric membrane which will then generate an electrical signal that can be treated by electronics to allow storage of the energy on a battery.

II. 1 Thermomechanical Conversion

The bimetal deviates or hits between two points or surfaces (with or without contact): a hot and a cold one. In our application, the bimetal (preformed to have two bistable states) will change between a "low" state where its central part will be in contact with the hot spot, and an "up" state where its central part will be in contact with the cold point. To retrieve a maximum of mechanical power, the bimetal must have a snap frequency as high as possible. To achieve this: 1/ the volume of the bimetal should be as low as possible to retrieve the temperature of the heat source (or the cold sink) as quickly as possible, while maintaining sufficient amplitude of deformation in order to hit the mechanical-to-electric transducer, 2/ the two temperatures of the bimetallic transition (up and down) have to be the closest, 3/ the number of bimetals must be high (matrix) and their miniaturization must be technically possible.

Each bimetal, oscillating between the hot and cold states, will hit or deform a piezoelectric membrane, whose signals can be processed by dedicated electronics to optimize the conversion of the signals generated.

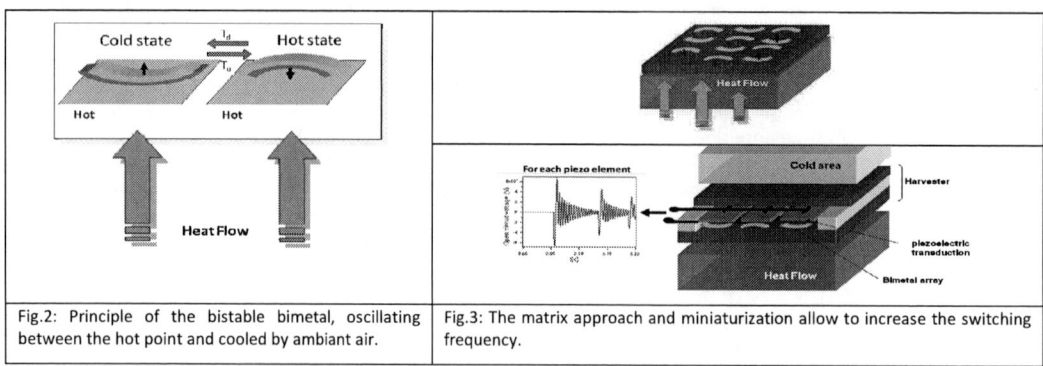

Fig.2: Principle of the bistable bimetal, oscillating between the hot point and cooled by ambient air.

Fig.3: The matrix approach and miniaturization allow to increase the switching frequency.

II.2 Thermofluidic conversion

The second principle of conversion of heat into mechanical impulses is based on the generation of cycles of evaporation and condensation of a working fluid. These cycles of evaporation and condensation cause variations of pressure and therefore deformation of a piezoelectric membrane that confines the working fluid. The liquid evaporates suddenly in a vaporization chamber when it comes in contact with the hot surface. This surface is at a temperature above the boiling point of the liquid. The sudden evaporation leads to a significant increase of pressure within the cavity.

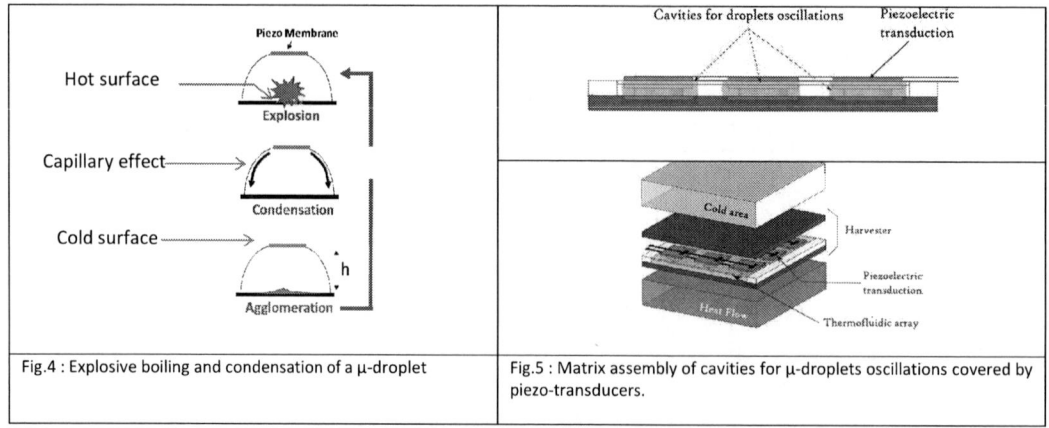

Fig.4 : Explosive boiling and condensation of a μ-droplet

Fig.5 : Matrix assembly of cavities for μ-droplets oscillations covered by piezo-transducers.

The generation of pressure created due to the evaporation forces the steam towards the cold surface. Subsequently, the vapor condenses and comes back by capillary forces or gravity to the hot surface. This evaporation/condensation cycle starts again when the droplet comes in contact with the hot surface. The originality of this energy recovery system is the first step in the conversion process, where heat is transformed into variation of pressure thanks to cycles of explosive evaporation. The effect is based on the rapid liquid phase change in areas of micro-confinement to optimize the explosive boiling. The device thus creates periodic pressure peaks from a constant temperature (or slowly varying) heat source. With small dimensions, the frequency of the mechanical pulses on the piezo-element will increase, and a relatively higher power can be expected. This variation in pressure deforms the piezoelectric membrane and induces voltage/current peaks at its terminals that are transported by connections to the outside. They can be processed by electronic circuits for energy recovery and be treated to be stored and/or used.

III. Proof of Concept: thermomechanical & thermofluidic macro-prototypes

Our innovative way of harvesting enables the fabrication of thin modules that work without a heat sink at temperatures close to ambient. The key point of the integration of this technology is the ability to keep an important gradient on a device body by smart control of the thermal flow. The first prototypes working with macro-scale elements have been fabricated to demonstrate the functionality of both approaches.

Figures 6 to 9 describe the macro-scale prototyping of the thermomecanical approach. In fig.6, the bimetal switches between a 55°C hot plate, shocks a piezo-transducer placed on top, and is cooled only with ambient air. A voltage of 7V was obtained on the piezo transducer at a frequency of 1Hz. In fig.7, ten modules have been connected and work together to load a storage capacitor.

The generated electric pulses can charge a storage unit through a dedicated harvesting circuit (fig. 8). The first centimeter scale prototypes built this way give over 11 µW of electrical power by using a bimetal that snaps at a frequency around 1 Hz (fig.9).

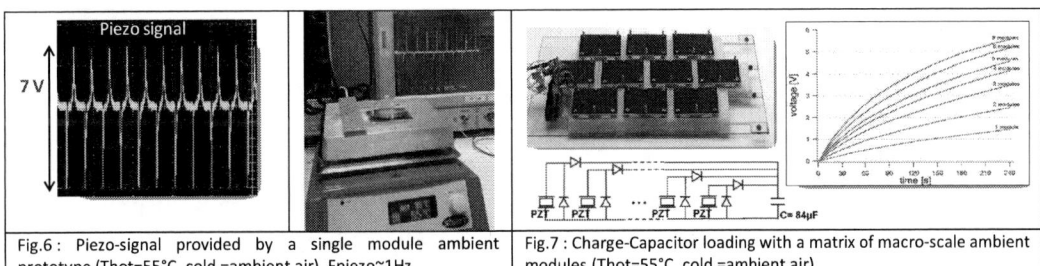

Fig.6 : Piezo-signal provided by a single module ambient prototype (Thot=55°C, cold =ambient air). Fpiezo~1Hz.	Fig.7 : Charge-Capacitor loading with a matrix of macro-scale ambient modules (Thot=55°C, cold =ambient air)

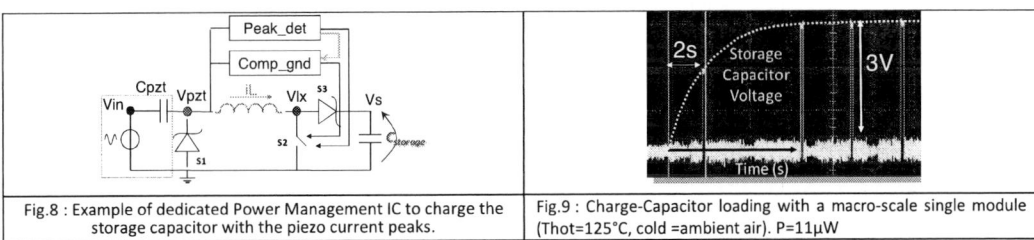

Fig.8 : Example of dedicated Power Management IC to charge the storage capacitor with the piezo current peaks.	Fig.9 : Charge-Capacitor loading with a macro-scale single module (Thot=125°C, cold =ambient air). P=11µW

The figures 10 and 11 are dedicated to the proof of concept of the thermofluidic approach. The cycle of the fluid can be implemented in two ways. In a first approach, the explosive boiling/condensation cycle is designed in a vertical configuration, with the condensation chamber placed above the explosion chamber (fig.10). Figure 11 describes an alternative approach where the explosion chamber and the condensation are placed at the same level, but the heat is distributed accordingly to their positions (fig.11).

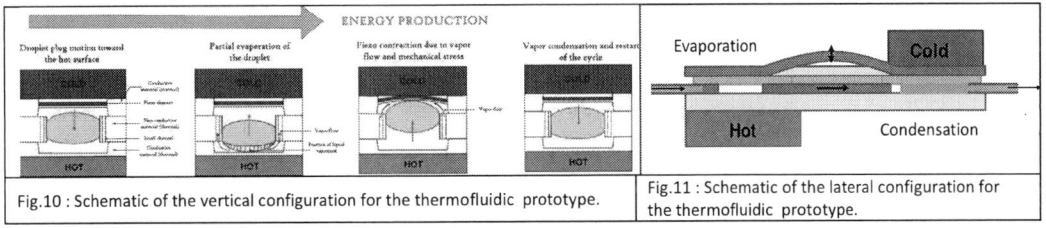

Fig.10 : Schematic of the vertical configuration for the thermofluidic prototype.	Fig.11 : Schematic of the lateral configuration for the thermofluidic prototype.

The figures 12 and 13 shows the first thermo-fluidic prototypes (vertical & lateral cycles) and the associated piezo-electric signals generated (Thot~130°C).

   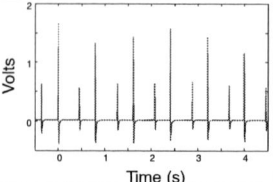

Fig.12: picture of the 2cm² vertical cycle prototype and associated signal provided by the piezo transducer.	Fig.13: picture of the 2cm² lateral cycle transparent prototype and associated signal provided by the piezo transducer.

The voltage variation is typical of what happens when the oscillation starts inside the macro-prototype (Figure 12). Indeed there is a transient behavior followed by "stable" oscillations. When the temperature is increased, the first "explosion" leads to a higher value of voltage (11 V), then the oscillation is damped and the voltage variation is comprised between 2 and 3 V.

IV. Scaling for Performances improvements

One of the keys to improve the power generated by the device is to increase the oscillation frequency of each bimetallic strip or of each boiling droplet. Laws modeling downscaling have been established for the bimetal elements and show that the miniaturization of the dimensions by a factor k increases the power generated by a factor k, thanks to the increased rate of heat transfer at small scale [O.Puscasu et al, submitted to SNW 2012].

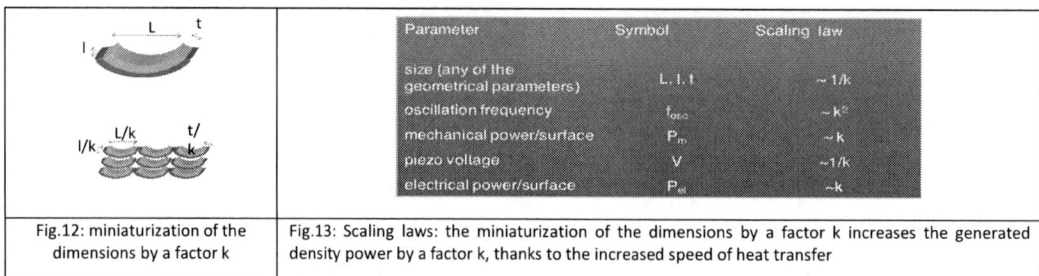

Fig.12: miniaturization of the dimensions by a factor k	Fig.13: Scaling laws: the miniaturization of the dimensions by a factor k increases the generated density power by a factor k, thanks to the increased speed of heat transfer

Thus, adaptation of the fabrication processes to produce reduced scale structures will be a key point for achieving higher levels of generated power.

V. A world of applications

The concepts presented here will be developped to allow the deployment of autonomous thermal energy recovery technologies, thanks to the conjunction with ultra-low power electronics. The evolution of the market for stand-alone systems is exponential, with many potential applications such as: Control of energy in buildings (air conditioning control, improving the distribution of energy), Power control and security systems in industrial environments (busbars, high voltage areas...), Control and security infrastructures (drinking water systems, sewage systems, gas distribution networks,...), Water, gas and electricity smartmeters, Control of buried electrical cables, Automotive (tire pressure sensors, wireless sensors ...), medical monitoring systems positioned on the body, cardio frequency meters, seismic sensors, alarm systems and more.

VI. Conclusions

Thanks to the conjunction of ultra-low power electronic development, the integration of autonomous sensors and electronics with ambient energy harvesting will be achievable. The application market is very wide, from environment and industrial sensors to medical portable applications; however the vision of the first and main application is a power supply for massively distributed sensor networks. To achieve this, innovative ways of thermal energy harvesting have been presented in this paper. Our new approaches are based on the interaction of a switching bimetal or explosive boiling μ-droplets with a piezoelectric transducer. The generated electric power density can be increased by miniaturizing each device, but an optimal working point should be found in order to have high power due to scaling and enough signal so as to overcome the threshold of the harvesting circuit. The concept is unique in the sense that it is based on a matrix structure of micro or mini energy nodes which will work together and which do not require the use of radiators, thanks to the controlled thermal resistance. This opens the door to new properties and features of the object, with better performance for low power applications.

[1] Hudak et al. Journal of Applied Physics 103, 101301 (2008), [2] Wittrick. W. H., Stability of a bimetallic disk, The Quarterly Journal of Mechanics and Applied Mathematics (1953), [3] S. Timoshenko, Analysis of bi-metal thermostats, J. Opt. Soc. Am. 11 (1925) 233-256, [4] Kanthal Thermostatic Bimetal Handbook, Kanthal AB., [5] Tadigadapa et al. Meas. Sci. Technol. 20 (2009) 092001, [6] Kandlikar,: Stabilization of flow boiling in microchannels using pressure drop elements and fabricated nucleation sites. Journal of Heat Transfer, 128:389–396, 2006., [7] Lu, C.-F. : Micro-droplet cooling apparatus, 2008., [8] Mateù, L. et Moll, F. : Review of energy harvesting techniques and applications for microelectronics. In SPIE, Sevilla, Spain, 2005., [9] Okuyama, K. et Mori, S. : Self-excited oscillating flow heat pipe, 2010.

New type steep-S device using the bipolar action

Digh Hisamoto, Shin-ichi Saito, Akio Shima, Hiroyuki Yoshimoto, and Kazuyoshi Torii
Central Research Laboratory, Hitachi, Ltd., Kokubunji, Tokyo 185-8601, Japan
Tel:+81-42-323-1111, E-mail: dai.hisamoto.pd@hitachi.com

Abstract

We have proposed an alternative approach for developing a steep subthreshold swing FET that is less than the theoretical diffusion-based limit of 60 mV/decade at room temperature. Instead of using a simple IGFET, we formed a complex device in a "single device" and worked it as a sub-circuit, which resulted in a steep subthreshold swing. We formed a tunnel junction in a drain diffusion layer of the MOSFET so that we could stuff a tunnel-injection bipolar, a resistor, and a MOSFET inside a single "scaled MOSFET". We used device simulation to clarify the concept of "device complex". Results showed a steep subthreshold swing even if the supply voltage was low (~0.2 V).

Introduction

Tunnel FETs have recently become a key candidate to overcome the lower limit of the subthreshold swing (SS) of MOSFETs, which is a crucial issue in scaling down both supply voltage and power consumption.

Most previous research has focused on reversing the bias condition that provides the high electric fields achieved in band-to-band tunneling and on enhancing these electric fields (Fig. 1). However, such approaches are not able to suppress the power consumption because high supply voltages are inevitable.

In this work, we developed a novel complex MOSFET (CxFET) that enables the use of multiple mechanisms (rather than a single mechanism) to achieve a steep SS. We used forward bias instead of reverse bias on the tunnel junction to ensure sufficient functionality under low supply voltages. In this work, we first clarified the concept of "device complex" with device simulation and then experimentally demonstrated that the proposed approach was able to achieve a steep SS of less than 60 mV/decade at room temperature.

Device Concept

The typical device structure of an N-channel CxFET and the equivalent circuit is shown in Fig. 2. The CxFET has a tunnel junction in the drain composed of a tunnel-injection bipolar transistor with a p-well substrate. Regarding the symbol of Esaki diode, the tunnel-injection bipolar was denoted in the equivalent circuit. The tunnel junction has two roles: (1) the hole injection under the low forward bias condition acts as an emitter base in a tunnel-injection bipolar transistor, and (2) the resistance of the tunnel junction and the MOSFET compose an inverter. This inverter generates a full-swing output (Vcc → 0 V) at around Vg = Vt. The inverter output directly drives the tunnel junction, which stimulates the tunnel-injection bipolar. The consequent accelerated tunnel injection provides a steep SS in the drain current of less than 60 mV/decade at room temperature.

Figure 3 shows simulated Id-Vgs characteristics and SS characteristics. A steep SS of less than 60 mV/decade was observed. The potential distribution for Vgs = −0.5 V and −0.3 V is shown in Fig. 4. For the 0.2 V change in the gate bias, a dramatic potential change of over 0.4 V appeared at the n^+ drain region, which caused a steep SS. The current flow at Vgs = −0.4 V in the device is shown in Fig. 5. The current at the channel surface mainly consisted of electrons, whereas the substrate current mainly consisted of holes, suggesting that tunnel injection occurred under the forward bias condition. The sub-circuit actions were what caused the steep SS.

Experiments

We fabricated CxFETs on 60-nm-thick SOI wafers so that we could control the junction precisely. The process flow was almost the same as in conventional NMOS except for p^{++} ion implantation. We formed the tunnel junction by laser spike annealing (LSA) after the p^{++} ion implantation (see the inset of Fig. 6).

Id-Vgs curves of a typical CxFET are shown in Fig. 6. A steep SS of less than 60 mV/decade was observed for over four decades in the subthreshold region. There were identical curves in the forward and reverse bias modes, which seems to suggest that the steep SS was not caused by the junction break-down mechanism. The swing values of the CxFET under low supply voltages are plotted in Fig. 7. Even for Vds = 0.2 V, an SS of below 60 mV/decade was observed while the off-state leakage current degraded the swing value.

Conclusions

We proposed "Device Complex" as a concept for developing devices with a steep SS and employed this concept to fabricate CxFETs using a tunnel junction formed in the drain node as the BJT and resistor. It is possible to use various combinations to form the device complex, so we expect the proposed approach to pave the way for the development of devices with a steep SS and low power consumption.

Reference:

[1] M.M. Atalla, et al., USPatent:3045129(1962). [2] S.R.Hofstein and G. Warfield, Trans. Electron Devices, vol.12, p.66, 1965. [3] E. Takeda, et al., IEDM, p.402, 1988. [4] C. Aydin, et al., Appl. Phys. lett., vol.84, p.1780, 2004. [5] W.Y. Choi, et al., Electron Device Lett., vol.28, p.743, 2007. [6] V. Nagavarapu, et al., Trans. Electron Devices, vol.55, p.1013, 2008. [7] K. Jeon, et al., VLSI Symp. p.121, 2010. [8] K.K. Bhuwalka, et al., Jpn. J. Appl. Phys., vol.43, p.4073, 2004. [9] S. Kim, et al., VLSI Symp. p.178, 2009. [10] J. Koga, et al., Electron Device Lett., vol.20, p.529, 1999. [11] D. Hisamoto, et al., IEDM, p. 233, 2011

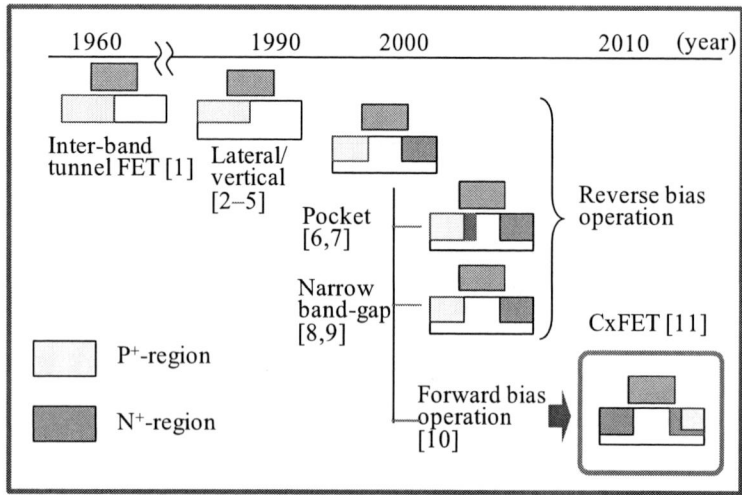

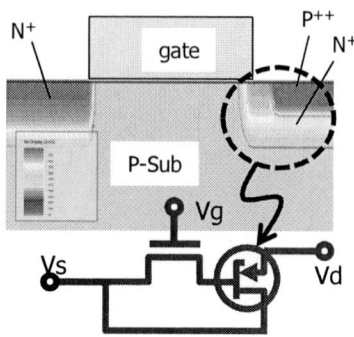

Fig. 1 Previous tunnel FET research and developments. To clarify the junction configurations, only N-type devices are shown.

Fig. 2 Simulated structure and equivalent circuit. Net dopant concentration profile is shown. Red and purple correspond to 10^{20} and 10^{16} cm^{-3}, respectively. The dotted circle is a tunnel injection BJT.

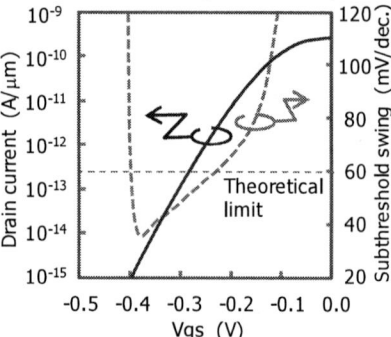

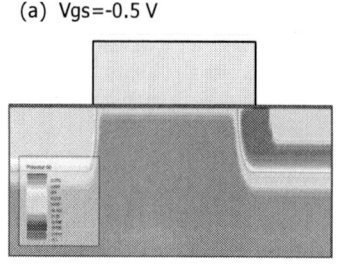

Fig. 3 Simulated Id-Vgs characteristics (left-hand y axis) and swing value (right-hand y axis). The thin dotted line shows the theoretical SS limit at room temperature.

Fig. 4 Simulated potential distribution for subthreshold operations of n-type CxFET. Vds/Vsub were 0.5 and 0 V, respectively. For n+-type diffusion area, red represents 0.5 V and yellow represents 0 V.

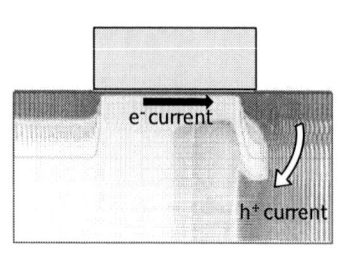

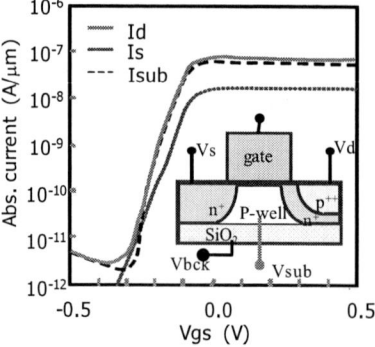

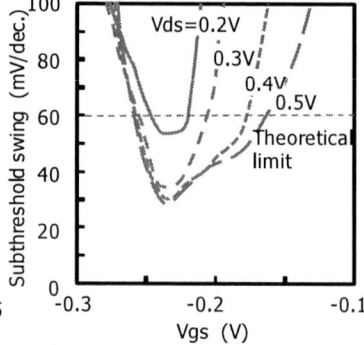

Fig. 5 Simulated current flow at Vgs = 0.4 V. The surface current was mainly composed of electrons and the body current was mainly composed of holes. Drain bias was 0.5 V.

Fig. 6 Fabricated device structure with 60-nm-thick SOI wafer (inset) and measured typical current characteristics with the gate bias as a variable. Vds = 0.6 V at room temperature.

Fig. 7 Subthreshold swing characteristics at room temperature under low drain bias conditions. Vds = 0.5 V down to 0.2 V.

978-1-4673-0996-7/12 $31.00 © 2012 IEEE

Experimental Demonstration of Temperature Stability of Si-Tunnel FET over Si-MOSFET

Shinji Migita, Koichi Fukuda, Yukinori Morita, and Hiroyuki Ota

Collaborative Research Team Green Nanoelectronics Center (GNC),
National Institute of Advanced Industrial Science and Technology (AIST), Tsukuba, Ibaraki 305-8569, Japan
e-mail: s-migita@aist.go.jp

Abstract

Temperature dependences of tunnel field-effect transistor (TFET) and MOSFET were experimentally compared on the same SOI wafer. Validity of the TFET result was corroborated by simulation. It is demonstrated that V_{TH} shift and off-current increment of Si-TFET with temperature were smaller in comparison with Si-MOSFET. Temperature stability of TFET is promising for ultra-low power VLSI.

Keywords: tunnel FET, MOSFET, SOI, temperature dependence, threshold voltage, and ultra-low power VLSI.

Introduction

TFET functions with the interband tunneling and has a potential to achieve steep sub-threshold swing less than the theoretical limit of conventional MOSFET (60mV/decade). This characteristic is promising for the development of low-power VLSI that functions by a deep sub-1 V supply voltage [1, 2].

As the supply voltage decreases to below 1 V, the tolerance of V_{TH} variation becomes extremely small. It is well known that V_{TH} variation originates from dopant fluctuation and variation of metal work function. In addition, V_{TH} changes with the operating temperature. According to the formula of band-to-band tunneling (BTBT) current [3], TFETs show temperature dependent electrical properties that are caused by the band gap narrowing of semiconductor with temperature. It is also confirmed in several experimental studies [4-8].

This work reports the comparative study of temperature dependences of TFETs and MOSFETs that were fabricated on the same SOI wafer. Temperature stability of Si-TFETs over Si-MOSFETs up to 400K was confirmed by smaller V_{TH} variation and smaller I_{OFF} increment.

Experimental and Simulation

TFETs and MOSFETs were fabricated on the same SOI wafer with the process flow shown in **Fig. 1**. Using photo masks, ion implantations for TFETs and MOSFETs were executed in a CMOS process manner. Physical analyses of the devices revealed that the heavy dose condition of arsenic ion in this experiment was overdone (**Fig. 2**). Defects in SOI layer could not be annihilated by the activation anneal, and the influence of these defects appeared in the electrical properties of TFET as the trap-assisted current. Reproducibility of electrical characteristics was inspected by measurement of more than 10 devices.

Performance of Si-TFETs was corroborated using TCAD simulator [9] with newly developed non-local band-to-band tunneling module [10]. The band gap narrowing effect is programmed to investigate the temperature dependence.

Results and Discussion

I_D-V_G and I_D-V_D characteristics show successful operation of Si-TFET (**Fig. 3**) and Si-MOSFET (**Fig. 4**). I_{OFF} of TFET is smaller than MOSFET by 5 orders, and I_{ON} is small by 2 orders. Gate leakage currents are small and negligible in the analyses of I_D in both devices.

Temperature dependences of TFET (**Fig. 5**) show the occurrence of trap-assisted tunneling (TAT) at low V_G region and the dominance of BTBT with the increment of V_G swing. Increment of I_{ON} in TFET with temperature originates from the band gap narrowing. In contrast I_{ON} of MOSFET decreases with temperature (**Fig. 6**), which is caused by the increment of phonon scattering in the channel. Both in TFET and MOSFET, the absolute value of V_{TH} decreases as the temperature increases. Temperature dependence of I_D-V_G characteristic of Si-TFET was reproduced in simulation (**Fig. 7**). Except for the existence of TAT in experimental Si-TFET, the overall performances are identical.

V_{TH} shifts and I_{ON}-I_{OFF} changes with temperature are summarized in **Fig. 8**. V_{TH} shift with temperature in Si-TFET is as small as half of Si-MOSFET (Fig. 8(a)). Although the band gap narrowing is the common origin of V_{TH} shift in both devices, the difference of current flow mechanism seems to bring the dissimilarity in the amount of V_{TH} shifts. I_{OFF} increment with temperature in Si-TFET is also smaller than that of Si-MOSFET. I_{ON} of Si-TFET increases with temperature opposite to the trend of Si-MOSFET. These results suggest that at higher environmental temperature Si-TFET can maintain low stand-by power and higher drive performance.

It must be investigated in future whether the V_{TH} shift of Si-TFET with temperature is small enough for the low-power device. Furthermore, studies using advanced channel materials, such as Ge and InGaAs, are also essential.

Conclusions

Temperature dependence of material properties induces change in device performances. As far as the V_{TH} shift and the I_{OFF} increment with temperature were examined, Si-TFET showed better temperature stability than Si-MOSFET. Temperature stability thus demonstrated would be a strong motivation of applying TFETs for low-power VLSI that requires strict V_{TH} control.

Acknowledgements

This research is granted by JSPS through FIRST Program initiated by CSTP. Device fabrication was supported by ICAN-AIST.

References

[1] A. Seabaugh and Q. Zhang, *Proc. IEEE* **98**, (2010) 2095.
[2] A. Ionescu and H. Riel, *Nature* **479**, (2011) 329.
[3] E.O. Kane, *J. Phys. Chem. Solids* **12**, (1959) 181.
[4] F. Mayer *et al.*, *IEDM Tech. Dig.* 2008, p.163.
[5] P.-F. Guo *et al.*, *IEEE Electron Device Lett.* **30**, (2009) 981.
[6] S. Mookerjea *et al.*, *IEDM Tech. Dig.* 2009, p.949.
[7] D. Leonelli *et al.*, *Jpn. J. Appl. Phys.* **50**, (2011) 04DC05.
[8] J. Wan, C.Le Royer, A. Zaslavsky, and S. Cristoloveanu, *Solid-State Electron.* **65-66**, (2011) 226.
[9] HyENEXSS™, ver. 5.5, Selete, 2011.
[10] K. Fukuda *et. al.*, Int. Sym. "Develop. Core Tech. Green Nanoelectronics" (Mar. 2012, Tokyo) p.72.

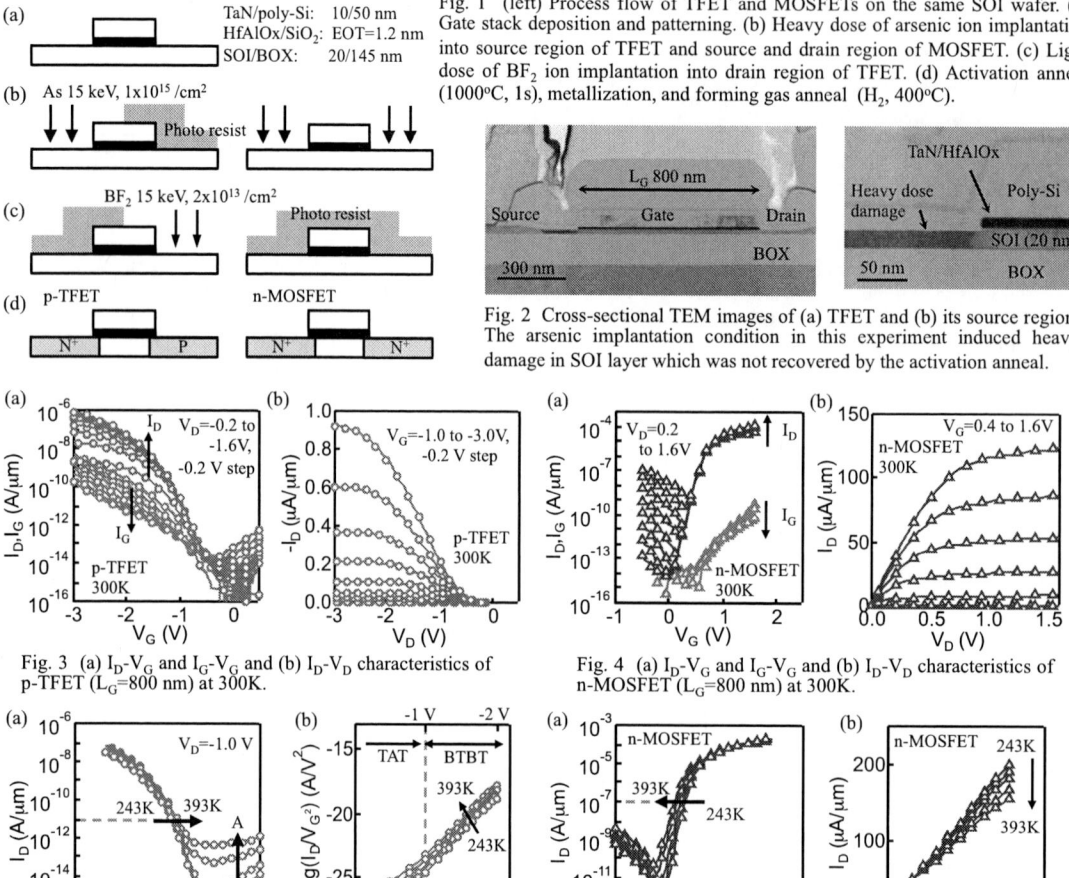

Fig. 1 (left) Process flow of TFET and MOSFETs on the same SOI wafer. (a) Gate stack deposition and patterning. (b) Heavy dose of arsenic ion implantation into source region of TFET and source and drain region of MOSFET. (c) Light dose of BF_2 ion implantation into drain region of TFET. (d) Activation anneal (1000°C, 1s), metallization, and forming gas anneal (H_2, 400°C).

Fig. 2 Cross-sectional TEM images of (a) TFET and (b) its source region. The arsenic implantation condition in this experiment induced heavy damage in SOI layer which was not recovered by the activation anneal.

Fig. 3 (a) I_D-V_G and I_G-V_G and (b) I_D-V_D characteristics of p-TFET (L_G=800 nm) at 300K.

Fig. 4 (a) I_D-V_G and I_G-V_G and (b) I_D-V_D characteristics of n-MOSFET (L_G=800 nm) at 300K.

Fig. 5 (a) Temperature dependences of I_D-V_G characteristics of p-TFET (V_D=-1 V). Increment of off-current with temperature (arrow A) indicates the trap-assisted tunneling (TAT). V_{TH} was defined at I_D=10^{-11} A/μm. (b) Log(I_D/V_G^2)–(1/V_G) plot of Fig. 5(a). Linear trend in the V_G range between -1 and -2 V proves the BTBT.

Fig. 6 (a) Temperature dependences of I_D-V_G characteristics of n-MOSFET (V_D=1 V). V_{TH} was defined at I_D=10^{-7} A/μm. (b) Linear plot of Fig. 5(a). Decrease of drive current with temperature is caused by the increment of phonon scattering at high temperatures.

Si Channel		Doping levels (/cm³)	
Thickness	10 nm	Source	n, 2×10^{20}
Length	100 nm	Channel	p, 1×10^{17}
Drain offset	50 nm	Drain	p, 1×10^{19}
EOT	1 nm		

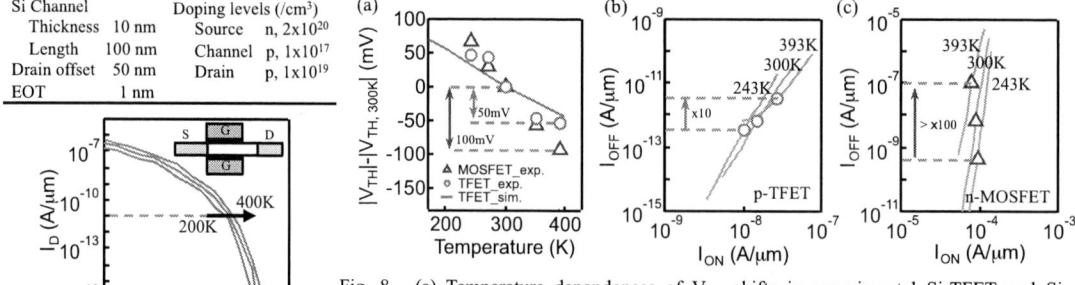

Fig. 7 Simulation parameters and results of temperature dependence of double-gate Si-FET. The V_{TH} shift with temperature is reproduced.

Fig. 8 (a) Temperature dependences of V_{TH} shifts in experimental Si-TFET and Si-MOSFET and simulation Si-TFET. V_{TH} values at 300K are referred as standards. V_{TH} shift of TFET with temperature is about half of MOSFET. (b) and (c) show I_{ON}-I_{OFF} change with temperature for experimental Si-TFET and Si-MOSFET. I_{OFF} increment in Si-TFET is as small as one order while that of MOSFET is larger than two orders. I_{ON} of Si-TFET increases at higher temperatures in contrast to MOSFET. Thus Si-TFET has better temperature stability over Si-MOSFET.

978-1-4673-0996-7/12 $31.00 © 2012 IEEE

Scale laws for enhanced power for MEMS based heat energy harvesting

O. Puscasu[1,2], S. Monfray[1], F. Boeuf[1], G. Savelli[3], F. Gaillard[3], D. Guyomar[2], T. Skotnicki[1]

[1]STMicroelectronics (Crolles 2) SAS, 850 rue Jean Monnet, 38920 Crolles Cedex, France
[2]LGEF, INSA Lyon, 8 Rue de la Physique 69621 Villeurbanne Cedex, France
[3]CEA Liten, 17 rue des martyrs 38054 Grenoble Cedex 9, France

Abstract

An innovative approach to thermal energy harvesting is presented. It consists of a two step conversion of heat into electricity. The new technique can be used for powering ultra-low power electronics and autonomous systems. One of the keys to improve the generated power density is downscaling of individual devices. Laws modeling downscaling have been established in this paper and show that the miniaturization of the devices by a factor k increases the generated power density by the same factor, due to the increased speed of heat transfer. The scaling laws predict increasing power gain when miniaturizing the devices with use of e.g. VLSI technologies. This can help in providing a strong alternative to Seebeck devices.

I. Introduction

Heat is one of the most abundant energy sources that can be converted into electricity in order to power circuits. Significant efforts have been made in order to harvest heat through the development of thermoelectric generators (TEGs), based on the Seebeck effect [1]. They require rare materials based on bismuth telluride or complex nanostructures to work at ambient temperature. These materials show low thermal insulating properties, which makes keeping temperature gradients difficult and the use of a heat sink necessary (fig. 1).

An innovative way of harvesting, that allows to avoid the above mentioned difficulties is presented in this work. It enables the fabrication of thin modules that work without a heat sink at temperatures close to ambient. The key point of the integration of this technology is the ability to keep an important gradient on a device body by intelligent control over the thermal flow (fig. 2).

II. Working principle and first results

An elementary device of this technology is based on a preformed bimetal (fig. 3) that snaps when it is being heated, and hits a piezoelectric, thus generating electric pulses (fig. 4). The bimetal then cools down by contact with the piezoelectric and snaps back, thus making a periodical movement possible. The generated electric pulses can charge a storage unit through a harvesting circuit (fig. 5).

The first centimeter scale prototypes built this way give over 10 μW of electrical power by using a bimetal that snaps at a frequency around 1 Hz, and a full wave harvesting circuit.

III. Thermo mechanical conversion by a bimetal

A thermal bimetal is a double layer consisting of a material of high coefficient of thermal expansion (CTE>10·10^{-6} K^{-1}) and a material of lower CTE. It is necessary to give the layer a concave shape (spherical cup, cylindrical shell or more complex) in order to make it snap, that is to move quickly from one stable position to another under the effect of rising or decreasing temperature [2].

When a bimetal is heated by contact with a hot source (fig 6), the average temperature with time is given by:

$$\overline{T}(t) = \frac{8}{\pi^2}\exp\left(-\pi^2 \alpha_T \frac{t}{L^2}\right)(T_c - T_h) + T_h \ (1)$$

with T_h- temperature of the hot source, T_c – temperature of the cold bimetal, α_T – average thermal diffusivity, L – heat propagation length.

The time lapse needed to reach a target average temperature (snap temperature) in the bimetal is proportional to the squared length:

$$t \sim L^2 \ (2)$$

This implies that the snap frequency is proportional to the inverse of the squared length:

$$f \sim \frac{1}{L^2} \ (3)$$

Thus, the reduction of the size of a bimetal by k would multiply the oscillation frequency by k^2. This fact is confirmed by experiments at macro scale, for sizes close to 1 cm (fig 7). Analytical expressions show that snap temperatures do not change with scale [2] (fig 8).

The elastic energy accumulated during the heating step, calculated according to the stress distribution in [3] (fig 9), will be partially converted into kinetic energy during the snap action:

$$E_{elast} = \frac{Y}{32}(\alpha_2 - \alpha_1)^2 \Delta T^2 V_b \sim E_k = \frac{\rho V_b \cdot v^2}{2} \ (4)$$

with Y – the Young modulus of the materials, α_2- coefficient of thermal expansion of the upper layer, α_1 – coefficient of thermal expansion of the down layer, ΔT – difference between snap temperature and the ambient, V_b – the bimetal volume, E_k – kinetic energy, ρ- density of the materials, v – maximal bimetal speed.

The kinetic energy will be transmitted to the piezoelectric and make it vibrate. The bimetal speed resulting from the former expression is independent of scale:

$$v \sim \sqrt{\frac{Y}{\rho}} \cdot (\alpha_2 - \alpha_1)\Delta T \ (5)$$

By replacing one large bimetal with several small ones occupying the same surface, with sizes divided by k (fig. 10), the total mass will be decreased by k, but the frequency of each element will be multiplied by k^2, bringing up the total transmitted mechanical power by k (Table 1):

$$\frac{P_{tot\,s}}{P_l} = \frac{N_s E_{ks} f_s}{E_{kl} f_l} = \frac{N_s \rho V_s \frac{v^2}{2}}{\rho V_l \frac{v^2}{2}} \frac{L_l^2}{L_s^2} = N_s \frac{V_s}{V_l} \frac{L_l^2}{L_s^2} = k = \sqrt{N_s} \ (6)$$

with $P_{tot\,s}$ – the total mechanical power transmitted by small bimetals, P_l – the mechanical power transmitted by a large bimetal, N_s is the number of bimetals and V is the volume of a bimetal, the index s refers to small bimetals, and the index l to the large one. The main parameter of the calculation is k – the scaling factor.

IV. Electromechanical conversion by the piezoelectric

The force developed by the bimetal for a given ΔT scales down as k^2 [4], which leads to a deflection in the piezoelectric divided by k and an average stress that does not vary with scale. The resulting piezoelectric voltage and induced electric energy will depend on the stress and the piezoelectric thickness [5]:

$$U = \frac{d_{31}}{\varepsilon}\cdot\sigma_L\cdot s \sim \frac{1}{k} \ (7); \ E_{el} = \frac{C_p U^2}{2} \ (8)$$

with all the constants and variables referring to the piezoelectric: U – voltage, d_{31} – bending charge constant, ε – dielectric constant, σ_L – longitudinal stress, s-thickness, E_{el} – electric energy on the piezoelectric, C_p – capacitance.

By combining the equations 3, 7, and 8 one can deduce that the electric power generated by several piezoelectric-bimetal components (P_{eltot}) will be k times larger compared to a single component occupying the same surface:

$$P_{eltot} = N_s \cdot E_{el} \cdot f \sim k \ (9)$$

This leads to considering device miniaturization and integration using VLSI techniques (fig. 11). Such an approach can help in obtaining superior power compared to the projections for the Seebeck modules.

Meanwhile the generated voltage will decrease with the thickness of the piezoelectric. It should be nonetheless kept high enough, until an optimal size is reached, in order to overcome the threshold of the harvesting circuit.

The heat transfer through the bimetals will be accelerated by decreasing the scale, so the degree of miniaturization is to be chosen depending on the available external thermal gradient.

V. Conclusion

An innovative way of thermal energy harvesting has been presented. It is based on the interaction of a thermal bimetal with a piezoelectric. The generated electric power density can be increased by a factor k when miniaturizing each device by the same factor k. An optimal working point should be found in order to have high power due to scaling and enough signal so as to overcome the threshold of the harvesting circuit. Such an approach can lead to superior power compared to the projections for the Seebeck modules.

[1] Hudak et al. Journal of Applied Physics 103, 101301 (2008)
[2] Wittrick. W. H., Stability of a bimetallic disk, The Quarterly Journal of Mechanics and Applied Mathematics (1953).
[3] S. Timoshenko, Analysis of bi-metal thermostats, J. Opt. Soc. Am. 11 (1925) 233-256
[4] Kanthal Thermostatic Bimetal Handbook, Kanthal AB.
[5] Tadigadapa et al. Meas. Sci. Technol. 20 (2009) 092001

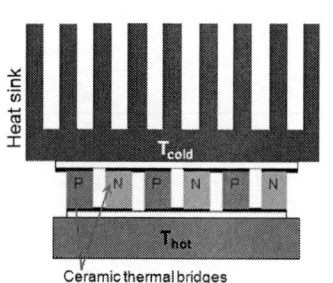

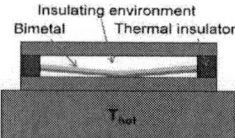

Figure 2. Module with thermal conductivity controlled by a bimetal

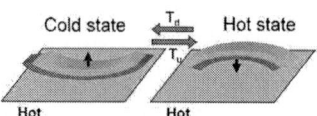

Figure 3. Cold and hot states of a bimetal

Figure 1. Seebeck module with heat sink

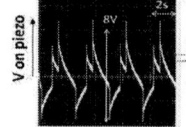

Figure 4. Example of measured signal for a bimetal and a piezoelectric working together, T_{up}= 122°C

Figure 5. Experimental charging of a 10 μF capacitor through a full wave harvesting circuit

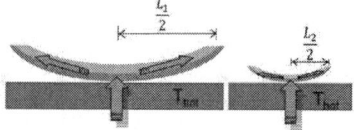

Figure 6. Bimetals of different lengths heated up. The arrows show the direction of the heat flux

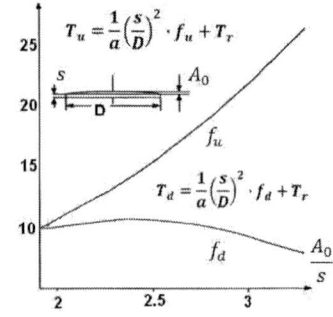

Figure 7. Snap temperature equations for a bimetal disk, according to Wittrick [2] (the constant a depends on the material, T_r is the room temperature)

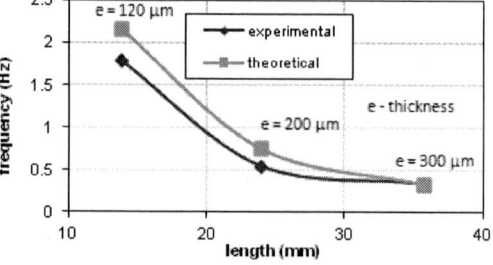

Figure 8. Evolution of the oscillation frequency of a bimetal with size

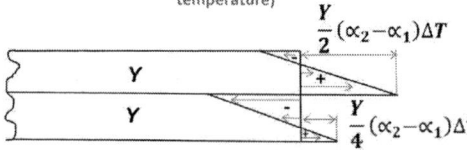

Figure 9. Stress distribution in a heated bimetal with two equal thickness layers, according to Timoshenko [3].

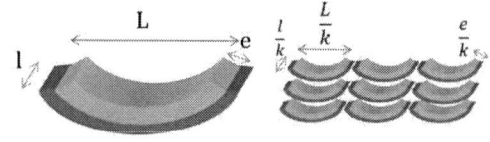

Figure 10. Replacement of a large bimetal by a series of small bimetals occupying the same surface

Figure 11. Design for micro scale integration of the devices and expected signal shape

Parameter	Symbol	Scaling law
size (any of the geometrical parameters)	L, l, t	" 1/k
oscillation frequency	f	" k²
mechanical power/surface	P_m	" k
transmitted thermal power/surface	P_t	"k
snap temperatures	T_u, T_l	same
bimetal force	F	"1/k²
piezo and bimetal deflection	A	"1/k
stress, strain inside piezo	σ_i, ε_i	same
piezo capacitance	C_p	"1/k
piezo voltage	U	"1/k
electrical power/surface	P_{elect}	"k

Table 1. Scaling laws round up for our innovative thermo mechanical harvester

Energy-Efficiency and Thermal Management in Nanoscale Devices

Albert D. Liao[1], Zhun-Yong Ong[2], Andrey Y. Serov[1], Feng Xiong[1,3], Eric Pop[1,3]

[1]Dept. of Electrical & Computer Engineering, [2]Dept. of Physics, [3]Beckman Institute
University of Illinois at Urbana-Champaign, Urbana, IL 61801, U.S.A.
Email: epop@illinois.edu

Abstract — **Power consumption and thermal management are significant challenges in electronics, from mobile devices to data centers. A fundamental examination of such aspects could lead to orders of magnitude improvements in energy efficiency. We present recent highlights from our work examining dissipation in nanoscale devices, at contacts, interfaces, and in novel materials. Advances include the use of high-thermal conductivity materials (graphene), low-power data storage (based on phase change rather than charge), and thermoelectric effects for highly localized cooling. Results suggest much room to improve power dissipation in nanoscale electronics, towards fundamental limits, through the co-design of geometry and materials.**

I. INTRODUCTION

Power dissipation and heat conduction in electronics begin at the level of nanoscale transistors and interconnects [1], as shown in Fig. 1. CMOS field-effect transistors (FETs) are the building blocks of modern electronics, typically based on Si technology with features down to ~25 nm [5]. However, scaling devices to sub-10 nm dimensions will likely involve new materials from groups III-V like GaAs, InAs or InSb [6], or group IV like Ge or C [7, 8], which have electronic properties (e.g. mobility) superior to Si. With the exception of C (as nanotubes or graphene), such materials all have lower thermal conductivity than Si [1].

Further scaling will also involve new geometries such as multi-gate (FinFET) devices [9], nanowires, or nanotubes [7]. In all cases, power dissipation begins with electrons or holes that are accelerated by high electric fields, of the order ~10 V/μm. The "hot" carriers scatter with the lattice, emitting acoustic (AP) and optical phonons (OPs) [10, 11]. Modern devices operate at voltages near ~1 V, strongly emitting OPs which have energies ~35 meV in Ge or GaAs, 60 meV in Si, 180 meV in CNTs and graphene [10, 11]. Typical OP lifetimes are of the order ~1 ps [12, 13], however the thermal time constants of CMOS devices are of the order ~10 ns due to the larger volumetric heat capacity [1]. It is worth noting that the cooling time of a single carbon nanotube (CNT) is ~0.1 ns [14], likely representing the smallest thermal time constant achievable in CMOS or CMOS-like devices.

II. DIFFUSIVE VS. QUASI-BALLISTIC DISSIPATION

The physical phenomena responsible for changes in device temperature during operation are the Joule and Peltier effects, as shown in Fig. 2. The Joule effect [1] occurs as charge carriers dissipate energy with the lattice, and is proportional to resistance and the square of the current ($\sim RI^2$). The Peltier effect [15] is proportional to the magnitude of the current through and the difference in Peltier coefficient at a junction of dissimilar materials, leading to either heating or cooling depending on the direction of current flow ($\sim \pm \Pi_{AB}I$).

Transport in nanoscale devices is typically either diffusive or quasi-ballistic. The length scale which determines this behavior is the principal inelastic scattering (energy relaxation) length, which is typically that of optical phonon emission, λ_{OP}. In diffusive transport (Fig 2a, $L \gg \lambda_{OP}$) power dissipation occurs through strong emission of OPs heating the device channel. In quasi-ballistic transport (Fig. 2b, $L \sim \lambda_{OP}$) there are few OPs emitted within the channel, and significant heating can occur at the device contacts.

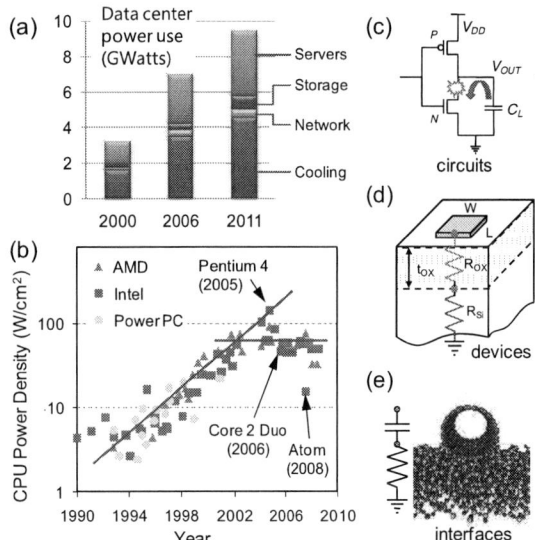

Fig. 1. (a) Power consumption of U.S. data centers has doubled almost every five years, with a significant portion dedicated to cooling (by comparison, electricity use in a small country like Hungary is ~4.5 GW) [1]. (b) Power consumption of CPUs rose exponentially until 2005, then flattened due to thermal dissipation challenges. (c-e) A comprehensive, fundamental examination of energy use in circuits, devices, contacts, and nanoscale interfaces could revolutionize the energy-efficiency of all electronics [1].

978-1-4673-0996-7/12 $31.00 © 2012 IEEE

III. EFFECTS OF CONTACTS AND INTERFACES

In addition to Joule heating, contacts can also lead to thermionic (TI) and thermoelectric (TE) effects. TI effects occur when a contact barrier acts as an energy filter, selectively allowing predominantly "hot" carriers to escape into the contact. Thus, the contact heats up, while the device slightly cools, as shown in Fig. 2c. TE effects occur at the contact between two dissimilar materials, where a difference in Seebeck coefficient leads to a cooling or heating effect, depending on the direction of the current flow. The latter has been recently observed at the contacts between graphene transistors and Pd contacts [4], and could be exploited in future devices to "shift" some of the heat normally dissipated within the channel into the larger metal contact.

The temperature difference and resistance to heat flow across device boundaries can be quantified by the thermal resistance per unit area, $R_B" = \Delta T/Q" \approx 2 \times 10^{-8}$ $m^2 KW^{-1}$ at room temperature [1, 16]. This is approximately equivalent to the thermal resistance of a ~25 nm layer of SiO_2 (ref. [1]), and could become a limiting dissipation bottleneck in highly scaled devices and interconnects [3]. For instance, based on calculations shown in Fig. 3, the boundary resistance is particularly dominant in sub~40 nm devices, together with the thermal resistance to heat flow into the contacts.

IV. DISSIPATION IN NANOSCALE DEVICES

The pathway of heat conduction between devices and environment has traditionally been considered key in determining the temperature [1] and ultimately the reliability of a device [17]. However, as devices are scaled down, four additional heat flow considerations must be taken into account. First, heat conduction into contacts plays a greater part, and the role of contact thermoelectric (Peltier) phenomena also becomes enhanced, as summarized in Section III [4]. Second, novel devices (e.g. FinFET, nanowire, CNT) contain many interfaces between the semiconductor body (Si, Ge, C, GaAs, etc.) where dissipation takes place, and the external heat sink. Thus, the associated thermal boundary resistances (R_B) must be carefully taken into account. Third, nanoscale devices are smaller than the bulk phonon mean free path, e.g. λ_{ph} ~ 100 nm in Si and comparable in most semiconductors. Thus, the thermal conductivity is lower than well-known bulk values, for instance that of a ~20 nm thin silicon-on-insulator (SOI) film is k_{SOI} ~ 25 $Wm^{-1}K^{-1}$ instead of the thermal conductivity of bulk Si, k_{Si} ~ 150 $Wm^{-1}K^{-1}$ [18]. Fourth, modern devices have electrical switching transients of the order 10-100 ps, ranging between those of carrier scattering events (0.1-1 ps) and the thermal relaxation of the lattice (~10 ns). Thus, nanoscale devices operate in a complex temporal fashion, where non-equilibrium situations can arise between electrons, OPs, and the lattice whose heat capacity is determined by APs [19].

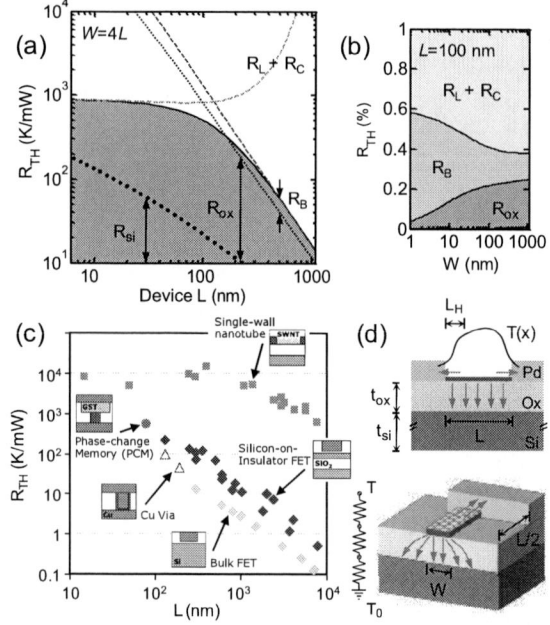

Fig. 3. (a) Scaling of thermal resistance (R_{TH}) of a thin-body SOI device or graphene-on-insulator (GOI) device as a function of channel length, based on model in Ref. [3]. (b) Relative contribution of heat flow along channel length and contacts ($R_L + R_C$), from thermal boundary resistance (R_B) and of the underlying oxide (R_{ox}), for a channel length L = 100 nm, as a function of channel width. (c) Experimental data [1] of thermal resistance for various devices, showing trends as modelled in (a). (d) Schematic of heat flow along device and into contacts (top) and into substrate (bottom) [3].

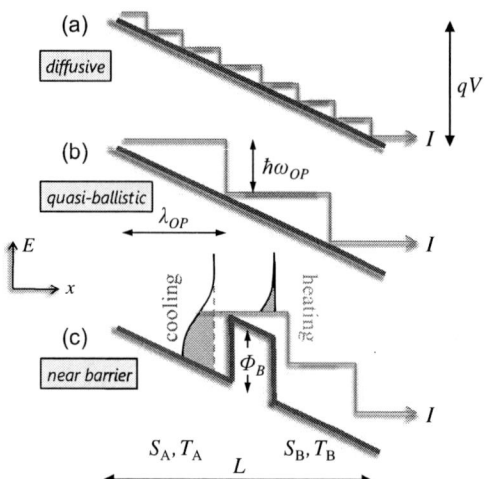

Fig. 2. Band diagrams [1] of energy dissipation in (a) device under diffusive transport ($L \gg \lambda_{OP}$, $qV \gg \hbar\omega_{OP}$), (b) quasi-ballistic device ($L \sim \lambda_{OP}$, $qV \sim \hbar\omega_{OP}$), and (c) near a barrier. The barrier can act as an energy filter, cooling the lattice to its left, heating to its right (thermionic effect). If a difference in temperature or Seebeck coefficient also exists, a Peltier power term appears, $P = (\Pi_A - \Pi_B)I$, where $\Pi_A = S_A T_A$ and S_A is the Seebeck coefficient of A, etc. [4].

Figure 3 illustrates the roles that various heat conduction mechanisms play in nanoscale devices. Numerical calculations in Fig. 3a-b are based on the models in Refs. [3, 20], corresponding to a graphene device on SiO_2 thickness $t_{ox} = 50$ nm, graphene-SiO_2 boundary resistance $R_B = 2 \times 10^{-8}$ m^{2}KW^{-1}, graphene thickness $t_g = 0.34$ nm and thermal conductivity $k_g = 150$ Wm^{-1}K^{-1}, consistent with graphene nanoribbons [3] or encased graphene [21]. Interestingly the numerical analysis is effectively identical to that for an ultra-thin SOI device with thickness $t_{SOI} = 5$ nm and thermal conductivity $k_{SOI} = 10$ Wm^{-1}K^{-1} [18] (same $t \cdot k$ product). Thus, the trends observed in the figure are also typical for other modern and future nanoscale transistors, including ultra-thin body SOI, multi-gate (FinFET), CNT or nanowire FETs. The results of Fig. 3a-b are consistent with experimental data summarized in Fig. 3c.

Figures 3a-b reveal several interesting trends as devices are scaled deeply into nanometer dimensions. First, the thermal resistance of large devices (up to 1 µm) remains limited by the "vertical" heat flow path which includes the boundary thermal resistance (R_B), the oxide (R_{ox}), and to a smaller degree the Si substrate (R_{si}). As devices are scaled below ~100 nm, the "lateral" heat flow path into the metal contacts ($R_L + R_C$) takes on a more important role, and can dominate for ~10 nm long FETs. The R_B with the substrate and R_C with the metal contacts also become increasingly important in sub-100 nm devices, almost regardless of the thermal properties of the device or substrate. Figure 3b also shows the percentage contribution of each thermal pathway as a function of device width, W (here for $L = 100$ nm). Dissipation from devices widths approaching ~1 nm is dominated by R_B with the substrate and R_C with the contacts, which has been experimentally confirmed for the case of CNTs [22, 23], as shown by the flattened R_{TH} (of CNTs) in Fig. 3c.

V. LOW-POWER MEMORY DEVICES

Power dissipation has also become a challenge for data storage, due to leakage in volatile memory and the large number of memory elements employed. In this context, phase change materials (PCMs) have attracted wide interest for memory and reprogrammable circuits with low voltage operation, fast switching, and high endurance [24]. PCMs are chalcogenides like $Ge_2Sb_2Te_5$ (GST) which have amorphous and crystalline phases with a large ratio (>100×) of electrical resistivity. The data in PCM memory are stored as changes in bit resistance, which can be reversibly switched with short voltage pulses and localized Joule heating. However, a drawback of PCMs has been their high programming current (>0.1 mA), as Joule heat must be delivered to a finite bit volume.

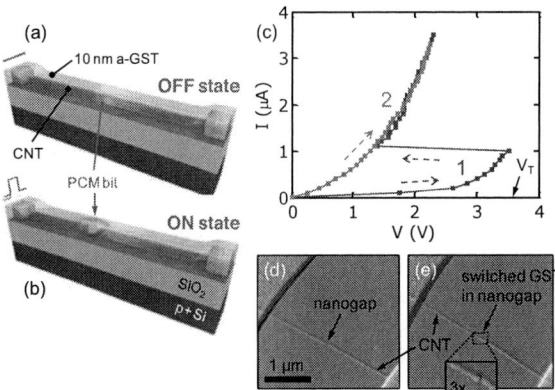

Fig. 4. (a) Schematic of low-power PCM memory device with carbon nanotube (CNT) electrodes [2]. The device is in its OFF state immediately after fabrication, with highly resistive amorphous $Ge_2Sb_2Te_5$ (GST) in the nanogap. (b) The device is switched to its ON state after an electric field in the nanogap changes the bit to its conductive crystalline phase. (c) Experimental current-voltage (*I-V*) of PCM device with CNT diameter ~3 nm, nanogap ~35 nm, and GST film thickness ~10 nm. Switching at ~1 µA, is ~100× lower power than previous state-of-the-art. (d) and (e) AFM images of the same device before and after switching [2].

To explore the fundamental scaling limits of PCM and design devices with ultra-low-power consumption, we have recently built devices with electrodes made from carbon nanotubes (CNTs) rather than metals [2, 25], as shown in Fig. 4. CNTs have higher electrical conductivity and can be more reliable than metals down to diameters of only ~2 nm; thus, they form electrodes contacting PCM bits of just tens of nm^3. We demonstrated that CNTs are compatible with GST deposition conditions, and that the contact resistance between the two materials is of good quality [25]. In the next step, we created nanogaps (~10 to 200 nm) in the middle of CNTs by "cutting" them with AFM tips or by electrical breakdown. Then we sputtered ~10-nm of amorphous GST to cover the device and fill the nanogaps (Fig. 4). This forms PCM devices with CNT electrodes and extremely small memory bits [2].

The CNT electrodes are very effective in addressing nanometer scale PCM bits, and thus the programming current and energy are scaled down significantly. To test initial memory switching, we source current and measure voltage across the devices, as shown in Fig. 4c. The amorphous bit displays threshold switching at a voltage V_T as is typical with GST, and a sharp transformation to a conductive phase under high E-field. Importantly, little voltage is dropped across the CNT electrodes, which are much more conductive (~50 kΩ) than the GST bit (~1-100 MΩ, depending on phase). Once threshold switching occurs, the bit crystallizes (SET transition) due to Joule heating. Testing hundreds of such devices [2] has uncovered trends

978-1-4673-0996-7/12 $31.00 © 2012 IEEE

suggesting that PCM memory could be scaled down to achieve sub-fJ per bit operation in sub-5-nm bits.

CONCLUSION

The fundamental limits of energy dissipation in nanoscale devices have not yet been reached. For instance, in PCMs these limits are set by phase transformations and heating at single nanometer dimensions. As both memory and logic devices are scaled down, novel materials are introduced, some with higher thermal conductivity than Si (e.g. CNTs), most with lower thermal conductivity. However, almost regardless of channel materials, for ultra-small devices the influence of thermal boundary resistance will be more important. Dissipation at contacts (particularly in quasi-ballistic channels) and contact thermoelectric effects will also play stronger roles, potentially allowing new mechanisms (e.g. highly localized, built-in Peltier cooling) to be leveraged for energy-efficient device operation.

ACKNOWLEDGEMENT

This work has been in part supported by the ONR, AFOSR, the Marco/MSD Focus Center (F.X.), and an NSF CAREER Award (E.P.). A.D.L acknowledges support from the NRI Coufal Fellowship.

REFERENCES

[1] E. Pop, "Energy dissipation and transport in nanoscale devices," *Nano Research*, vol. 3, pp. 147-169, 2010.

[2] F. Xiong, *et al.*, "Low-Power Switching of Phase-Change Materials with Carbon Nanotube Electrodes," *Science*, vol. 332, pp. 568-570, 2011.

[3] A. D. Liao, *et al.*, "Thermally Limited Current Carrying Ability of Graphene Nanoribbons," *Phys. Rev. Lett.*, vol. 106, p. 256801, 2011.

[4] K. L. Grosse, *et al.*, "Nanoscale Joule heating, Peltier cooling and current crowding at graphene-metal contacts," *Nat Nano*, vol. 6, pp. 287-290, 2011.

[5] R. Chau, *et al.*, "Integrated nanoelectronics for the future," *Nat Mater*, vol. 6, pp. 810-812, 2007.

[6] J. A. del Alamo, "Nanometre-scale electronics with III-V compound semiconductors," *Nature*, vol. 479, pp. 317-323, 2011.

[7] P. Avouris and R. Martel, "Progress in Carbon Nanotube Electronics and Photonics," *MRS Bulletin*, vol. 35, pp. 306-313, 2010.

[8] K. Saraswat, *et al.*, "High performance germanium MOSFETs," *Materials Science and Engineering: B*, vol. 135, pp. 242-249, 2006.

[9] N. Collaert, *et al.*, "Multi-gate devices for the 32 nm technology node and beyond," *Solid-State Electronics*, vol. 52, pp. 1291-1296, 2008.

[10] E. Pop, *et al.*, "Monte Carlo simulation of Joule heating in bulk and strained silicon," *Appl. Phys. Lett.*, vol. 86, p. 082101, 2005.

[11] V. Perebeinos and P. Avouris, "Inelastic scattering and current saturation in graphene," *Phys. Rev. B*, vol. 81, p. 195442, 2010.

[12] K. Kang, *et al.*, "Lifetimes of optical phonons in graphene and graphite by time-resolved incoherent anti-Stokes Raman scattering," *Phys. Rev. B*, vol. 81, p. 165405, 2010.

[13] Z.-Y. Ong, *et al.*, "Reduction of phonon lifetimes and thermal conductivity of a carbon nanotube on amorphous silica," *Phys. Rev. B*, vol. 84, p. 165418, 2011.

[14] Z.-Y. Ong and E. Pop, "Molecular dynamics simulation of thermal boundary conductance between carbon nanotubes and SiO_2," *Phys. Rev. B*, vol. 81, p. 155408, 2010.

[15] F. J. DiSalvo, "Thermoelectric Cooling and Power Generation," *Science*, vol. 285, pp. 703-706, 1999.

[16] Y. K. Koh, *et al.*, "Heat Conduction across Monolayer and Few-Layer Graphenes," *Nano Letters*, vol. 10, pp. 4363-4368, 2010.

[17] D. K. Schroder and J. A. Babcock, "Negative bias temperature instability: Road to cross in deep submicron silicon semiconductor manufacturing," *J. Appl. Phys.*, vol. 94, pp. 1-18, 2003.

[18] W. Liu, *et al.*, "Modeling and Data for Thermal Conductivity of Ultrathin Single-Crystal SOI Layers at High Temperature," *IEEE Trans. Electron Devices*, vol. 53, pp. 1868-1876, 2006.

[19] S. Sinha, *et al.*, "Non-Equilibrium Phonon Distributions in Sub-100 nm Silicon Transistors," *J. Heat Transfer*, vol. 128, pp. 638-647, 2006.

[20] M.-H. Bae, *et al.*, "Scaling of High-Field Transport and Localized Heating in Graphene Transistors," *ACS Nano*, vol. 5, pp. 7936-7944, 2011.

[21] W. Jang, *et al.*, "Thickness-Dependent Thermal Conductivity of Encased Graphene and Ultrathin Graphite," *Nano Letters*, vol. 10, pp. 3909-3913, 2010.

[22] A. Liao, *et al.*, "Thermal dissipation and variability in electrical breakdown of carbon nanotube devices," *Phys. Rev. B*, vol. 82, p. 205406, 2010.

[23] E. Pop, "The role of electrical and thermal contact resistance for Joule breakdown of single-wall carbon nanotubes," *Nanotechnology*, vol. 19, p. 295202, 2008.

[24] G. W. Burr, *et al.*, "Phase change memory technology," *Journal of Vacuum Science & Technology B*, vol. 28, pp. 223-262, 2010.

[25] F. Xiong, *et al.*, "Inducing chalcogenide phase change with ultra-narrow carbon nanotube heaters," *Appl. Phys. Lett.*, vol. 95, p. 243103, 2009.

Comparative Study of Tri-Gate- and Double-Gate-Type Poly-Si Fin-Channel Split-Gate Flash Memories

Y. X. Liu[1], T. Kamei[2], T. Matsukawa[1], K. Endo[1], S. O'uchi[1], J. Tsukada[1], H. Yamauchi[1], Y. Ishikawa[1],
T. Hayashida[2], K. Sakamoto[1], A. Ogura[2], and M. Masahara[1, 2]

[1]National Institute of Advanced Industrial Science and Technology (AIST), [2] Meiji University.

Tsukuba Central 2, 1-1-1 Umezono, Tsukuba-shi, Ibaraki 305-8568, Japan, Phone: +81-298-61-3417, E-mail: yx-liu@aist.go.jp

Abstract

The tri-gate (TG)- and double-gate (DG)-type poly-Si fin-channel split-gate flash memories with a thin n^+-poly-Si floating-gate (FG) have successfully been fabricated, and their electrical characteristics including the variations of threshold voltage (V_t) and S-slope have been comparatively investigated. It was experimentally found that better short-channel effect (SCE) immunity, smaller V_t variations, and a higher program speed are obtained in the TG-type flash memories than in the DG-type memories. Moreover, it was also confirmed that over-erase is effectively suppressed by split-gate structure.

Introduction

It is well recognized that three-dimensional (3D) channel devices, such as fin-type double-gate (DG) and tri-gate (TG) devices provide excellent short-channel effect (SCE) immunity owing to the strong controllability of channel potential by the multiple gates. Therefore, recently, fin-like poly-Si channels have actively been used in the fabrication of high-density and low-coast stacked NAND flash memories [1, 2]. However, most of the developed memory cell transistors have an oxide-nitride-oxide (ONO) charge-trapping layer, and the variations of V_t and S-slope in the cell transistors have not been investigated sufficiently. Very recently, we have reported poly-Si fin-channel stack-gate flash memory [3] and crystal-Si fin-channel split-gate flash memory with highly suppressed over-erase [4].

As a further study, in this work, we fabricate TG- and DG-type poly-Si fin-channel split-gate flash memories with a thin n^+-poly-Si FG, and investigate the gate structure dependence on the variability of V_t and S-slope before and after a program/erase (P/E) cycle.

Device Fabrication

Figure 1 shows the schematic 3D device structure of the poly-Si fin-channel split-gate flash memory and its cross-sectional view. It is clear that the FG length (L_{FG}) is smaller than the control-gate (CG) length (L_{CG}). Figure 2 shows the abbreviated device fabrication process flow. First, (100)-oriented bulk Si wafers were thermally oxidized and an 80-nm-thick non-doped poly-Si layer was deposited at 620 °C by LPCVD. After thermal oxidation, fin-channels were fabricated by electron-beam (EB) lithography and RIE. The SiO_2 fin hard-mask was etched by RIE for TG-type, but it was remained for DG-type devices. Then, an 8-nm-thick tunnel oxide (T_{ox}) was formed, followed by the deposition of a 30-nm-thick n^+-poly-Si layer as the FG material. After the FG formation, an 18-nm-thick TEOS-SiO_2 layer was deposited as the interpoly dielectric (IPD) layer. As the CG material, a 20-nm-thick PVD-TiN layer and a 100-nm-thick n^+-poly-Si layer were used. After the CG formation, ion implantation (I/I), RTA for dopant activation and metallization were performed.

Results and discussion

Figure 3 shows the SEM and cross-sectional STEM images of the fabricated TG- and DG-type poly-Si fin-channel split-gate flash memories. It is clear from Fig.3 (a) that a 148-nm FG is formed partially on a fin-channel, i.e., $L_{FG} < L_{CG}$. This fact is further confirmed by comparison of Fig. 3(c) and 3(d). Moreover, it is also clearly confirmed that no SiO_2 hard-mask exists on top of poly-Si fin-channel in the TG-type as shown in Fig. 3(c), but it still remains in the DG-type as shown in Fig. 3(b).

Figures 4(a) and 4(b) show the initial I_d-V_g characteristics of the fabricated TG- and DG-type devices with the same L_{CG} of 256 nm, respectively. The V_t values were evaluated at a constant drain current of $I_d = 1 \times 10^{-8}$ A. The measured statistical variations of the V_t and S-slope are shown in Fig. 5 and Fig. 6, respectively. Note that smaller V_t variations and better S-slope are obtained in the TG-type than in the DG-type. It is also clear from Fig. 7 that both of σV_t and $<S>$ are improved by using TG-structure due to the additional top gate and recessed bottom SiO_2 region as shown in Fig. 3(c), which strengthen the controllability of the channel potential and increase the coupling ratio of the FG to CG. The measured I_d-V_g characteristics of the fabricated TG- and DG-type devices after one P/E cycle are shown in Fig. 8. The statistical V_t variations after one P/E cycle are evaluated as shown in Fig. 9, and the σV_t and $<S>$ values are summarized in Fig. 10. Note that a significant improvement in σV_t and $<S>$ is obtained in the TG-type as compared with DG-type. The higher program speed and a larger memory widow are also observed in TG-type than DG-type as shown in Fig. 12. Moreover, it is noteworthy that almost the same I_d-V_g curve and almost a constant V_t value are obtained at erase time > 1 ms, as shown in Fig. 11(b) and Fig. 12, respectively. This indicates that split-gate is very effective to suppress over-erase.

Conclusion

The TG- and DG-type poly-Si fin-channel flash memories have successfully been fabricated and a quantitative comparison of their electrical characteristics has been made. It was experimentally confirmed that TG-type shows better SCE immunity, smaller σV_t values and a higher program speed than DG-type. It was also confirmed that over-erase is effectively suppressed by split-gate.

References

[1] H. Tanaka et al., Symp. VLSI Tech., p.14 (2007).

[2] H.-T. Lue et al., Symp. VLSI Tech., p. 131 (2010).

[3] Y. X. Liu et al., ESSDERC Tech Dig., p. 203 (2011).

[4] T. Kamei et al., IEEE EDL Vol. 33, No. 3, p. 345 (2012).

Acknowledgement

This work was supported in part by the Nanotechnology Project of NEDO Japan.

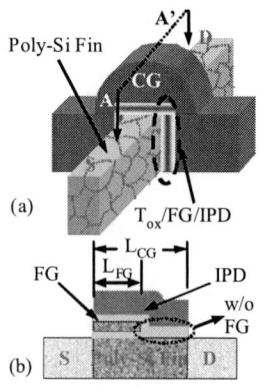

Fig. 1. (a) Schematic 3D diagram of poly-Si fin-channel split-gate flash memory, (b) cross-sectional view of the device along line A-A'.

- (100)-orientd bulk Si wafers
- Thermal oxidation (SiO_2 = 170-nm)
- Non-doped poly-Si = 80-nm by LPCVD
- Thermal oxidation (SiO_2 = 50 nm)
- Fin-channels by EB-lithography & RIE
- Fin hard-mask RIE for tri-gate (TG)
- Tunnel oxide formation, T_{ox} = 8.0-nm
- n^+-poly-Si = 30-nm deposition for FG
- IPD (TEOS-SiO_2 = 18-nm)
- 20-nm TiN/100-nm n^+-poly-Si for CG
- Extension I/I (D = 4x10^{14} cm^{-2}, θ = 60°)
- SD I/I (D = 1.5x10^{15} cm^{-2}, θ = 7°)
- Dopant activation RTA (T = 830 ^{o}C, 2s)
- Metallization & PMA (450 ^{o}C, 30 min)

Fig. 2. Device fabrication process flow for the poly-Si fin-channel split-gate flash memory. Double-gate (DG) and tri-gate (TG) type flash memories were fabricated.

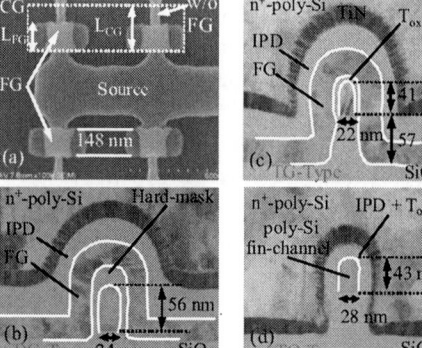

Fig. 3. (a) SEM image of the poly-Si fin-chanenl flash memory after FG formation, (b) cross-sectional STEM image of DG-type, (c) and (d) cross-sectional STEM images of TG-type at with and without FG regions.

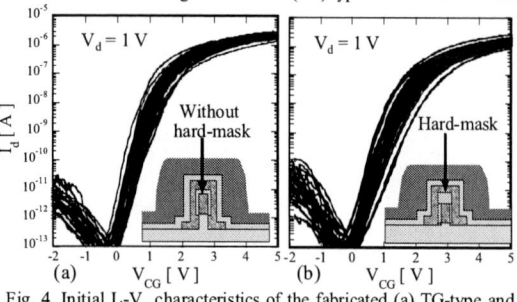

Fig. 4. Initial I_d-V_g characteristics of the fabricated (a) TG-type and (b) DG-type poly-Si fin-channel split-gate flash memory cell transistors with the same L_{CG} of 256 nm.

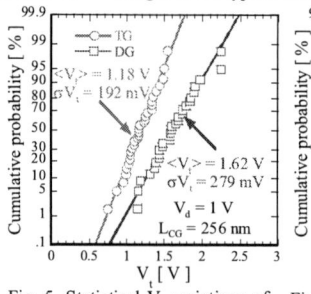

Fig. 5. Statistical V_t variations of the fabricated TG- and DG-type flash memory cell transistors.

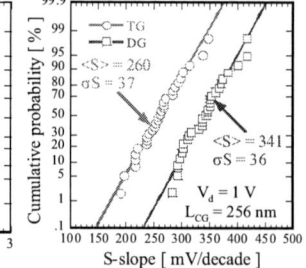

Fig. 6. Statistical S-slope variations of the fabricated TG-and DG-type flash memory cell transistors.

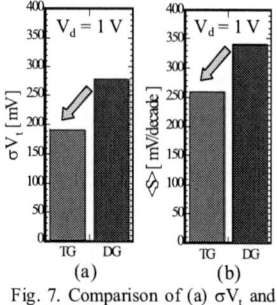

Fig. 7. Comparison of (a) σV_t and (b) <S> values of the TG- and DG-type devices with L_{CG} = 256 nm.

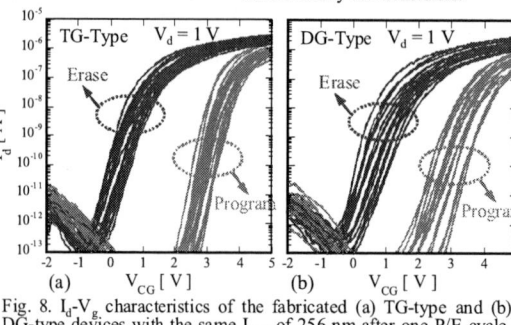

Fig. 8. I_d-V_g characteristics of the fabricated (a) TG-type and (b) DG-type devices with the same L_{CG} of 256 nm after one P/E cycle. P/E conditions: V_{CG} = +10/-10 V, V_d = V_s = 0V, t = 50 ms.

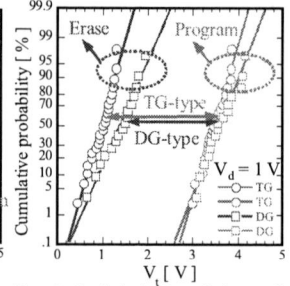

Fig. 9. Statistical V_t variations of the TG- and DG-type devices with L_{CG} = 256 nm after one P/E cycle.

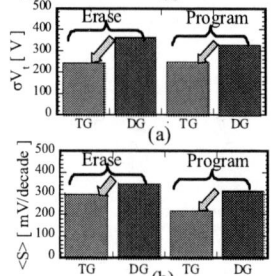

Fig. 10. Comparison of (a) σV_t and (b) <S> values of the TG- and DG-type devices after one P/E cycle.

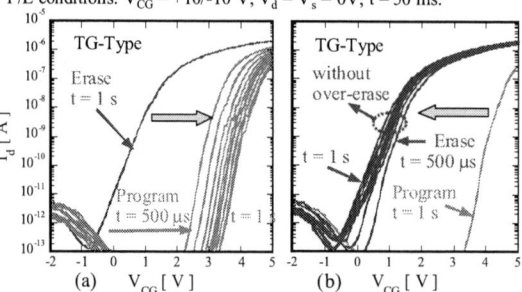

Fig. 11. I_d-V_g characteristics of the fabricated TG-type split-gate flash memory with L_{CG} = 256 nm at different (a) program and (b) erase pulse times. P/E bias conditions: V_{CG} = +10/-10 V, V_d = V_s = 0V.

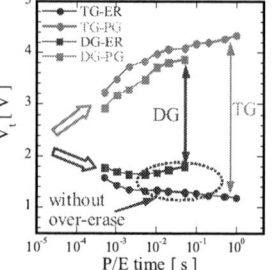

Fig. 12. P/E characteristics of the fabricated TG- and DG-type devices with L_{CG} = 256 nm.

Variation-Aware Study of BJT-based Capacitorless DRAM Cell Scaling Limit

Min Hee Cho, Wookhyun Kwon, Nuo Xu, and Tsu-Jae King Liu

Department of Electrical Engineering and Computer Sciences, University of California, Berkeley, CA 94720 USA

*Phone: +1-510-326-8858, Fax: +1-510-642-2739, E-mail: cmh12@eecs.berkeley.edu

Abstract

The scaling limit of the BJT-based capacitorless DRAM cell is investigated via 3-D process and device simulations, accounting for systematic and random sources of variation. The cell design and operating voltages are optimized at each gate length, following a constant electric field methodology. Retention time decreases with gate length, so that the scaling limit is expected to be 16.5 nm or 13 nm, depending on the application.

I. Introduction

The original capacitorless (single-transistor) dynamic random access memory (DRAM) cell design was proposed to achieve high-density storage with a standard silicon-on-insulator (SOI) CMOS process [1], and is scalable to approximately 25 nm channel length [2]. An improved BJT-based capacitorless DRAM cell design was recently proposed to provide for higher sensing margin and longer retention time [3]. In this work, the scaling limit of the BJT-based cell design is studied via 3-D process and device simulations [4], accounting for process-induced variations.

II. 25 nm Reference Cell Design and Scaling Constraints

The layout dimensions of the reference cell (with 25 nm gate length, L_g) were selected based on 22 nm SOI CMOS technology [5, 6]. Default cell operating voltages are summarized in Table I. Due to reliability considerations, the gate oxide (SiO_2) thickness (T_{ox}) and gate-sidewall spacer width (W_{spacer}) cannot be scaled down with L_g and hence are fixed at 3 nm and 21 nm, respectively [6]. Although a thinner body is beneficial for suppressing variations [7], retention time falls to zero if the body is too thin (< 8 nm) to adequately store charge (Figs. 1 and 2). The body thickness (T_{Si}) is therefore fixed at 9 nm in this study, to allow for ±1 nm variation.

III. Optimization of Cell Operating Voltages

Band-to-band tunneling (BTBT) limits retention time [6] so it is important to avoid increasing the peak electric field in the Hold state as L_g is scaled down. The cell operating voltages are adjusted together with L_g to maintain a constant peak electric field, and to maximize the retention time (Table II and Fig. 3). Fig. 4 shows that the nominal retention time decreases with scaling, falling below 10 ms at 9 nm L_g.

IV. Impact of Variations and L_g Scaling Limit

Systematic variations (normal distributions) in T_{Si}, buried oxide thickness (T_{BOX}), T_{ox}, and W_{spacer} are considered (Table III). Random dopant fluctuation (RDF) effects are investigated via Kinetic Monte Carlo (KMC) simulation (Fig. 5) and found to affect the local electric field and thereby the impact ionization rate and BTBT, and

hence the sensing current (Fig. 6 (a)) [11]. To gauge the influence of each variation source, the concept of Sigma Sensitivity (SS) [12] is used: SS is defined to be the deviation (from the nominal value) in Read current that results from a standard-deviation change in the parameter of interest, keeping all other parameters fixed, and is plotted in Figs. 6 (b) and (c) for Read 1 and Read 0 currents, respectively.

The signal sense margin (SSM) [13] is defined as

$$SSM = <\Delta I_{sensing}> - \alpha \times (\sigma_{Read\,0} + \sigma_{Read\,1}) \quad (1)$$

where $<\Delta I_{sensing}>$ is the median sensing margin (Read 1 current − Read 0 current), and α is set to 4.5 assuming 64 redundancies for a 16 Mbit array. The total standard deviation in Read 0 current [14] is calculated as:

$$\sigma_{Read\,0} \approx \sqrt{\begin{array}{c} \left(\sigma_{Read\,0,RDF}\right)^2 + \left(\sigma_{Read\,0,T_{OX}}\right)^2 + \left(\sigma_{Read\,0,T_{Si}}\right)^2 \\ + \left(\sigma_{Read\,0,T_{BOX}}\right)^2 + \left(\sigma_{Read\,0,W_{Spacer}}\right)^2 \end{array}} \quad (2)$$

Fig. 6 (d) shows how SSM and the nominal sensing current margin ($\Delta I_{sensing}$) each depend on the Hold time. Without accounting for variations, the retention time is overestimated to be 0.423 s (the Hold time at which $\Delta I_{sensing}$ falls below 60 µA/µm). Process-induced variations effectively reduce the retention time by ~68%, to 0.135 s (the Hold time at which SSM falls below 0).

The corresponding results in Fig. 7 for 12 nm L_g clearly show the increased impact of variations at shorter gate length, as SSM becomes negative by 1 µs Hold time. From the plot of variation-aware retention time (Hold time at SSM = 0) vs. L_g in Fig. 8, it can be seen that the minimum L_g is ~16.5 nm for stand-alone DRAM applications (64 ms retention time [15]). For embedded DRAM applications (1 ms retention time [16]), the minimum L_g is ~13 nm.

Acknowledgements

This work was supported in part by the Center for Circuit & System Solutions (C2S2) Focus Center, one of six research centers funded under the Focus Center Research Program, a Semiconductor Research Corporation program. M.H. Cho also appreciates the support of Samsung Electronics.

References [1] H.-J. Wann, *IEDM* 1993 [2] N. Butt, *TED*, 2007. [3] S. Okhonin, *IEDM*, 2007. [4] Sentaurus manual, 2010. [5] C. Shin, *SOI Conf*, 2009. [6] M.H. Cho, *SSDM*, 2011. [7] T. Ohtou, *EDL*, 2007. [8] ITRS 2010. [9] O. Faynot, *IEDM*, 2010. [10] W. Schwarzenbach, *Electrochem. Soc.* 2011. [11] H. Furuhashi *SOI Conf*, 2008. [12] M. H. Cho, *EDL*, 2010 [13] F. Matsuoka, *IEDM*, 2007. [14] C. Shin, Ph.D thesis, UC Berkeley, 2011. [15] JEDEC spec, 2012. [16] C. S. Wang, TSMC article (available online).

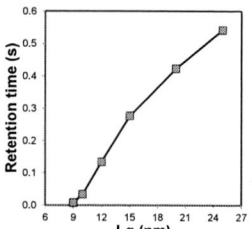

Fig. 1. Nominal retention behavior, for cells with different body thicknesses.

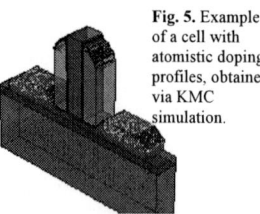

Fig. 4. Nominal retention time *vs.* gate length.

Fig. 5. Example of a cell with atomistic doping profiles, obtained via KMC simulation.

Table I. Cell operating voltages (in Volts) for the reference cell design (L_g = 25 nm)

	Write 1	Write 0	Hold	Read
Front Gate Voltage, Vgs	-1	0	-1.6	-1
Back Gate Voltage, Vbg	2.5	2.5	2.5	2.5
Drain Voltage, Vds	1.7	-0.5	0	1.4
Source Voltage, Vs	0	0	0	0

Fig. 2. Device cross-sections showing hole density contour plots during a Hold operation, for two different body thicknesses. $L_g = 25$ nm.

Table II. Optimized values of back-gate bias (Vbg), gate voltage during Hold operation (Vghold), and drain voltage during Read operation (Vdr) [6].

Lg (nm)	Vbg (V)	Vghold (V)	Vdr (V)
9	2.3	-1.7	1
10	2.4	-1.7	1
12	2.5	-1.7	1
15	2.5	-1.6	1.1
20	2.5	-1.6	1.3
25	2.5	-1.6	1.4

Table III. Parameter variations considered.

	Body thickness (T_Si)	BOX thickness (T_BOX)	Gate oxide thickness (T_OX)	Spacer width (W_spacer)
Nominal Value	9 nm	10 nm	3 nm	21 nm
Standard Dev.	1σ = 1.6 Å	1σ = 1.6 Å	1σ = 0.4 Å	1σ = 3.5 Å
References	[8, 9]	[8, 10]	[8]	3σ = 5%
LL (-4.5σ)	8.28 nm	9.28 nm	2.82 nm	19.43 nm
UL (+4.5σ)	9.72 nm	10.72 nm	3.18 nm	22.58 nm

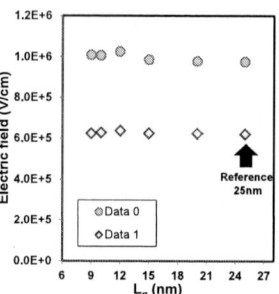

Fig. 3. Peak electric field within the body region during a Hold operation, for "1" and "0" data states. The cell operating voltages are adjusted to maintain the same peak field (to within 5%) for each gate length.

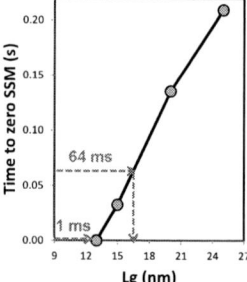

Fig. 8. Time to zero SSM *vs.* Lg. Scaling limit is 13 nm for embedded DRAM applications (1 ms retention time) or 16.5 nm for stand-alone DRAM applications (64 ms retention time).

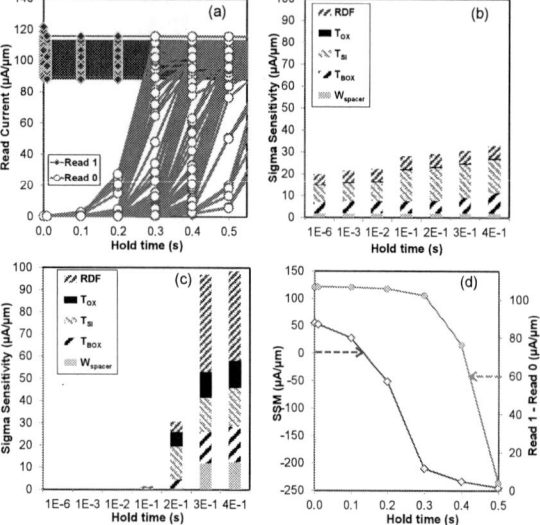

Fig. 6. Simulation results for a cell with 20 nm L_g: (a) Impact of RDF (100 cases) (b) Sigma sensitivity *vs.* Hold time for Read 1 and (c) for Read 0. (d) Signal sense margin (SSM) and nominal sensing current margin as a function of Hold time.

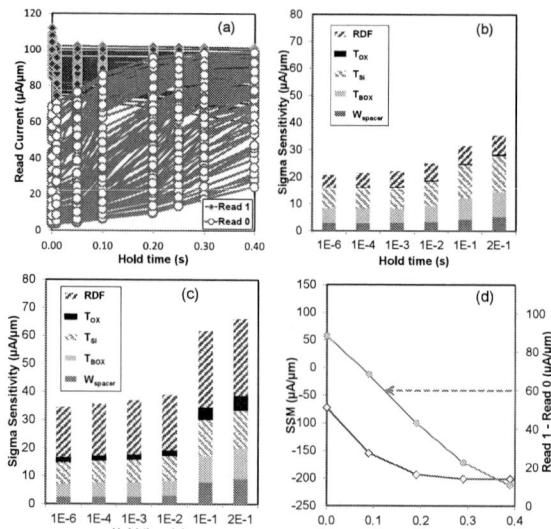

Fig. 7. Simulation results for a cell with 12 nm L_g: (a) Impact of RDF (100 cases) (b) Sigma sensitivity *vs.* Hold time for Read 1 and (c) for Read 0. (d) Signal sense margin (SSM) and nominal sensing current margin as a function of Hold time.

978-1-4673-0996-7/12 $31.00 © 2012 IEEE

Investigation into the Effect of the Variation of Gate Dimensions on Program Characteristics in 3D NAND Flash Array

Joo Yun Seo, Yoon Kim, Se Hwan Park, Wandong Kim, Do-Bin Kim, Jong-Ho Lee, Hyungcheol Shin, and Byung-Gook Park*

Inter-university Semiconductor Research Center (ISRC) and
School of Electrical Engineering and Computer Science, Seoul National University,
San 56-1, Sillim-dong, Gwanak-gu, Seoul 151-742, Republic of Korea.
Tel.: +82-2-880-7279, Fax: +82-2-882-4658, E-mail address: jooyun@snu.ac.kr

Abstract — **In 3D stacked NAND flash memory, the number of stacked layers tends to increase for high density storage capacity. With the increase of the height of devices, it is important to achieve a good vertical etch profile by which word line (WL) gate dimensions are affected. In this paper, we investigate the effect of the variation of gate dimensions on the program characteristics in 3D NAND flash memory array by using TCAD simulation. Also, we compare the cell characteristics of NAND flash with different structures, gate-all-around (GAA) and double gate (DG). (keywords: 3D stacked NAND flash)**

I. INTRODUCTION

As the quest for higher storage capability continues, NAND flash memory has become the most attractive storage device. However, the limitation of photo-lithography technology hinders further scaling of NAND flash memory. Consequently, many groups have proposed three-dimensional (3-D) stackable memory devices as one of the most promising solutions [1]-[3]. Fig. 1 presents one of 3D stacked NAND flash memory structures. Moreover, there is a number of fabrication methods designed to expand storage capacity. In any case, to get higher density of storage capacity, it is inevitable to increase the number of stacked layers. In the highly stacked 3D array, a slight deviance from a right angle of the etch slope results in drastic change of the dimensions among the gates. This paper reports how the dimensions of the gates affect the program characteristics in 3D NAND flash array architectures.

II. RESULTS AND DISCUSSION

We adopted the gate-all-around (GAA) structure and the double gate (DG) structure to perform the simulation. Three flash cells are connected in series, and the diameter of the nanowire and body thickness of the double gate structure is 20 nm. Flash cells have charge trap layers and the thickness of tunnel oxide/nitride/blocking oxide layer is 3/6/6 nm, respectively. Between the cell gates, virtual source and drain region is formed by the fringing field of the gates

[4]. With the same pitch size (100 nm), the gate length (L_g) is varied from 40 nm to 70 nm. Also, the length between adjacent channels in the same layer is defined as L_{gap}, which is from 30 nm to 50 nm. As described in Fig. 2, the gates in the upper layer have shorter gate length, longer word line gap, and longer L_{gap}.

In the case of the GAA structure, the variation of L_g does not affect the I-V characteristic at the initial state. Even though there is a slight difference in ΔV_T, the level of bit-line current remains constant (Fig. 3). In addition, the program characteristic of the GAA structure is rarely changed by the variation of L_{gap} whether the gate length is 40nm or 70nm (Fig. 4).

However, the cell characteristic of the DG structure is vulnerable to the change in the gate dimension. When L_g is under 50nm, program characteristic cannot be obtained. As shown in Fig. 5, there is a notable reduction in the level of bit line current as L_g decreases. It is because the fringing field is not sufficient enough to induce the inversion layer between gates as the distance between gates gets longer, which leads to the low electron density in the virtual source and drain region. This means that DG NAND flash memory cells located in the upper layer do not guarantee stable program characteristics. In other words, DG structure in 3D NAND flash memory is unreliable due to its unstable cell characteristics with the variation of gate dimensions. To guarantee the stable cell characteristics of DG NAND flash memory, the gap between WL gates should be smaller than 50 nm. Fig. 6 indicates the maximum height (H_{max}) of total stacked layers that DG NAND can have as a function of the etch slope. H_{max} is the height when the WL gap is 50nm at the top of gates with the assumption that the WL gap is 10nm at the bottom of gates. The results indicate that the etch slope limits the total height of stacked layers when DG NAND is adopted for 3D array architecture.

II. CONCLUSION

This paper reports the effect of the gate dimension on a 3D NAND flash memory array by using TCAD simulation. NAND flash memory cells featuring a

GAA structure are less sensitive to the variation of the gate dimensions than cells featuring a DG structure. Therefore, GAA structure is suitable to be adopted in 3D stacked NAND flash memory for ultra-high-density storage devices.

ACKNOWLEDGMENT

This work was supported by the IT R&D program of MKE/KEIT (10035320, Development of novel 3D stacked devices and core materials for the next generation flash memory).

REFERENCES

[1] J. Choi et al. "3D approaches for non-volatile memory," VLSI Tech. Dig., pp. 178-179, 2011.

[2] H. T. Lue et al. "A highly scalable 8-layer 3D vertical-gate(VG) TFT NAND flash using junction-free buried channel BE-SONOS device," VLSI Tech. Dig., pp. 131-132, 2010.

[3] J. Y. Seo , Y. Kim, S.-H. Park,W. Kim, and B.-G. Park, "Compact Bit-line stacked array," IEEK summer conference, pp. 358-359, 2011.

[4] W. Kim et al. "Arch NAND flash memory array with improved virtual source/drain performance", IEEE Electron Device Lett, vol. 31, 12, pp. 1374–1376, 2010.

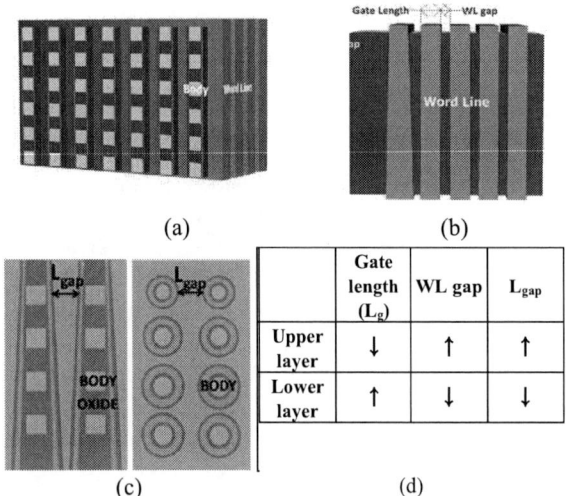

Fig. 1. Bird's eye view of 3D stacked NAND Flash memory

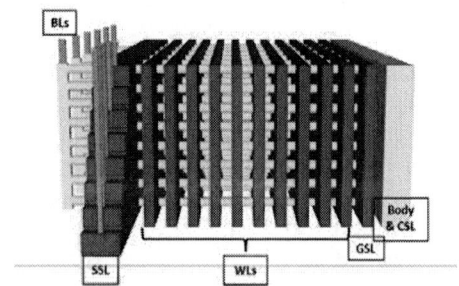

	Gate length (L_g)	WL gap	L_{gap}
Upper layer	↓	↑	↑
Lower layer	↑	↓	↓

(d)

Fig. 2. (a) Double Gate stacked NAND array (b) Gate length variation with etch slope (c) Definition of L_{gap} in double gate and GAA structure (d) Tendency of L_g, WL gap, and L_{gap} in different layer

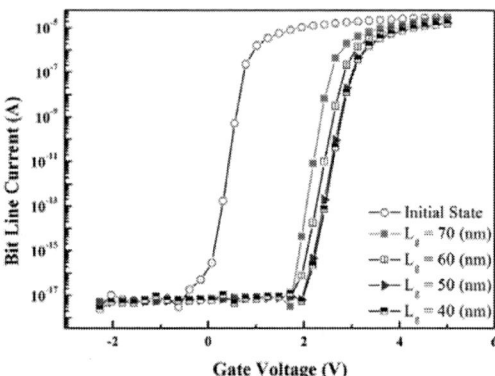

Fig. 3. Program characteristics of GAA NAND flash array with L_g variation

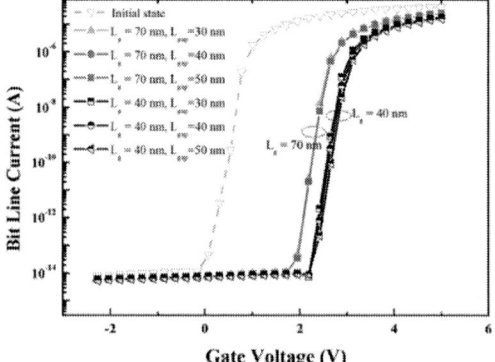

Fig. 4. Program characteristics of GAA NAND flash array with L_{gap} variation

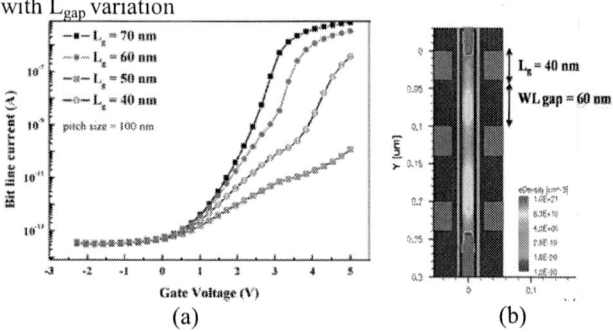

Fig. 5. (a) I-V characteristics of fresh cells with double gate structure (L_g is from 40 nm to 70 nm) (b) Electron density when gate voltage is 5 V (L_g = 40 nm).

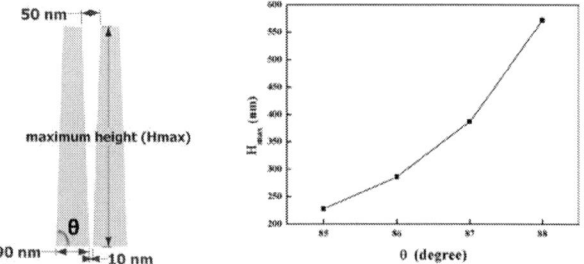

Fig. 6. The maximum height of total stacked layers as a function of etch slope

978-1-4673-0996-7/12 $31.00 © 2012 IEEE

A novel Gate-All-Around Ultra-Thin p-channel Poly-Si TFT Functioning as Transistor and Flash Memory with Silicon Nanocrystals

Hung-Bin Chen[1*], Shih-Han Lin[2], Jia-Jiun Wu[1], Yung-Chun Wu[2], and Chun-Yen Chang[1]

[1]Institute of Electronics, National Chiao Tung University, Taiwan, Republic of China

[2]Department of Engineering and System Science, National Tsing Hua University, Hsinchu, Taiwan, Republic of China

*Corresponding author. Tel: 886-3-571-2121 ext: 52981; Email: chenlays@hotmail.com

Abstract

A novel gate-all-around ultra-thin p-channel poly-Si TFT functioning as transistor and flash memory with silicon nanocrystals have been successfully demonstrated. The process is simple and mask free. For the 3-nm-thick channel devices, the S.S. of 88 mV/dec and I_{on}/I_{off} ratio of more than 10^8 can be achieved. Extreme low applied voltage for band-to-band-tunneling-induced hot electron injection tunneling (BBHE) operation and excellent retention are proposed.

Introduction

Recently, poly-Si TFTs have received a considerable attention in vertical stacking to increase device density [1]. Flash memory, a nonvolatile memory (NVM), is extensively adopted in portable products owing to its nonvolatile feature and low cost [2]. Ultra-thin-body (UTB) channel TFT is promising candidates for future CMOS devices and applications due to the immunity to the short-channel effects and I_{OFF} reduction [3]. The gate-all-around (GAA) structure, which the gate can control four edges of the active region, performs the best gate control ability [4] to enhance the performance of the TFTs and NVM. Moreover, the GAA structure enhances the electric field in tunneling oxide and reduces the electric field in blocking oxide, so the program and erase (P/E) speed are enhanced [5]; however, high electric field region also causes the reliability issue. Nanocrystals (NCs) as the charge storage layer have been investigated for improving reliability. Si-NCs promote high P/E efficiency owing to their narrow bandgap, deep conduction band, and absence of metal contamination [6][7].

Device Fabrication

The key process of the UTB TFTs is shown in Fig.1. A 50 nm undoped amorphous silicon (a-Si) layer was deposited on a 400nm thermal oxide layer. Next, a-Si layer was solid-phase recrystallized (SPC) at 600 ^{0}C for 24 hr in N_2 ambient. The active layers were defined as nanowires (NWs) by e-beam lithography [Fig. 1a and 1b] and mesa-etched by time-controlled wet etching of the buried oxide to release the poly-Si bodies [Fig. 1c]. After suspended NWs were formed, 22-nm-thick layers of thermal oxide were deposited as gate oxide and the self-aligned raised S/D are formed in the meanwhile [Fig. 1d]. A hybrid Si_3N_4(3nm)/ Si-NC(2nm)/ Si_3N_4(3nm) trap layer is deposited by LPCVD, followed by a furnace annealing at 1050 ^{0}C for 30 min. Then, 20-nm-thick *tetra-ethyl-ortho-silicate* (TEOS) oxide was deposited as blocking oxide. Subsequently, 200-nm-thick poly-Si layer was deposited and patterned as gate electrode. After S/D formation, it was activated by rapid thermal annealing at 1050^{0}C for 1 sec. Finally, passivation and metallization were performed. The 3-D schematic structure of the GAA-UTB RSDNW TFTs is shown in the Fig. 1e. The SEM picture of active pattern is shown in fig. 1f and 1g. The conventional top gate planar-TFTs were fabricated to serve as control samples. Fig. 2 shows cross-sectional TEM image of the GAA-UTB NW TFTs and a 5-nm-thick channel with single crystalline is obtained [Fig. 2c]. Fig. 3 shows cross-sectional TEM image of the single channel (SC) TFTs. The Ω-gate covering the 3-nm-thick channel is observed at the edge of the SC. Figure 3d presents plane-view TEM image of the Si-NCs, embedded in the Si_3N_4 layer.

Result and Discussion

Id-Vg and Id-Vd characteristics of GAA-UTB RSDNW-TFTs, Ω-gate UTB SC TFTs and planar-TFTs are shown in Fig. 4. The GAA-UTB TFTs (W/L= 0.1μm×10/0.5μm) and Ω-gate UTB TFTs (W/L= 1μm/0.5μm) shows the steeper subthreshold swing (SS), higher I_{on}/I_{off} ratio and smaller DIBL effect than planar TFTs (W/L= 1μm/1μm) due to the thin body of TFT structure and excellent controllability of the gate over the channel which suppresses short channel effect. The Ω-gate UTB TFTs shows the better on-resistance than GAA-UTB ones due to the thicker channel. A single crystal-like cross-sectional image is obtained in the ultra-thin channel [Fig. 2c and 3b]. Thus, the total amount of defects in the channel can be reduced to improve device performance. Fig. 5 exhibits the temperature dependence on Id-Vg characteristics. The planar TFT is sensitive to temperature. It could be attributed two reasons. The one is the stronger phonon scattering to carries transport far away center of channel [8]. The other is energy quantization of electrons in the ultra-thin channel strongly suppresses the temperature dependence of threshold voltage [9]. The GAA-UTB TFTs shows the less increase of the drain current. Fig. 6 shows the Id-Vg curve of the GAA-UTB NVM and Ω-gate UTB NVM in fresh and programmed states. The programming of GAA NVM by a 12 V pulse in 10^{-3} s at gate and -12 V pulse in 10^{-3} s at drain (BBHE) achieves a 1.9 V of a threshold voltage shift (ΔVth). The programming of Ω-gate NVM by a 5 V pulse in 1 s at gate and -5 V pulse in 1 s at drain achieves a 2 V of ΔVth. The Ω-gate NVM shows the extreme low voltage operation on thick gate oxide. It could be attributed the thinner channel would help the BBHE operation. The electrons from the valance band tunneling near center of the channel easily accelerate for gaining more energy and are injected into the trapping layer. Fig. 7 and Fig. 8 show the program/erase speed. The BBHE operation is an efficient operation than FN one [10]. Fig. 9 shows the retention characteristics at 85°C. Thanks to the good charge confinement property of NCs, the stored charge does not loss too much.

Conclusion

A novel GAA TFTs and NVM with ultra-thin p-channel and self-aligned raised S/D structure have been successfully demonstrated. The UTB devices have a steeper S.S., higher I_{on}/I_{off} ratio, smaller DIBL and larger driving current than planar TFTs. The GAA and Ω-gate UTB structure with Si NCs NVM performs great P/E speed and good reliability. The process of the ultra-thin channel is simple with mask free and highly compatible with the current flash process, which is highly promised for the future 3D stacked high-density applications.

Reference

[1] S. J. Choi *et al.*, VLSI Tech. Dig., p.111, 2010. [2] S. M. Jung, et al., IEDM Tech. Dig., p. 503, 2006. [3] B. Kim *et al.*, IRPS, p. 126, 2011. [4] N. Singh, et al., ED, 55, p. 3107, 2008. [5] K. H. Yeo, et al., VLSI Tech., p.138, 2008. [6] T. Y. Chiang, et al., EDL., 29, p. 1148, 2008. [7] H. B. Chen *et al.*, EDL, 33, p. 537, 2012. [8] C. W. Lee *et al.*, ED, 57, p. 620, 2010. [9] Y. Omura *et al.*, SSE, 48, p.1661, 2004 [10] M. T. Wu et al., ED, 54, p. 699, 2007.

978-1-4673-0996-7/12 $31.00 © 2012 IEEE

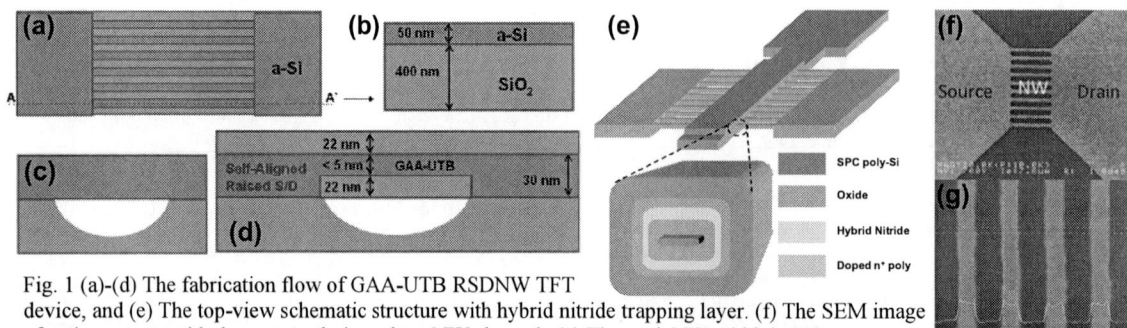

Fig. 1 (a)-(d) The fabrication flow of GAA-UTB RSDNW TFT device, and (e) The top-view schematic structure with hybrid nitride trapping layer. (f) The SEM image of active pattern with the source, drain and ten NW channels (g) The each NW width is 119 nm.

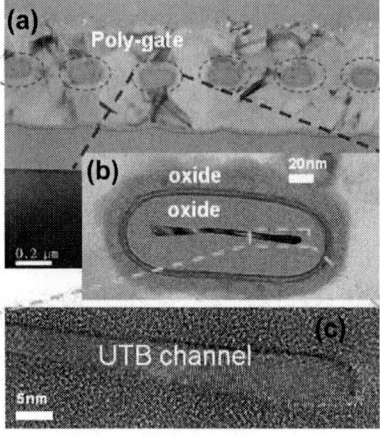

Fig. 2 (a)-(c) A single crystal-like ultra-thin channel is obtained.

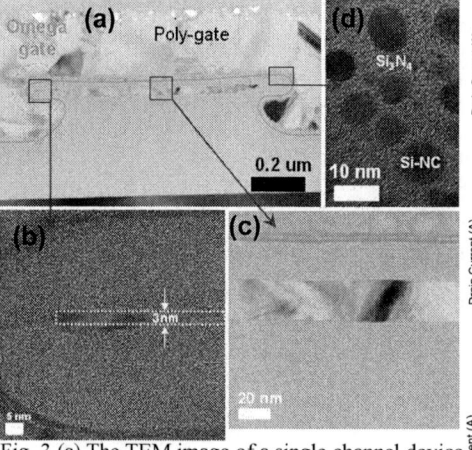

Fig. 3 (a) The TEM image of a single channel device (b)-(c) The UTB is 3nm(corner) and 25nm(middle). (d) Top-view of Si-NCs on the control wafer.

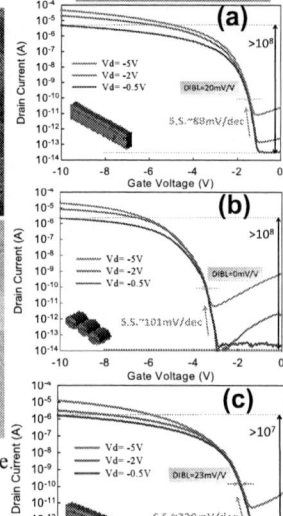

Fig. 4 (a)-(c) Id-Vg and (d) Id-Vd of Omega gate SC, GAA NW and planar TFTs.

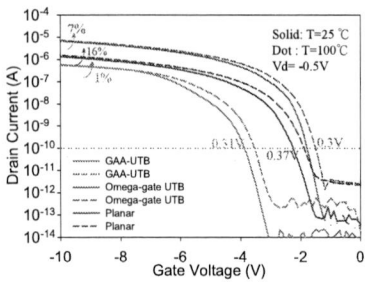

Fig. 5 Temperature dependence on Id-Vg of Omega-gate SC, GAA NW and planar TFTs.

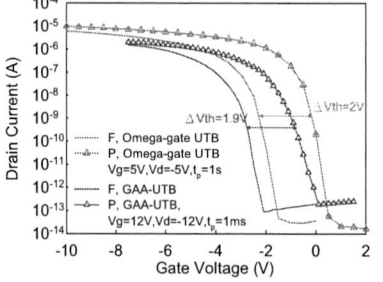

Fig. 6 Id-Vg of fresh and programmed states of Omega-gate SC and GAA NW devices

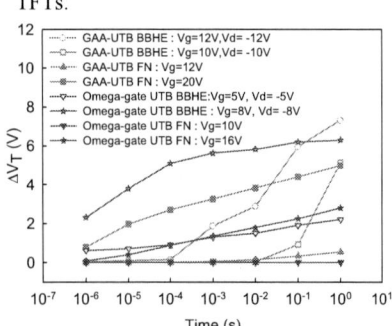

Fig. 7 Program characteristics of GAA NW and Omega-gate SC under various BBHE and FN bias conditions.

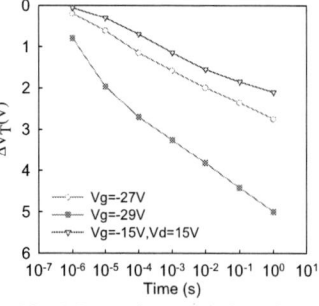

Fig. 8 Erase characteristics of GAA NW and Omega-gate SC.

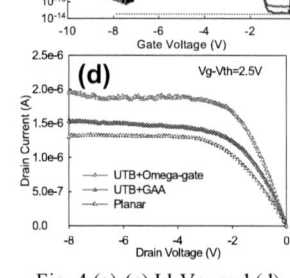

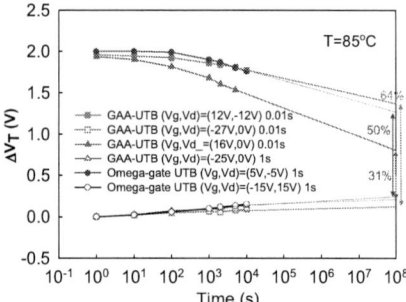

Fig. 9 Retention characteristics at 85 °C of GAA NW and Omega-gate SC.

978-1-4673-0996-7/12 $31.00 © 2012 IEEE

Graphene for More Moore and More Than Moore Applications

M.C. Lemme, S. Vaziri, A.D. Smith, J. Li, S. Rodriguez, A. Rusu, M. Ostling

KTH Royal Institute of Technology, Electrum 229, 16440 Kista, Sweden

phone: +46-8-790-4351, email: lemme@kth.se

Graphene has caught the attention of the electronic device community as a potential future option for More Moore and More Than Moore devices and applications. This is owed to its remarkable material properties, which include ballistic conductance over several hundred nanometers or charge carrier mobilities of several 100.000 cm^2/Vs in pristine graphene. Furthermore, standard CMOS technology may be applied to graphene in order to make devices. Integrated graphene devices, however, are performance limited by scattering due to defects in the graphene and its dielectric environment [1, 2] and high contact resistance [3, 4]. In addition, graphene has no energy band gap (**Figure 1**) and hence graphene MOSFETs (GFETs) cannot be switched off, but instead show ambipolar behaviour [5]. This has steered interest away from logic to analog radio frequency (RF) applications [6, 7]. This talk will systematically compare the expected RF performance of realistic GFETs with current silicon CMOS technology [8]. GFETs slightly lag behind in maximum cut-off frequency $F_{T,max}$ (**Figure 2**) up to a carrier mobility of 3000 cm^2/Vs, where they can achieve similar RF performance as 65nm silicon FETs. While a strongly nonlinear voltage-dependent gate capacitance inherently limits performance, other parasitics such as contact resistance are expected to be optimized as GFET process technology improves.

Some advantages offered by graphene may be exploited in novel device designs. One example, a graphene-based hot electron transistor will be discussed in the talk. This "graphene-base" transistor (GBT) [9], can potentially provide very low off currents and current saturation in the output characteristics, two of the main issues associated with conventional GFETs. The GBT is based on a vertical arrangement of emitter, base, and collector contacts (**Figure 3**). In the off-state, charge carriers face a dielectric barrier. In the on-state, the emitter-base diode injects hot electrons across the base into the conducting band of the insulator separating the base from the collector (BCI). The GBT allows minimizing I_{off} by proper BCI design and shows current saturation when the output voltage exceeds the value necessary to remove the tunneling barrier of the BCI. The GBT is expected to provide THz operation and may also have potential for logic integration.

The unique band structure of graphene makes it an interesting optoelectronic material. The linear dispersion relation from 0 up to ±1eV translates to broadband photodetection from UV to THz frequencies, while the high mobility enables fast detectors [10]. The talk will include results from gated graphene devices that allow tuning the graphene channel from bipolar to unipolar and thus a gate-activated photoresponse in the visible lght spectrum [11]. This can be explained with a model of the photothermal effect: elevated temperatures at the p-n-junction induce thermoelectric currents. The possible extension into far-IR / terahertz radiation, the high conductivity of graphene could lead to high-speed broadband bolometers with submicrometer pixilation.

Mechanical properties of graphene include an exceptional Young's modulus of ~1 TPa [12] and elastic stretchability of up to 20%. As the band structure is changed when graphene is strained, graphene could be a potential material for electro-mechanical transducers. We propose a device structure suitable to detect strain in suspended graphene films via electrical measurements [13]. A schematic of a possible layout is shown in **Figure 4**. Electrical data obtained from such structures will be discussed in the talk.

Finally, graphene research includes large-scale and high-throughput solution based processing for low-cost electronics. An efficient method is the direct exfoliation of graphite into single-layer or few-layer graphene in certain solvents. We have proposes a simple yet general route to prepare high-concentration surfactant-free graphene liquid dispersions [14]. The method is based on a distillation-assisted solvent exchange technique that leads to graphene dispersions with few-layer graphene flakes without severe defects. The dispersions reach graphene concentrations as high as 0.39 mg/mL and are stable for at least 10 h without any surfactant/polymer stabilization. The technique can be used to fabricate transparent conductive thin films (**Figure 5**).

Acknowledgement
The authors would like to thank A. Delin (KTH), F. Niklaus (KTH), A. Fischer (KTH), E. Alarcon (UPC Barcelona), W. Mehr (IHP Germany), G. Lupina (IHP), G. Lippert (IHP) and J. Dabrowski (IHP) for fruitful collaboration and discussions.

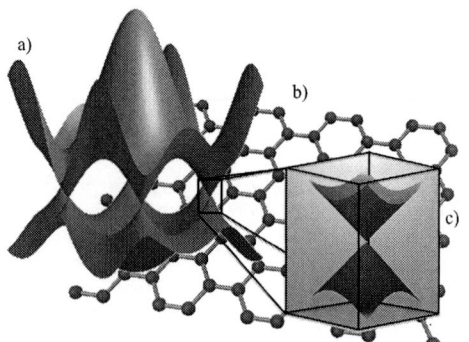

Figure 1: Band structure of graphene emphasizing the linear dispersion relation at the k-points and the absence of an energy band gap.

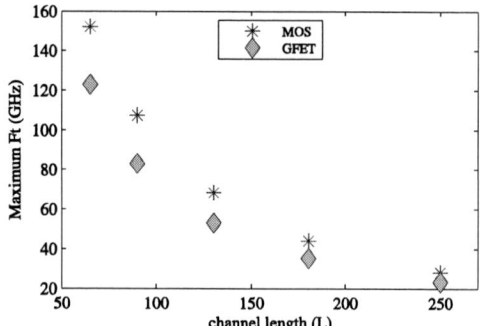

Figure 2: Simulated $F_{T,max}$ for silicon and graphene MOSFETs with a mobility of $\mu = 2.500$ cm^2/Vs.

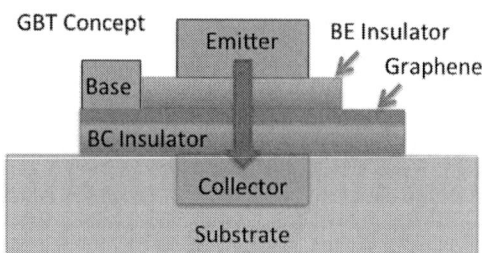

Figure 3: Schematic of the Graphene-Base Transistor (GBT) device concept. In the On-State, hot carriers are injected from the emitter into the conduction band of the base-collector dielectric.

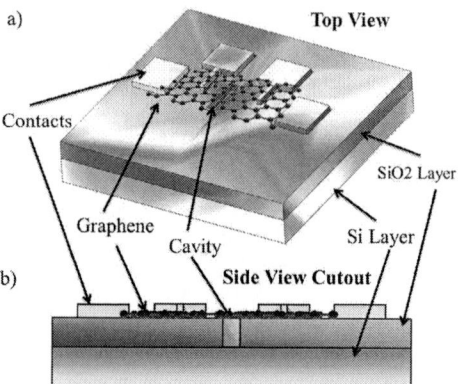

Figure 4: Schematic of a possible layout for graphene based electro-mechanical transducers.

Figure 5: Photograph of two conductive graphene thin films on glass slides on top of a paper with the KTH logo demonstrating the transparency of the graphene films.

References

[1] M. C. Lemme, T. J. Echtermeyer, M. Baus, B. N. Szafranek, J. Bolten, M. Schmidt, T. Wahlbrink, and H. Kurz, "Mobility in graphene double gate field effect transistors," *Solid-State Electronics,* vol. 52, pp. 514-518, Apr 2008.

[2] S. Adam, E. H. Hwang, V. M. Galitski, and S. Das Sarma, "A self-consistent theory for graphene transport," *Proceedings of the National Academy of Sciences,* vol. 104, pp. 18392-18397, November 20, 2007 2007.

[3] S. Vaziri, M. Ostling, and M. C. Lemme, "A Hysteresis-Free High-k Dielectric and Contact Resistance Considerations for Graphene Field Effect Transistors," *ECS Transactions,* vol. 41, pp. 165-171, 2011.

[4] F. Xia, V. Perebeinos, Y.-m. Lin, Y. Wu, and P. Avouris, "The origins and limits of metal-graphene junction resistance," *Nat Nano,* vol. 6, pp. 179-184, 2011.

[5] M. C. Lemme, T. J. Echtermeyer, M. Baus, and H. Kurz, "A Graphene Field-Effect Device," *Electron Device Letters, IEEE,* vol. 28, pp. 282-284, 2007.

[6] I. Meric, N. Baklitskaya, P. Kim, and K. L. Shepard, "RF performance of top-gated, zero-bandgap graphene field-effect transistors," in *Electron Devices Meeting, 2008. IEDM 2008. IEEE International*, 2008, pp. 1-4.

[7] Y. Q. Wu, Y.-M. Lin, K. A. Jenkins, J. A. Ott, C. Dimitrakopoulos, D. B. Farmer, F. Xia, A. Grill, D. A. Antoniadis, and P. Avouris, "RF performance of short channel graphene field-effect transistor," in *Electron Devices Meeting (IEDM), 2010 IEEE International* San Francisco, 2010, pp. 9.6.1 - 9.6.3.

[8] S. V. S. Rodriguez, M. Ostling, A. Rusu, E. Alarcon, M.C. Lemme, "RF Performance Projections of Graphene FETs vs. Silicon MOSFETs," *arXiv:1110.0978v1,* 2011.

[9] W. Mehr, J. C. Scheytt, J. Dabrowski, G. Lippert, Y.-H. Xie, M. C. Lemme, M. Ostling, and G. Lupina, "Vertical Transistor with a Graphene Base," *IEEE Electron Device Letters,* vol. 33, 2012.

[10] T. Mueller, F. Xia, and P. Avouris, "Graphene photodetectors for high-speed optical communications," *Nat Photon,* vol. 4, pp. 297-301, 2010.

[11] M. C. Lemme, F. H. L. Koppens, A. L. Falk, M. S. Rudner, H. Park, L. S. Levitov, and C. M. Marcus, "Gate-Activated Photoresponse in a Graphene p n Junction," *Nano Letters,* vol. 11, pp. 4134–4137, 2011.

[12] C. Lee, X. Wei, J. W. Kysar, and J. Hone, "Measurement of the Elastic Properties and Intrinsic Strength of Monolayer Graphene," *Science,* vol. 321, pp. 385-388, July 18, 2008 2008.

[13] A. D. Smith, S. Vaziri, A. Delin, M. Östling, and M. C. Lemme, "Strain Engineering in Suspended Graphene Devices for Pressure Sensor Applications," in *13th International Conference on Ultimate Integration on Silicon (ULIS)* Grenoble, France, 2012.

[14] J. Li, F. Ye, S. Vaziri, M. Muhammed, M. C. Lemme, and M. Östling, "A Simple Route towards High-Concentration Surfactant-Free Graphene Dispersions," *Carbon,* 2012.

High Performance Ω-Gate Ge FinFET Featuring Low Temperature Si₂H₆ Passivation and Implantless Schottky-Barrier NiGe Metallic Source/Drain

Bin Liu,[1] Xiao Gong,[1] Genquan Han,[1] Phyllis Shi Ya Lim,[1] Yi Tong,[1] Qian Zhou,[1] Yue Yang,[1] Nicolas Daval,[2] Matthieu Pulido,[2] Daniel Delprat,[2] Bich-Yen Nguyen,[2] and Yee-Chia Yeo.[1],*

[1] Department of Electrical and Computer Engineering, National University of Singapore (NUS), Singapore 117576.
[2] Soitec, Parc technologique des fontaines, F-38190 Bernin, France.
*Phone: +65-6516-2298, Fax: +65-6779-1103, Email: yeo@ieee.org

Abstract

We report the first Ω-gate Germanium (Ge) p-channel FinFET with low-temperature Si₂H₆ passivation and implantless Schottky-barrier nickel germanide (NiGe) metallic Source/Drain, formed on high-quality GeOI substrates using sub-400 °C process modules. As compared with reported multi-gate (MuG) Ge devices in which the Ge channels were formed by top-down approaches, the Ge FinFETs in this work have a record high on-state current I_{ON} of ~494 μA/μm at V_{GS} - V_{TH} = -1 V and V_{DS} = -1 V. A high I_{ON}/I_{OFF} ratio of more than 3×10^4 and a high peak saturation transconductance $G_{MSatMax}$ of ~540 μS/μm were achieved.

I. Introduction

Germanium is a promising alternative channel material for sub-14 nm CMOS technology due to its high electron and hole carrier mobilities. High performance Ge devices have been fabricated on Ge bulk, as well as Germanium-On-Insulator (GeOI) substrates [1]-[10]. The FinFET or multi-gate (MuG) device architecture has excellent control of short channel effects and has been adopted at the 22 nm node and beyond. To achieve higher drive current and better short-channel control, Ge MuGFETs, including Gate-All-Around (GAA) devices, have been realized by various techniques in the past few years [5]-[10].

In this paper, we report high performance Ge channel Ω-gate FinFETs with low temperature Si₂H₆ passivated gate stack and implantless self-aligned metallic NiGe S/D (Fig. 1). At similar gate overdrive and drain voltage, the devices exhibit a record high I_{ON} of ~494 μA/μm, as compared with other Ge MuGFETs whose channels were formed by top-down fabrication techniques [7]-[10]. High transconductance and decent subthreshold swing SS were also achieved.

II. Germanium FinFET: Integration of Sub-400 °C Modules

The key process flow to fabricate Ge FinFETs in this work is shown in Fig. 2. High quality 8-inch GeOI wafers were formed by Smart Cut™ technology. TEM images of GeOI used are shown in Fig. 3. After n-well formation, Ge fins were defined by electron beam lithography (EBL) and dry etch.

A cyclic DHF (1:50) and DI water clean was performed. The samples were then loaded into an ultra-high vacuum (UHV) tool for further surface cleaning by SF₆ plasma. SF₆ plasma cleaning not only removes the native oxide but also removes some Ge which could have been damaged during fin etch. Sub-400 °C in situ Si₂H₆ passivation was then performed in the tool to form a high-quality Si passivation layer, followed by gate stack [~4 nm HfO₂/TaN] formation and patterning.

Metal gate spacer (formed after normal gate etch process usually used for planar devices) was removed using a recipe with high TaN/HfO₂ etch selectivity. This is required for effective integration of NiGe metallic S/D. Lastly, a sub-400 °C process formed the self-aligned NiGe metallic S/D.

A TEM image demonstrating a FinFET with Ω-shaped metal gate is shown in Fig. 4(a). The fin height H_{FIN} is ~23 nm, and W_{FIN} is ~60 nm. The rounded corners of the fin were achieved by SF₆ plasma cleaning in the UHV system. Cross-sectional schematics across and along the gate of a FinFET are shown in Fig. 1 (b) and (c), respectively. Fig. 4 (b) is a HRTEM image of the Si₂H₆ passivated Ge (100) surface, showing a SiO₂/Si interlayer beneath the TaN/HfO₂ gate stack. Devices with fin widths W_{FIN} down to ~60 nm and L_G from ~140 nm to ~380 nm were fabricated in this work.

III. Results and Discussion

A. Electrical Characteristics

Low gate leakage current density of ~1.5×10^{-6} A/cm² was obtained from the Si₂H₆ passivated gate stack at a gate voltage V_G of -1 V, as shown from Fig. 5. I_S-V_{GS} and $|I_D|$-V_{GS} characteristics of a Ge FinFET with a L_G of ~380 nm and a W_{FIN} of ~70 nm are shown in Fig. 6 (left axis). The total effective channel width of this device W_{EFF} is ~150 nm. W_{EFF} is calculated as W_{FIN} + 2 H_{FIN} + 2 $W_{FIN,Btm}$. The Source-to-Drain direction is along <110>. A I_{DLin}/I_{MIN} or I_{ON}/I_{OFF} ratio of ~2×10^4 was achieved at V_{DS} = -50 mV, with I_{DLin} taken at V_{GS} - V_{TH} = -1 V and I_{MIN}

taken as the lowest drain current. A decent SS of ~158 mV/dec. was observed on this device. DIBL of this device is 355 mV/V. The relatively large DIBL may be due to the backside negative interface charges near Ge and SiO₂ interface [11]-[12]. G_M-V_{GS} (Fig. 6, right axis) characteristics show peak linear transconductance $G_{MLinMax}$ of 31 μS/μm and $G_{MSatMax}$ of 348 μS/μm at V_{DS} = -50 mV and -1 V, respectively. $|I_D|$-V_{DS} plots (Fig. 7) of the same Ge FinFET demonstrates a high I_{ON} of 318 μA/μm taken at a gate overdrive V_{GS} - V_{TH} = -1 V and V_{DS} = -1 V.

Fig. 8 shows the current vs. V_{GS} (left axis) and G_M vs. V_{GS} (right axis) characteristics of a shorter channel device with a L_G of ~160 nm and a W_{FIN} of ~70 nm. Low I_{MIN} (I_{OFF}) of 1.66 nA/μm is demonstrated at V_{DS} = -50 mV. SS is ~220 mV/decade. I_{ON}/I_{OFF} is above 3×10^4 at V_{DS} = -50 mV. $G_{MLinMax}$ of 60 μS/μm and $G_{MSatMax}$ of 532 μS/μm at V_{DS} = -50 mV and -1 V, respectively, were achieved. At V_{GS} - V_{TH} = -1 V and V_{DS} = -1 V, this device demonstrates an even higher I_{ON} of 494 μA/μm, as shown in the $|I_D|$-V_{DS} plots in Fig. 9, which is the highest among all Ge MuGFETs fabricated using top-down approaches [7]-[10].

B. Series Resistance Reduction

Fig. 10 (a) shows the total resistance R_{Total}-L_G plot at V_{DS} = 50 mV for devices with a W_{FIN} of 70 nm. The extrapolated Source-Drain resistance R_{SD} is ~2680 Ω, which is ~400 Ω·μm when normalized by W_{EFF}. Fig 10 (b) is R_{Total}-$|V_{GS}|$ plot of a device with a L_G of ~230 nm. Fig. 11 (a) shows a SEM image of the FinFET after TaN spacer etch, and Fig. 11 (b) is a zoomed-in SEM of the gate region. Removal of TaN spacer allows NiGe to be formed on the Ge fin sidewall, increasing the NiGe contact area [as illustrated by Fig. 11 (c)-(d)] and contributing to the low R_{SD} observed.

C. Statistics and Scaling of Performance and Geometry

Fig. 12 shows the statistical cumulative plots of SS and I_{OFF} at V_{DS} = -50 mV, showing decent median SS of ~187 mV/dec. and I_{OFF} of ~1.2 nA/μm for FinFETs with metallic S/D. Fig. 13 shows linear drain current I_{DLin}-L_G (left axis) and $G_{MLinMax}$-L_G (right axis) characteristics for device with W_{FIN} of ~70 nm, with I_{DLin} taken at V_{GS} - V_{TH} = -1 V and V_{DS} = -50 mV. Similarly, Fig. 14 shows normalized I_{ON} and $G_{MSatMax}$ for devices with different L_G. It is clearly seen that current and transconductance increase as L_G decreases for the Ge FinFETs.

D. Benchmarking

Table I. compares FinFETs in this work with reported Ge MuGFETs [5]-[10]. All current values are normalized by W_{EFF} or perimeter for a fair comparison. A record high I_{ON} of 494 μA/μm is achieved in this work for Ge MuGFETs in which the Ge fins/wires are fabricated by a top-down fabrication process [7]-[10]. With proper S/D doping and passivation of backside of the Ge layer [11]-[12], better short channel control could be achieved.

IV. Conclusion

In this paper, we demonstrate high performance Ω-gate Ge FinFETs fabricated using sub-400 °C process modules. The integration flow features low temperature Si₂H₆ passivated gate stack and implantless self-aligned metallic NiGe S/D. This is the first time that implantless self-aligned Schottky-barrier NiGe S/D was applied to 3D Ge MuGFETs. Comparing with other Ge MuGFETs with channel formed by top-down fabrication processes, the devices in this work have a record high I_{ON} of ~494 μA/μm. The device also exhibits high I_{ON}/I_{OFF} ratio of more than 3×10^4, high $G_{MSatMax}$ of ~540 μS/μm, and low R_{SD} of ~400 Ω·μm.

References

[1] P. Zimmerman et al., IEDM 2006, p.655.
[2] J. Mitard et al., VLSI Symp. 2009, p.82.
[3] R. Pillarisetty et al., IEDM 2010, p.6.7.1.
[4] L. Hutin et al., EDL 31, p.234, 2010.
[5] L. Zhang et al., Nano Lett. 6, p.2785, 2006.
[6] J. Xiang et al., Nature 441, p.489, 2006.
[7] J. Feng et al., EDL 28, p.637, 2007.
[8] S.-H. Hsu et al., IEDM 2011, p.35.2.1.
[9] J. Feng et al., EDL, 29, p. 805, 2008.
[10] J.W. Peng et al., IEDM 2009, p.38.2.1.
[11] P. Tsipas et al., APL 94, p. 012114, 2009.
[12] K. Romanjek et al., Microelectron. Eng. 86, pp. 1585, 2009.

978-1-4673-0996-7/12 $31.00 © 2012 IEEE

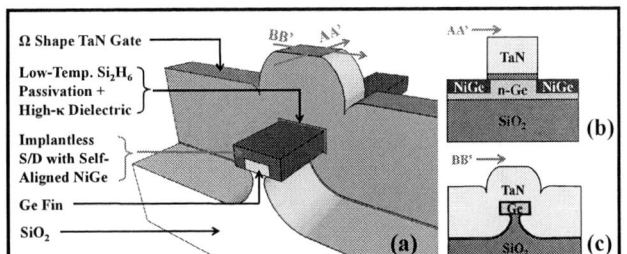
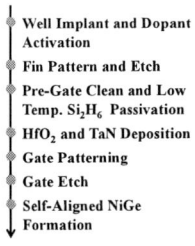
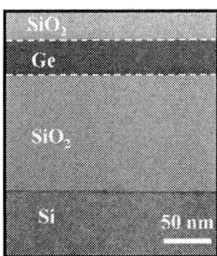

Fig. 1. (a) 3D Schematic of the Ω gate Ge FinFETs fabricated in this work, demonstrating some of the key features of the devices. (b) and (c) are cross sessional schematics along AA' (cross the gate) and BB' (along the gate) directions as indicated in Fig. 1 (a).

Fig. 2. Process flow to fabricate Ge FinFETs with metal S/D.

Fig. 3. (a) TEM of high quality GeOI wafer used in this work.

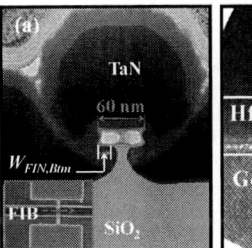
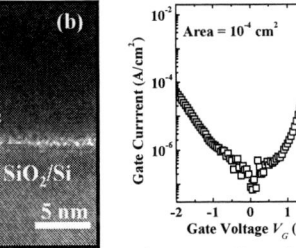
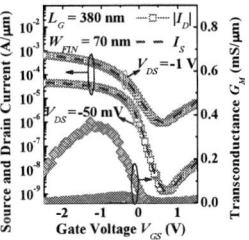
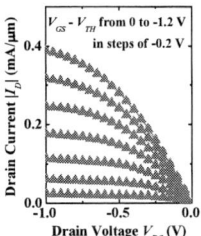

Fig. 4. (a) TEM image of the Ω gate Ge FinFET. Ge fins with W_{FIN} down to 60 nm were fabricated. H_{FIN} is ~23 nm. The total effective gate width W_{EFF} of is defined as W_{FIN} + 2 H_{FIN} +2 $W_{FIN,Btm}$. The rounded corners of the fin were achieved by SF_6 plasma cleaning in the UHV system. (b) HRTEM of gate stack formed on bulk (100) Ge substrate, demonstrating high quality SiO_2/Si passivation.

Fig. 5. Gate leakage current vs. gate voltage, showing low leakage of ~1.5 μA/cm² at V_G = -1 V. The gate area is 10^{-4} cm².

Fig. 6. Current-V_{GS} (left axis) and G_M-V_{GS} (right axis) characteristics of a Ge FinFET with a L_G of ~380 nm and a W_{FIN} of ~70 nm or a W_{EFF} of ~150 nm. A $G_{MSatMax}$ of 350 μS/μm was obtained.

Fig. 7. $|I_D|$-V_{DS} plots of the same Ge FinFET as shown in Fig. 6. At V_{GS} - V_{TH} = -1 V and V_{DS} = -1 V, I_{ON} is 318 μA/μm.

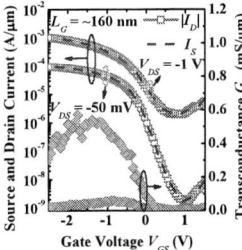
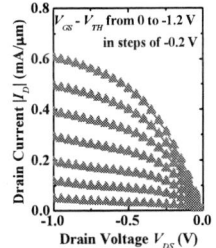
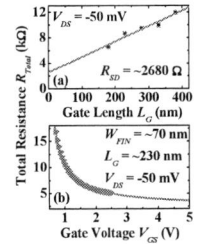
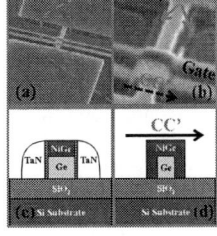
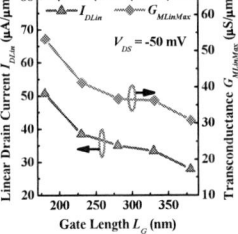

Fig. 8. Current-V_{GS} (left axis) and G_M-V_{GS} (right axis) characteristics of a Ge FinFET with a L_G of ~160 nm and a W_{FIN} of ~70 nm. A $G_{MSatMax}$ of 540 μS/μm was achieved at V_{DS} = -1 V.

Fig. 9. $|I_{DS}|$-V_{DS} plots of the same Ge FinFET as shown in Fig. 8. A high normalized I_{ON} of 494 μA/μm was obtained at V_{GS} - V_{TH} = -1 V and V_{DS} = -1 V.

Fig. 10. (a) R_{Total}-L_G plot for device with W_{FIN} of 70 nm. Extrapolated R_{SD} is ~2680 Ω, i.e. 400 Ω·μm. (b) R_{Total}-$|V_{GS}|$ plot of a device with a L_G of ~230 nm.

Fig. 11. (a) A SEM image of the FinFET after TaN spacer etch. (b) A zoomed-in SEM of the gate region. (c) and (d) are cross sessional schematics of NiGe formation on the fin region without and with TaN spacer etched.

Fig. 12. Cumulative plots of SS (left) and I_{OFF} (right) at V_{DS} = -50 mV. I_{OFF} was taken as the lowest point from $|I_D|$-V_{GS} curves at V_{DS} = -50 mV. Median SS of 187 mV/dec. and I_{OFF} of ~1 nA/μm were observed.

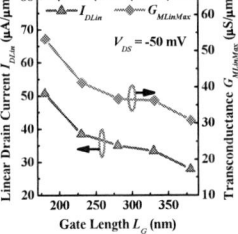

Fig. 13. I_{DLin} vs. L_G (left) and $G_{MLinMax}$ vs. L_G (right) characteristics of devices with fin width of ~70 nm.

Fig. 14. I_{ON}-L_G (left) and $G_{MSatMax}$-L_G (right) characteristics of devices with fin width of ~70 nm.

Table I. Comparison of the Ω gate FinFETs in this work with previously reported MuGFETs. Record high I_{ON} normalized by W_{EFF} is achieved for Ge MuGFETs of which the Ge fins/wires are fabricated by top-down fabrications [7]-[10].

	Ref. 5	Ref. 6	Ref. 7	Ref. 8	Ref. 9	Ref. 10	This work	
Structure	GAA (CVD)	GAA (CVD)	FinFET	GAA	GAA	GAA	Ω Gate FinFET	Ω Gate FinFET
Gate Dielectric	Si/4 nm Al_2O_3	Si/4nm HfO_2	SiON	4nm GeO_2/3nm Al_2O_3	GeON/8.1 nm Al_2O_3	Si/11 nm HfO_2	Si/4nm HfO_2	Si/4nm HfO_2
Fin Width/Wire Diameter (nm)	~20	18	130	52	200	35	70	70
L_G (nm)	3000	190	4500	183	1300	200	380	160
S/D	Ti	Ni	Boron + NiGe	Boron + NiGe	Boron + NiGe	Boron + NiGe	NiGe Only	NiGe Only
V_{DS} (V)	-1.5	-1	-1	-1	-1.1	-1	-1	-1
V_{GS}-V_{TH} (V)	-1.5*	-0.7	-2	-2	-1*	-1	-1	-1
I_{ON} (μA/μm)	~65*	2100# ~668^	~10*	232	~50*	604# ~192^	318	494
SS (mV/dec)	120	100	750*	130	71	160	158	220

CVD: Ge nanowire channels were synthesized by CVD (bottom-up) technique
*: Estimated from reference publications; #: Normalized by Diameter; ^: Normalized by Perimeter

978-1-4673-0996-7/12 $31.00 © 2012 IEEE   27

High-performance pMOSFETs with High-k Gate Dielectric and Dislocation-free Epitaxial Si/Ge Super-lattice Channel

Li-Jung Liu, Kuei-Shu Chang-Liao*, Chung-Hao Fu, Hsiao-Chi Hsieh, Chun-Chang Lu, Tien-Ko Wang, P.Y. Gu[a], and M.J. Tsai[a]

Department of Engineering and System Science, National Tsing Hua University, Hsinchu 30013, Taiwan, R.O.C.
[a] Electronics and Opto-Electronics Research Laboratories, Industrial Technology Research Institute, Hsinchu, Taiwan, ROC
* Tel: (886)-(3)-5742674, Fax: (886)-(3)-5720724, Email: lkschang@ess.nthu.edu.tw

Abstract

The pMOSFET device with a novel Si/Ge super-lattice (SL) channel is proposed in this work. Experimental results show that the electrical characteristics can be obviously improved by SL virtual substrate. The peak hole mobility of pMOSFET device with SL is enhanced to twice as high as that with Si one. The on-off ratio of Id-Vg curve is beyond 8 orders, and the EOT value of gate dielectric can be ~ 1 nm. The source/drain activation temperature at 650 ℃ is especially suitable for high-k gate dielectric process.

Introduction

As the gate oxide thickness of MOS devices becomes thinner, the high-k gate dielectric has been implemented to replace SiO_2 or SiON for nano-scale MOS device applications. However, a reduction in channel mobility is also encountered. A promising technique to solve this issue is to alternate Si channel with high mobility material, like Ge, which can offer two times higher electron mobility and four times higher hole mobility than Si [1]. Besides, the mobility improvement of Ge MOS can be increased by compressive strain [2]. Nevertheless, the Ge and SiGe MOSFET have several concerns: the cost of pure Ge wafer is higher, the method of Ge grown on Si is complicated [3], Ge up-diffusion after high temperature annealing resulting in degradation of device performance [4], and the mobility improvement is not enough for SiGe channel with low Ge content. In this work, a dislocation-free epitaxial super-lattice (SL) channel with Si/Ge period stacks is proposed for MOS device to achieve a high mobility and also maintain a good quality of gate dielectric.

Experiment

MOSFETs with Si, $Si_{0.7}Ge_{0.3}$, and Si/Ge super-lattice channel were fabricated on n-type Si (100) wafer, as shown in Fig. 1. A 2 nm thick HfO_2 were deposited by an atomic layer deposition (ALD). Then, a 50 nm thick TaN film was deposited by a sputtering to serve as the metal gate. For dopant activation, Si, $Si_{70}Ge_{30}$, and Si/Ge super-lattice channel were through 950, 800, 650 ℃ for 30 s in N_2, respectively.

Results

Fig. 2 shows the schematic cross-section of SL channel with repeating 11~20 periods of the Si 8 Å/Ge 4~8 Å stacks. Fig. 3 shows the HR-TEM images of the SL virtual structures with (a) Si 8 Å/Ge 4 Å and (b) Si 8 Å/Ge 8 Å stacks. Fig. 4 shows the HR-XRD rocking curve spectra of the SL structures, which reveal that the SL structures are epitaxially grown along Si (001) plane. The R_{rms} values from the 10 μm x 10 μm AFM scans of the 8/4 and 8/8 SL structures are 1.54 Å and 1.35 Å, as shown in Fig. 5 respectively. The roughness of the SL channel with an annealing at 900 ℃ for 30s in N_2 is found to be 1.2 Å, indicating the good thermal stability of the SL structure. Fig. 6 shows the HR-XRD reciprocal space mapping (RSM) along (a)

Si (100) and (b) Si (224) planes, indicating the strain status of the Ge layers in the SL structure are fully-strained, which is very helpful for the enhancement of the carrier mobility [2]. Fig. 7 (a) shows the comparison of EOT and leakage current density for MOS devices with Si, SiGe, and SL channels after a PDA at 700 ℃, indicating the Ge up-diffusion may be minor. Fig. 7 (b) shows the current versus bias voltage for S/D junctions in MOSFET with Si, SiGe, and SL channels, respectively, after various activation temperatures. Since the SL sample has lower S/D activation temperature (650 ℃), it is more suitable to integrate high-k dielectric process. Fig. 8 shows (a) the C-V hysteresis and (b) Gm*EOT*L/W versus gate voltages of devices with Si, SiGe, and SL channels, indicating good interface quality with employing SL channel. Fig. 9 shows (a) Id–Vg and (b) Id–Vd characteristics of MOSFETs with Si, SiGe, and SL channel. Good MOSFET characteristics are observed for SL sample, such as a relatively high drive current, an acceptable subthreshold swing (SS) of 84 mV/dec, a lower I_{OFF} leakage of 5×10^{-12} A, and large on-off ratio (over 8 orders). It can be attributed to higher mobility, minor Ge up-diffusion, good S/D junction, and a compressive stress with fully-strain in Ge layer achieved by SL channel. Fig. 10 shows hole mobility as a function of the effective electric field for MOSFETs with Si, SiGe, and SL channel. The peak hole mobility of the device with SL channel is about 150 $cm^2/(V \cdot s)$, which is enhanced by about 100 % and 30 % as compared to that with Si and SiGe, respectively. Results in this work are benchmarked with some reported ones as listed in Table 1. Compare to the SL sample, most other devices have larger EOT values; for those MOSFET devices with small EOT values and high Ge content, the mobilities are similar to the SL sample.

Conclusions

This work demonstrated a high-performance MOSFET with epitaxial SL on Si substrate. SL channel is characterized to have extremely low surface roughness, dislocation-free structure and fully-strained Ge layers. This design achieves good performance in terms of high mobility and on-off ratio. The temperature of S/D annealing at 650 ℃ is fully compatible with current VLSI fabrication method.

References

[1] Y. H. Wu et al., IEEE EDL, vol. 30(1), p. 72, 2009.
[2] K. Sawano et al., APL, vol. 95, p. 122109, 2009.
[3] A. Nayfeh et al., IEEE EDL, vol. 26(5), p. 311, 2005.
[4] J. Huang et al., Symp. VLSI Tech., p. 82, 2008.
[5] J. Oh et al., Symp. VLSI-TSA Tech., p. 40, 2008.
[6] H. Y. Yu et al., IEEE EDL, vol. 30(6), p. 675, 2009.
[7] S. H. Lee et al., Symp. VLSI Tech., p. 74, 2009.
[8] J. Mitard et al., Tech. Dig. IEDM, p. 10.6.1, 2010.

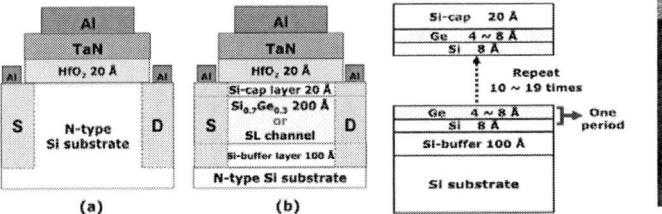
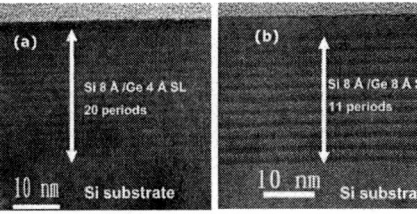
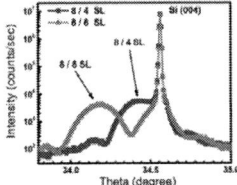

Fig. 1 The schematic cross-sections of pMOSFET devices with (a) Si and (b) SiGe or Si/Ge super-lattice (SL) channel.

Fig. 2 The schematic cross-section of SL channel with different thicknesses of Ge layers.

Fig. 3 The HR-TEM images for the as-deposited (a) 8/4 and (b) 8/8 SL structure. Clear interfaces between Si and Ge indicate good epitaxial structure of SL channel.

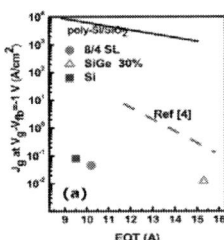
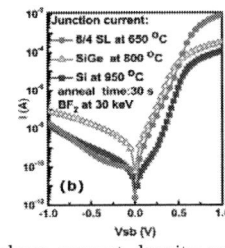
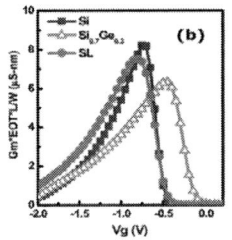

Fig. 4 The HR-XRD rocking curve spectra of 8/4 and 8/8 SL structures along Si (004) plane.

Fig. 5 Three-dimension topographic images of (a) 8/4 and (b) 8/8 SL structures from the 10 μm x 10 μm AFM scans.

Fig. 6 The HRXRD-RSM images of 8/8 SL structures along (a) Si (004) and (b) Si (224) planes. The images show the fully-strain status of the Ge layers in the SL structure.

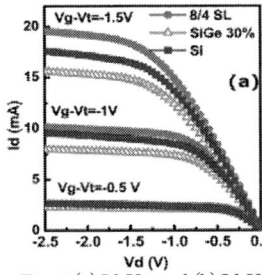
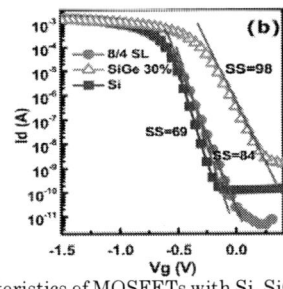
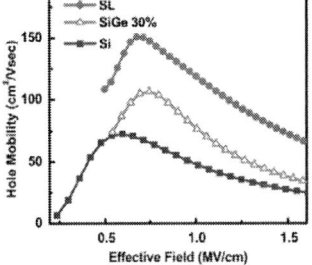

Fig. 7 (a) Comparison of leakage current density and EOT for MOS devices. (b) The current for S/D junction in MOSFET devices with Si, SiGe, and SL channel after various S/D activation temperatures.

Fig. 8 (a) Capacitance-voltage hysteresis and (b) Gm*EOT*L/W versus gate voltage for devices with Si, SiGe, and SL channel.

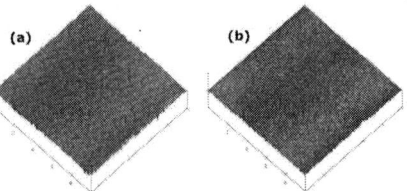
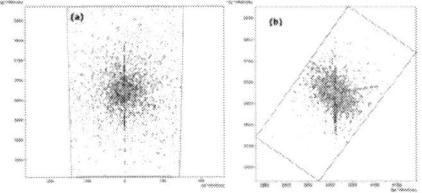

Fig. 9 (a) Id-Vg and (b) Id-Vd characteristics of MOSFETs with Si, SiGe, and SL channels, indicating good electrical characteristics of the device with SL channel.

Fig. 10 Hole mobility as a function of the effective electric field for MOSFETs with Si, SiGe, and SL channels.

Table 1 Comparison of main electrical parameters about MOSFET with SiGe or Ge virtual channels

Channel structure	Max-$\mu_{h\ (cm^2 V^{-1} s^{-1})}$	High-k dielectric	EOT (Å)
8/4 SiGe Super-lattice [This work]	150	2nm HfO$_2$	10
Epitaxial pure Ge [5]	170	5 nm HfO$_2$	-
Epitaxial Pure Ge [6]	400	GeO$_2$+4.5 nm Al$_2$O$_3$	-
Solid phase epitaxy Ge [1]	350	SiO$_2$	>25
Epitaxial Si$_{0.25}$Ge$_{0.75}$ [7]	140	HfSiOx	10
Epitaxial Si$_{0.5}$Ge$_{0.5}$ [7]	180	HfSiOx (flash anneal)	10
Epitaxial Si$_{0.45}$Ge$_{0.55}$ [8]	140	-	8.5

978-1-4673-0996-7/12 $31.00 © 2012 IEEE

Counter Dipole Layer Formation in SiO_2/High-k/SiO_2/Si Gate Stacks

S. Hibino, T. Nishimura, K. Nagashio, K. Kita and A. Toriumi

Department of Materials Engineering, The University of Tokyo

7-3-1 Hongo, Tokyo 113-8656, Japan

toriumi@material.t.u-tokyo.ac.jp ; phone +81-3-5841-7120

Abstract – **This paper presents experimental results of the counter dipole formation in SiO_2/high-k (Al_2O_3 and Y_2O_3)/SiO_2/Si gate stacks for the first time. The results definitely support the high-k/SiO_2 interface dipole layer formation in metal/high-k gate CMOS.**

1. Background and Objective

Several experimental results have already demonstrated that there should be a dipole layer at high-k/SiO_2 interface [1-3]. The dipole formation mechanisms have also been proposed [4-7]. Whatever the model is, the dipole effect is practically used for V_{TH} tuning in the band-edge metal/high-k gate stack CMOS. An optimum condition might be found by changing the dielectric material, and/or process conditions, but there is still a missing fact in the high-k/SiO_2 dipole formation experiment. It is the dipole cancelling effect (the counter dipole formation) in SiO_2/high-k/SiO_2/Si gate stacks, where no dipole effect should be found in that system. The objective of this paper is to demonstrate the counter dipole effect experimentally.

2. Results and Discussion

The multi-layer dielectric film stacks were prepared by PLD (pulsed laser deposition) in high-vacuum (10^{-6} Pa) on SiO_2/p-Si, followed by the low temperature PDA (post-deposition annealing) in N_2+0.1%O_2 at 600°C. The deposited layer thickness was adjusted by the laser pulse count and finally checked by TEM. Au was deposited by vacuum evaporation for the gate electrode.

V_{FB} shifts as compared with the SiO_2/Si gate stack expected from the phenomenological model [4] are shown in **Fig. 1**. The dipole directions in the present experiment results for Y_2O_3/ SiO_2 and Al_2O_3/SiO_2 shown in **Fig. 2** are quite consistent with the model. Next, the triple layer dielectric film stacks (high-k_1/high-k_2/SiO_2/Si) were prepared by changing the bottom high-k_2 film thickness under the fixed top layer high-k_1 film on SiO_2/Si. Measured V_{FB} values are plotted as a function of the bottom high-k_2 layer thickness in **Fig. 3**. The results show that V_{FB} is definitely determined by the bottom high-k/SiO_2 interfaces in both Au/**Y_2O_3/Al_2O_3/SiO_2**/Si and Au/**Al_2O_3/Y_2O_3/ SiO_2**/Si stacks. They are exactly the same

as those in case of ALD grown dielectric stacks reported previously [2]. When the top high-k layer thickness was changed, V_{FB} was not shifted very much. Thus, the present results indicate the existence of the bottom high-k/SiO_2 interface dipole in case of PLD-grown high-k gate stacks.

Most importantly, **SiO_2/Al_2O_3/SiO_2/Si** and **SiO_2/Y_2O_3/SiO_2/Si** stacks were prepared to investigate whether the top interface dipoles were formed or not. The top SiO_2 layer (3~4 nm) was grown by PLD of Si in O_2 (~1 Pa). The XTEM image is shown for Au/SiO_2/Al_2O_3/SiO_2/Si in **Fig. 4**. V_{FB} was measured by changing the top-SiO_2 thickness on the fixed high-k film. **Fig. 5** clearly shows that in both Au/SiO_2/Al_2O_3/SiO_2/Si and Au/SiO_2/Y_2O_3/SiO_2/Si gate stacks, V_{FB} values approach almost the same V_{FB} in Au/SiO_2/Si gate stack within a couple of nm-thick transition layer, as the SiO_2 thickness is increased. The results are exactly what we have expected from the schematic picture in **Fig. 6**, in which the bottom dipole is perfectly cancelled by the top one with the opposite direction from the bottom one. To our knowledge, this is the first demonstration of the counter dipole effect in the high-k gate stacks.

Although the dipole formation model is quite sure only from the results in Fig. 2 and 3, the present results make this model in high-k gate stacks more sound.

3. Conclusion

We have demonstrated the counter dipole effect in SiO_2/high-k/SiO_2/Si gate stacks, in addition to the confirmation of the dipole layer formation in high-k/SiO_2/Si gate stacks grown by PLD method. Thus, it is rigorously concluded that the dipole layer at high-k/SiO_2 interface is the dominant origin for the V_{FB} shift in high-k gate stacks.

Acknowledgement: This work was partly supported by a Grant-in-Aid for Scientific Research (S) by JSPS.

References

[1] Y. Yamamoto et al., Jpn. J. Appl. Phys. Pt.1, **46**, 7251(2007).

[2] K. Iwamoto et al., Appl. Phys. Lett. **92**, 132907 (2008).

[3] Y. Kamimuta et al., IEDM 2007, p.341.

[4] K. Kita and A. Toriumi, Appl. Phys. Lett. **94**, 32902 (2009).

[5] O. Sharia et al., Phys. Rev. **B77**, 085326 (2008).

[6] H. Jagannathan, ECS Trans. **16** (5) 19 (2008).

[7] A. Toriumi and K. Kita, ECS Trans. **19**(1), 243 (2009).

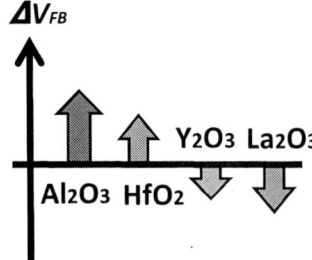

Fig. 1. V_{FB} shifts expected in high-k/SiO$_2$/Si gate stacks. As compared with the V_{FB} in SiO$_2$/Si MOS capacitor, dipole direction in terms of V_{FB} shift is schematically indicated. The opposite direction of Al$_2$O$_3$/SiO$_2$/Si from Y$_2$O$_3$/SiO$_2$/Si gate stacks is expected [4].

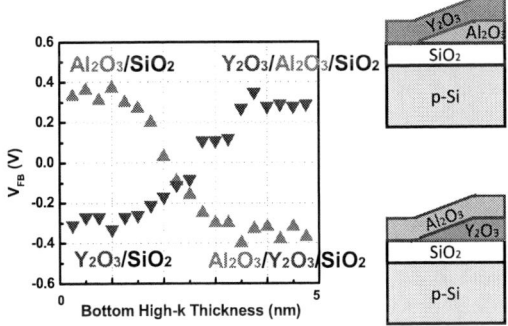

Fig. 3. The bottom high-k layer thickness dependences of V_{FB} in triple layer dielectric film gate stacks. The bottom layer thickness was systematically changed by using the movable mask in PLD chamber. The relative thickness was estimated by the laser pulse count, and the end-point thickness of the bottom high-k thickness was accurately estimated. V_{FB} shift is determined by the bottom interface at high-k/SiO$_2$. The V_{FB} transition layer thickness is around 2 nm.

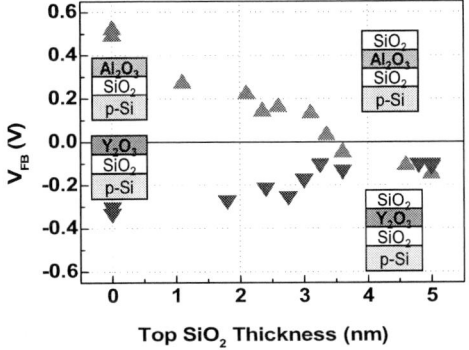

Fig. 5. The top SiO$_2$ thickness dependences of V_{FB} in SiO$_2$/high-k/SiO$_2$/Si gate stacks. The initial gate stacks (without the top SiO$_2$ layer) show the dipole shift of V_{FB}, while both V_{FB} values are approaching the initial V_{FB} in SiO$_2$/Si gate stack with increasing the top SiO$_2$ thickness. This is the direct evidence of the counter dipole effect (the dipole cancelation effect) in SiO$_2$/high-k/SiO$_2$/Si gate stacks.

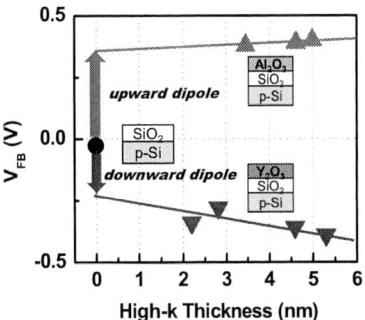

Fig. 2. Top high-k layer thickness dependences of V_{FB} on SiO$_2$/Si. Compared with the V_{FB} (●) of SiO$_2$/Si MOS capacitor, clear dipole effects are experimentally observed from the extrapolated values in high-k thickness dependences. The opposite direction of extrapolated V_{FB} values between Al$_2$O$_3$/SiO$_2$/Si (▲) and Y$_2$O$_3$/SiO$_2$/Si (▼) is quite consistent with the proposed model in Fig. 1.

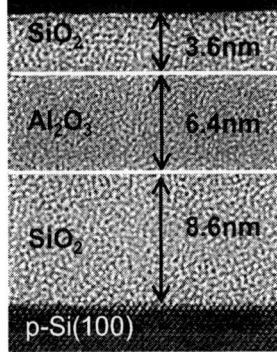

Fig. 4. The cross-sectional TEM image of SiO$_2$/Al$_2$O$_3$/SiO$_2$/Si stack. Both interfaces (top and bottom) are clearly identified. White lines are for the eye-guide.

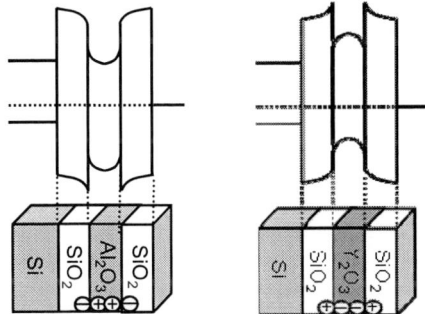

Fig. 6. Schematic pictures for explaining the counter dipole effect in SiO$_2$/high-k/SiO$_2$/Si gate stacks. It is understandable that V_{FB} is recovered to the value in SiO$_2$/Si gate stacks, even if the dipole direction in Al$_2$O$_3$ and Y$_2$O$_3$ is opposite from each other.

978-1-4673-0996-7/12 $31.00 © 2012 IEEE

Simultaneous Carrier Transport Enhancement and Variability Reduction in Si MOSFETs by Insertion of Partial Monolayers of Oxygen

R.J. Mears, [1]N. Xu, [1]N. Damrongplasit, H. Takeuchi, R.J. Stephenson,
N.W. Cody, A. Yiptong, X. Huang, M. Hytha and [1]T.-J. King-Liu

Mears Technologies, [1]EECS Department, University of California, Berkeley
189 Wells Avenue, Newton MA 02459 USA Tel +1 617 219 0600   E-mail hideki.takeuchi@mearstechnologies.com

Abstract

We demonstrate simultaneous NMOS and PMOS high-field mobility enhancement and variability reduction by inserting partial monolayers of oxygen during silicon epitaxy of the channel layer.

Introduction

The benefits of doping profile engineering to create super-steep retrograde and pulse profiles to enhance the scalability of the MOSFET are well known [1-3]. Reductions in short-channel effects and variability with such profiles has been demonstrated by multiple authors. However, it has also been shown that such profiles degrade carrier transport because of the higher body effect [4].

In this paper we demonstrate a new technique to enhance high-field mobility by sub-band engineering. The mobility enhancement is shown to arise from "quasi-confinement" due to partial monolayers of oxygen within the channel layer, which can selectively increase both the energy separation and spatial separation of the carrier sub-bands. Hence reduced scattering or lowering of effective mass through sub-band re-population can be achieved, for significant high-field mobility improvement for both NMOS and PMOS devices. Furthermore, a reduction in diffusion of electrical dopants through the channel layer provides for ultra-steep retrograde and pulse doping profiles with gradients as steep as 3.3nm/decade, even after typical post-STI anneals. Thus simultaneous carrier transport enhancement and variability reduction is achieved.

Channel Layer Enhancement Technique

Partial monolayers of oxygen are inserted during silicon epitaxy as shown in Figure 1. The oxygen atoms are introduced interstitially so that they cause minimal disruption to the silicon lattice and do not hinder subsequent silicon epitaxy. They are designed to cause multiple local perturbations of potential so that the channel carriers experience a confinement effect while the wavefunctions can still exist continuously through the material. We refer to this as "quasi-confinement".

Simulation of the carrier sub-band structure and mobility enhancement was carried out using a state-of-the-art Poisson-Schrödinger self-consistent simulator [5]. For example, as shown in Figures 2 and 3, the positions of the oxygen partial monolayers can be tailored to significantly enhance the ground state population of the low effective mass Δ-2 sub-band at the expense of the Δ-4 sub-band, and almost double the energy separation so that inter-band scattering is reduced. Carrier mobility enhancement of 30% or more is predicted for both NMOS and PMOS devices.

In addition, by simulation and experiment it has been found that electrical dopants (both donors and acceptors) preferentially situate themselves close to a partial monolayer. This is a result of local charge modulation induced by the oxygen partial monolayer. Additionally, the existence of dopant lattice sites with lower energy in the proximity of the partial monolayer impacts the lattice-hopping activation energy locally. Channel dopants implanted below the layers are attracted to the lower-most partial monolayer but are then impeded from further diffusion upward through the channel layer, so that ultra-steep retrograde profiles may be designed as shown in Figure 4.

Results and Discussion

The mobility enhancement achieved with oxygen partial monolayers is experimentally verified to be distinct from the effect of the improved doping profile, as shown in Figure 5. The control device with a retrograde channel doping profile has a higher low-field mobility because it has lower Vt; its high-field mobility is the same as the control device with uniform channel doping, as expected. By contrast, the device with oxygen partial monolayers has significantly improved transport properties due to the quasi-confinement effect. Additional simulations suggest that even greater degrees of mobility enhancement (~100%) are possible using this new technique.

The improved doping profile achieved with inserted oxygen partial monolayers is beneficial for reducing variability, by >50%, in Figure 6. 6-T SRAM cell simulations suggest significant yield benefit at Vdd=1.0V for the 28nm CMOS technology node (Figures 7-10).

Conclusion

A new technique for simultaneously enhancing carrier transport and reducing variability in NMOS and PMOS devices by inserting partial monolayers of oxygen during channel silicon epitaxy is demonstrated. This technique is compatible with and complementary to other approaches for enhancing MOSFET performance and scalability, and hence can facilitate scaling to the end of the roadmap.

REFERENCES

[1] K. Miyamoto, A. Strojwas, E. Hosomi, M. Ooida, H. Ezawa, M. Fukuda, Y. Matsubara, and K. Numata, "Novel circuit design and process technology for leading-edge products", in Symp. on VLSI Tech Dig., June 2010, pp.141-142.

[2] R-H. Yan, A. Ourmazd, and K. F. Lee, "Scaling the Si MOSFET: from bulk to SOI to bulk", IEEE Trans on Elect. Dev., vol.39, no.7, pp.1704-1710, 1992.

[3] A. Asenov et al., "Simulation of Intrinsic Parameter Fluctuations in Decananometer and Nanometer-Scale MOSFETs," IEEE Trans. Electron Devices, vol. 50, pp. 1837-1852, 2003.

[4] T. Skotnicki, "Advanced architectures for 0.18-0.12um CMOS generations", in ESSDERC Tech. Dig., Sept. 1996, pp.505-514.

[5] N. Xu, B. Ho, F. Andrieu, L. Smith, B.-Y. Nguyen, O. Weber, T. Poiroux, O. Faynot, T.-J. K. Liu, "Carrier Mobility Enhancement via Strain Engineering in Future Thin Body Transistors," IEEE Electron Device Letters, vol. 33, pp. 318-320, 2012.

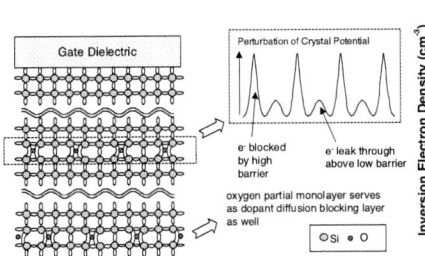

Fig.1 Introduction of partial monolayers of oxygen leading to formation of quasi-confinement potential. Oxygen partial monolayers serve as the dopant diffusion blocking layer as well.

Fig.2 Simulated electron sub-band population change due to insertion of oxygen partial monolayers.

Fig.3 Simulated electron sub-band energy change due to insertion of oxygen partial monolayers.

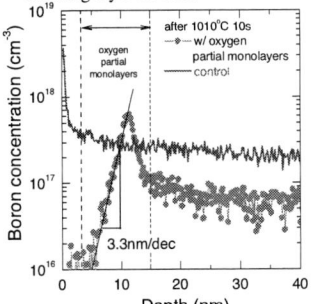

Fig.4 Ultra-steep retrograde boron profile is achieved with the oxygen partial monolayers after gate oxidation and 1010°C 10s RTA.

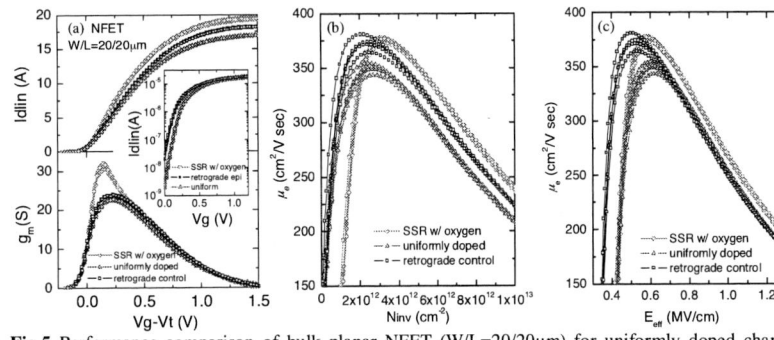

Fig.5 Performance comparison of bulk planar NFET (W/L=20/20μm) for uniformly doped channel, retrograde epi channel, and super-steep retrograde epi channel with inserted oxygen partial monolayers: (a) linear drain current and transconductance vs. gate overdrive (inset: turn-on characteristics); (b) inversion charge density vs. electron mobility; (c) effective field vs. electron mobility. Whereas enhancement of retrograde channel epi control is simply due to low Vt [(a) inset], the SSR channel with oxygen partial monolayers has improved transport properties [(b), (c)].

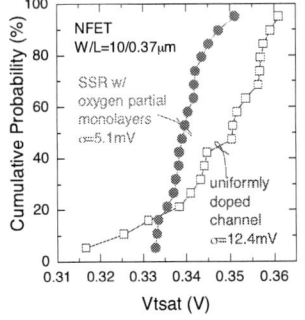

Fig.6 Measured Vt cumulative probability plots for uniformly doped silicon channel (control) and super-steep retrograde epi channel with inserted oxygen partial monolayers.

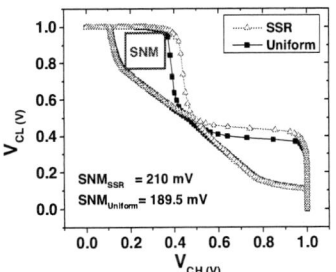

Fig.7 Comparison of simulated butterfly curves for 28nm CMOS 6-T SRAM cell: poly-Si/SiON (EOT=1.6nm); L=28nm; $W_{PU}/W_{PD}/W_{PG}$= 50nm/80nm/50nm. SNM is improved with SSR channel.

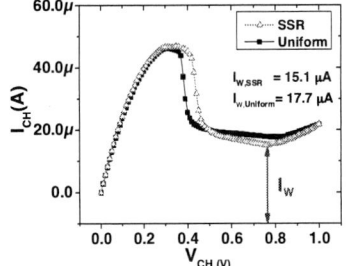

Fig.8 Comparison of simulated 6-T SRAM cell N-curves. The SSR channel shows slightly worse writeability due to non-optimized cell alpha ratio..

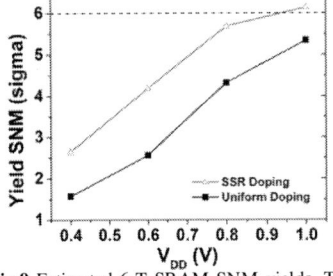

Fig.9 Estimated 6-T SRAM SNM yields. The 6σ yield criterion can be met at Vmin = 0.94 V for SSR, but not for uniform doping below V_{DD}=1.0V.

Fig.10 Estimated 6-T SRAM write yields. Due to non-optimized transistor sizing, SSR write yields are slightly degraded but do not limit Vmin scaling.

Transport in Graphene on Boron Nitride

D. K. Ferry[*]

School of Electrical, Computer, and Energy Engineering
Arizona State University, Tempe, AZ 85287-5706

Graphene has become of great interest in recent years for its unique band structure and prospective importance in both microwave and logic devices. Recently, the use of a boron nitride layer between the graphene and the silicon dioxide substrate has shown enhanced mobilities due to displacing the disorder charge, typical on the oxide, further from the graphene material.[1,2] On the other hand, like the oxide, boron nitride has polar optical modes which can interact with the carriers in graphene to lower their mobility. We have used an ensemble Monte Carlo (EMC) technique to study the transport in graphene on a boron nitride layer. Scattering by the intrinsic phonons of graphene,[3] as well as by the flexural modes of the rippled layer, and the remote polar mode of boron nitride has been included. The flexural modes are described by the model of Castro et al.[4] While the EMC uses the simple Dirac band structure, coupling constants for the intrinsic phonon modes are taken by fitting to scattering rates determined from first-principles calculations.[5] We find that, at low temperatures, the mobility is dominated primarily by the intrinsic graphene phonons and the flexural modes. This arises as the interfacial polar mode of boron nitride lies at an energy of 200 meV, which is largely too high to interact well with the majority of the carriers in graphene. On the other hand, at room temperature, the mobility begins to be dominated by the remote polar mode of the boron nitride. Nevertheless, the prospects of reaching a high velocity, needed for device performance particularly at microwave frequencies, remains very good.

---

[1] C. R. Dean et al., Nature Nanotech. **5**, 722 (2010).

[2] W. Gannett et al., Appl. Phys. Lett. **98**, 242105 (2011).

[3] R. S. Shishir et al., J. Comp. Electron. **8**, 43 (2009).

[4] E. V. Castro et al., Phys. Rev. Lett. **105**, 266601 (2010).

[5] S. Aboud and M. V. Fischetti, *private communication.*

[*] ferry@asu.edu, 480-965-2570, FAX 480-965-8058

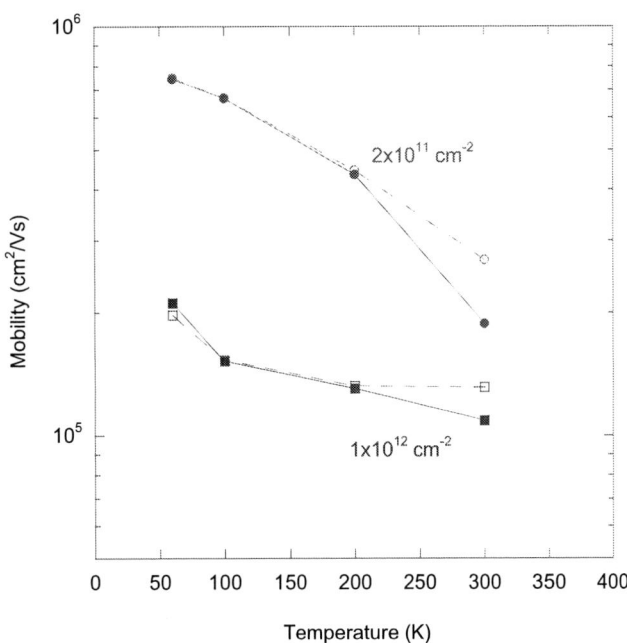

Fig. 1 The mobility of graphene on boron nitride. The dashed curve represents the case in which the remote polar modes of boron nitride are ignored. These modes dominate the mobility at room temperature.

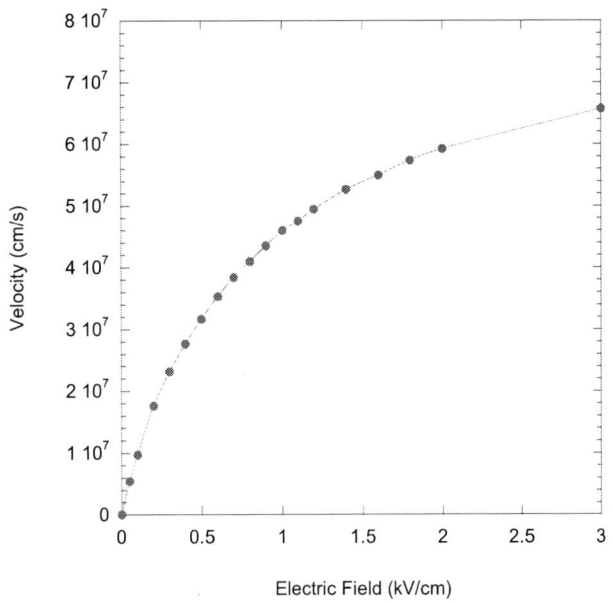

Fig. 2 The high field velocity in graphene on boron nitride at a carrier density of 10^{12} cm^{-2}. No real saturation appears in this curve.

Magnetic Tunnel Junction for Magnetoresistive Random Access Memory and Beyond

Hideo Ohno[1,2,3]

[1] Center for Spintronics Integrated Systems, Tohoku University, 2-1-1 Katahira, Aoba-ku, Sendai 980-8577, Japan
Phone: +81-22-217-5553 E-mail: ohno@riec.tohoku.ac.jp
[2] WPI-Advanced Institute for Materials Research, Tohoku University, Sendai, Japan
[3] Laboratory for Nanoelectronics and Spintronics, Research Institute of Electrical Communication, Tohoku Univeristy, Sendai 980-8577, Japan

1. Introduction

Magnetoresistive random access memory (MRAM) using magnetic tunnel junction (MTJ) combined with spin transfer torque (STT) write is capable of fast-read/write with high endurance, compatible with CMOS back-end-of-line (BEOL), and has high potential for scalability. Recent progress in perpendicular magnetic-easy axis MTJs shows that high performance MTJs can be made of MgO-CoFeB system that produces high tunnel magnetoresistance at room temperature, which are suitable for RAM applications. In addition to STT-MRAM applications, these high performance MTJs and its variants open a route to nonvolatile CMOS logic employing nonvolatile logic-in-memory architecture. Here, I review the current status of MTJ technology along with logic implementation of these nonvolatile devices.

2. MTJ Technology

MTJ needs to satisfy the following requirements simultaneously [1]: (1) small size *i.e.* scalability (F nm), (2) low current for STT switching ($I_{C0} \leq F$ µA), (3) high tunnel magnetoresistance (TMR) ratio (>100%), where the ratio is defined as $(R_{AP}-R_P)/R_P$, where R_P and R_{AP} are resistance at parallel magnetization configuration and at antiparallel configuration, respectively (4) high thermal stability factor of recording layer $\Delta = E/k_B T$, where E is energy barrier separating the two configurations, finally (5) capability to withstand annealing temperature of 350 °C to 400 °C required for standard CMOS processing. In 2010, a 40 nm perpendicular-MTJ (magnetization of the free and fixed layers is pointing perpendicular to the substrate) was demonstrated utilizing perpendicular anisotropy at the interface of MgO-CoFeB, which satisfies nearly all the requirements [2]. Later, it was shown that there is a length scale involved above which "nucleation" of magnetization reversal takes place in the recording layer of an MTJ. This length scale is approximately 40 nm in the case of MgO-CoFeB MTJ [3]. Because the upper limit of Δ is given by the nucleation phenomenon, one needs to increase the thickness of the recording layer to increase Δ [4]. By developing material technology for higher interface anisotropy to allow thicker recording layer and for reduced damping constant α to realize low switching current, high performance MTJ at further reduced dimension will be in sight.

3. Nonvolatile CMOS Logic with MTJs

Following the seminal work on 4 kbit spin RAM [5], a number of groups have demonstrated higher capacity of STT-MRAM [6], reaching 64 Mbit in the year 2010 [7]. With this technology in hand, embedded DRAM/SRAM replacement is also being pursued. Here, the 350 to 400 °C annealing capability is critical, because one cannot alter optimized CMOS logic process. Feature size requirement is more relaxed in this front. For high speed operation, a 3-terminal device that separates the write path from the read path is another possibility. Recently 16 kbit content addressable memory with 5 ns search time has been developed using 3-terminal domain wall motion device [8]. Nonvolatile logic-in-memory architecture has also been demonstrated; a 6T-2MTJ nonvolatile 2 kbit ternary content addressable memory that can reduce search mode power to 1/30 of the comparable CMOS realization [9]. It is also noteworthy that a 600 MHz nonvolatile latch is developed utilizing an incubation time of an MTJ [10].

4. Summary

I have reviewed current status of MTJ and how it can be used in memories and logic circuits, referring to some of our recent implementations. The ultimate scalability of MTJ technology will be determined by both materials involved and processing technology. It is difficult to foresee how far in dimension one can go at this point. But we should be able to learn from the materials science for hard disk media that can realize high Δ at dimensions less than 10nm and is continuing to develop a patterned one.

Acknowledgements

I thank my collaborators, particularly Professors T. Endoh, T. Hanyu, N. Kasai and S. Ikeda. A part of the work described here is supported by the FIRST program "Research and Development of Ultra-low Power Spintronics-based VLSIs" from JSPS.

References

[1] H. Ohno *et al.*, IEDM (invited) 2010.
[2] S. Ikeda *et al.*, Nature Materials, **9**, 721 (2010).
[3] H. Sato *et al.*, Appl. Phys. Lett. **99**, 042501 (2011).

[4] H. Sato *et al.*, IEEE Magn. Lett. **3**, 3000204 (2012)

[5] M. Hosomi *et al.*, IEDM 2005.

[6] T. Kawahara *et al.*, ISSCC 2007.

[7] K. Tsuchida *et al.*, ISSCC 2010.

[8] R. Nebashi *et al.*, VLSI Circuit Symposium 2011.

[9] S. Matsunaga *et al.*, VLSI Circuit Symposium 2011

[10] T. Endoh et al., IEDM 2011.

Systolic Architectures and Applications for Nanomagnet Logic

M. Niemier[#], X. Ju[†], M. Becherer[†], G. Csaba[#], X.S. Hu[#], D. Schmitt-Landsiedel[†], P. Lugli[†], W. Porod[#]

†Technical University of Munich, Munich, Germany, # University of Notre Dame, Notre Dame, IN, USA

Corresponding: 384 Fitzpatrick Hall, Notre Dame, IN 46556, USA, (574) 631-3858, (574) 631-9260 (fax), mniemier@nd.edu

Abstract — **Most NML research has studied small magnet ensembles for interconnect or isolated gates. We discuss how NML might be used to *process information*, as well as suitable system architecture-to-device architecture mappings. A case study for pattern matching hardware is presented.**

I. INTRODUCTION

Most work with nanomagnet logic (NML) has focused on devices that couple *in-plane* (Fig. 1a). With clock overheads, projections indicate that **iNML** circuits could be more energy efficient than low-power CMOS equivalents without performance losses [1]. Copper wires clad with ferromagnetic material on the sides and bottom, have been fabricated and used to re-evaluate line and gate ensembles [2]. Fig. 2 illustrates how larger systems would be clocked. Structures to move information between electrical and magnetic domains have also been proposed and simulated. NML devices can also be realized from multi-layered materials with perpendicular anisotropy such as Co/Pt (Fig. 1b). Irradiation of individual dots enables non-reciprocal propagation and defines dataflow directionality. Thus, **oNML** devices can be clocked with a homogeneous and *global* (out-of-plane) magnetic field – which simplifies drive circuitry. During each sinusoidal field cycle, switching events occur only if a neighbor is in a parallel, metastable state (see Fig. 3).

II. INML VS. ONML

When looking toward circuit architectures, the *oNML* device architecture could offer distinct advantages. **(1)** With iNML, signal routing requires data movement between anti-ferromagnetic (AF) and ferromagnetic (F) interconnect. This can be problematic when an input signal must fan out and serve as an input to more than one other gate. oNML devices always couple AF to neighboring devices, and can be resized such that neighboring devices have different footprints to simplify routing (Fig. 4). **(2)** Provided circuit inputs do not change, the potential for data races associated with different signal arrival times [3] at an iNML gate would be eliminated in oNML circuits (Fig. 5). **(3)** With oNML, FIB irradiation can define dataflow directionality *without* multi-phase clock schemes that are required for iNML – useful architecturally for realizing bidirectional dataflow. Bidirectional dataflow is difficult to achieve with an iNML line clock (Fig. 6).

(4) oNML ensembles can be pipelined at the device level independent of clock feature sizes.

III. A CASE STUDY: SYSTOLIC PATTERN MATCHING

We now consider oNML-based systolic [4] pattern matching (PM) hardware. (PM is commonplace in data mining, genomics, intrusion detection, etc.). The fundamental processing element (PE) in a systolic PM circuit appears in Fig. 7a. Individual bits of a data stream serve as one input to an XNOR gate. The other input to the XNOR is a bit (w_i) of a pattern of interest. If the bits match, the output of the XNOR gate is '1'; otherwise a '0' results. This output then becomes one input to the AND gate – the output of which captures the global history of multiple, concatenated PEs (see Fig. 7b). A schematic of four, concatenated, oNML, systolic PM PEs appears in Fig. 7c. After the pipeline is filled, a new match check will exit every $4T_{pulse}$ time units (to satisfy architectural timing constraints). Per micromagnetic simulations, the circuit functioned properly when subjected to an out-of-plane field where H_{pulse} was approximately ±70 mT, and T_{pulse} was 15 ns.

IV. COMPARISONS TO CMOS

With oNML, a new sequence would be tested every 60 ns. This number could be halved with interleaved data streams, *and* T_{pulse} could be reduced to just 2-3 ns [5]. The energy associated with magnet switching events during the re-evaluation of a single PE is ~45 aJ. We anticipate using microcoils to generate out-of-plane fields to produce $\pm H_{pulse}$. The energy from a clock will be a function of: (i) the area of a given PE amortized over microcoil area (~1 μm^2 vs. ~0.01-1 mm^2), (ii) the quality factor Q – a high Q inductor may be used in an LC tank circuit, and a large percentage of the clock-field generating energy can be recycled for subsequent cycles, and (iii) H_{pulse} requirements (per [6], fields of ~10 mT are possible with Co/Ni layers). *Initial* oNML projections for a PM PE without clock overhead are still 100X lower than CMOS designs at iso-performance (Fig. 7d). Additional study is warranted.

IV. REFERENCES

1. Dingler, A., et al., *NANOARCH*, (2009), 21-26.
2. Alam, M.T., et al. *IEEE TNANO*, *11*(2). 273-86, 2012.
3. Carlton, D.B., et al. *Nano Letters*, *8* (12). 4173-78, 2008.
4. Kung, H.T., *Computer*, *15* (1). 37-46, 1982.
5. Barman, A., et al. *JAP*, *101* (9), 109D102-103, 2007.
6. Ju, X., et al. *IEEE TNANO*, *11* (1). 97-104, 2012.

978-1-4673-0996-7/12 $31.00 © 2012 IEEE

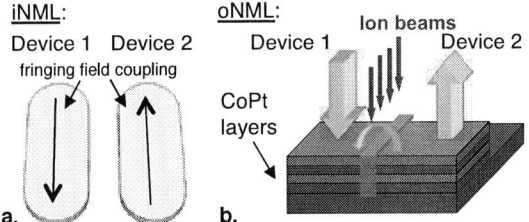

Figure 1: (a) iNML; **(b)** oNML devices can be made by etching, lift-off, focused ion beam (FIB) irradiation, or combinations thereof. Local manipulation of magnetic properties by FIB, changing multilayer composition, etc. enables a large design space.

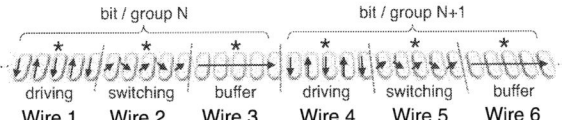

Figure 2: How a line clock would control larger iNML ensembles. Magnets over a given wire could (i) rest in a ground state to drive an adjacent group, (ii) be placed into a 0° state for re-evaluation and to define dataflow directionality, or (iii) relax into a new ground state to transmit information.

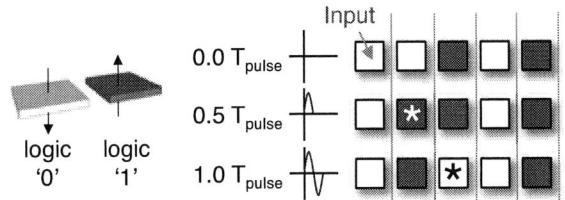

Figure 3: How oNML devices could be clocked: an AF line is used as an example. We assume dots are irradiated on the left edges. These dataflow patterns (as well as a majority gate) have been experimentally demonstrated (INTERMAG 2012).

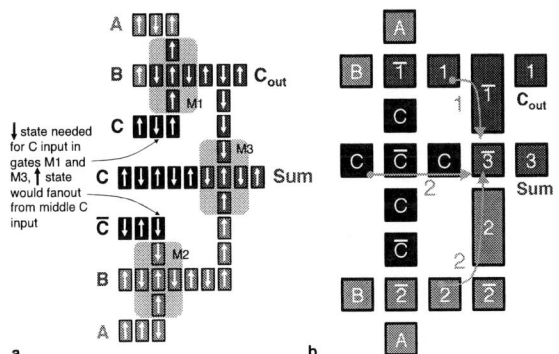

Figure 4: Signal routing in iNML, oNML adders; **(a)** AF-to-F transitions in iNML can be problematic if the original signal and a complement are required as gate inputs; **(b)** the ability to selectively re-size oNML devices eases signal routing complexity.

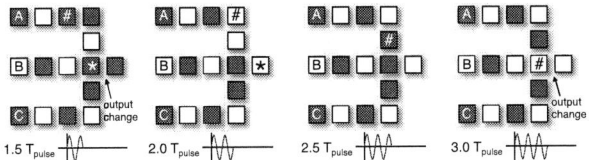

Figure 5: Provided inputs to an oNML gate do not change, data races in majority gates can be avoided. See multiple output changes.

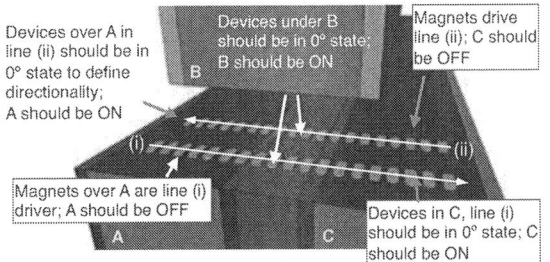

Figure 6: For iNML, assuming a line clock, three clock wires *cannot* facilitate local dataflow in multiple directions as different clock line excitation patterns are needed. While these could be addressed with an electric field based clock (via multiferroics or magnetostriction) more device contacts would be required.

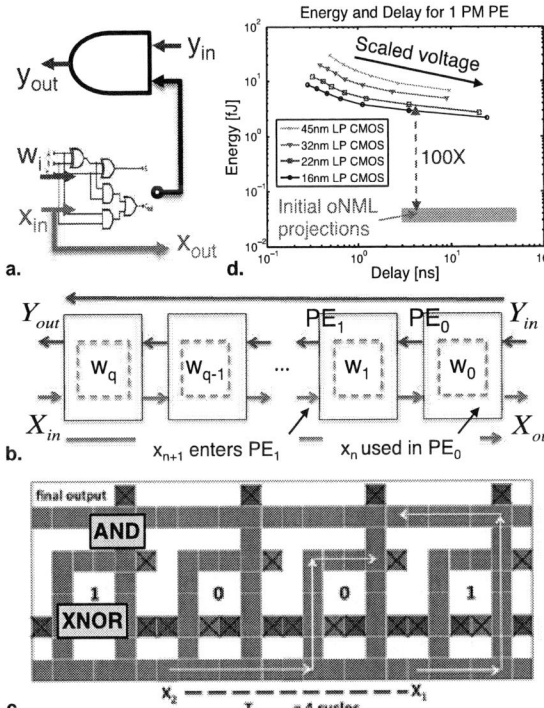

Figure 7: (a) Logic for systolic PM PE. Bits can be stored directly at the gate with no static power loss and be programmed via STT; **(b)** Concatenated PEs to look for N-bit patterns in streaming data. If the output of any XNOR gate is 0, the accumulated global history is set to 0, signifying there is no match in input stream bits x_m ... x_{m+n}. When the shifted global history reaches the leftmost PE, if the initial logic '1' has been preserved, a match is detected. **(c)** Schematic of four, concatenated oNML PEs. By resizing select dots, the critical paths through PEs can be balanced to meet architectural timing requirements. In simulations, square shaped devices were assumed to have 100×100 nm^2 footprints, while elongated dots have 210×100 nm^2 footprints. The distance between dots is 10 nm; **(d)** In CMOS, common hardware-based solutions for finding patterns in data leverage reprogrammable hardware or memory-based finite state machines. To make fair comparisons to oNML-based solutions given the extraneous overheads of memory references, etc., we chose to model a CMOS-based systolic array – where a PE like that shown in Fig. 7a was considered in SPICE using ASU predictive models. Projections for energy and delay for a single PE, at four different technology nodes, and with different supply voltages appear.

978-1-4673-0996-7/12 $31.00 © 2012 IEEE   39

Analysis of static noise margin and power-gating efficiency of a new nonvolatile SRAM cell using pseudo-spin-MOSFETs

Yusuke Shuto[1,3], Shuu'ichirou Yamamoto[2,3], and Satoshi Sugahara[1,3]

[1]ISEL, Tokyo Inst. Tech, Yokohama, Japan, [2]Dept. Info. Proccessing, Tokyo Inst. Tech., Japan, [3]CREST, JST, Japan.

E-mail: shuto@isl.titech.ac.jp

Abstract

Static noise margins (SNMs) and power-gating efficiency were computationally analyzed for our proposed nonvolatile SRAM (NV-SRAM) cell based on pseudo-spin-MOSFET (PS-MOSFET) architecture using spin-transfer-torque MTJs (STT-MTJs). The NV-SRAM cell has the same SNMs as an optimized 6T-SRAM cell. SNMs for other recently-proposed NV-SRAM cells using STT-MTJs were also evaluated, and we showed that their SNMs were deteriorated owing to the effect of the constituent STT-MTJs. Break-even time (BET) and power efficiency were analyzed for the NV-SRAM cell using PS-MOSFETs. The BET can be successfully minimized by controlling the bias of the cell. The average power dissipation can be effectively reduced by power-gating (PG) executions, and the further reduction is made possible by introducing a sleep mode.

Introduction

PG is the most attractive architecture to reduce static power dissipation in advanced CMOS logic systems, such as microprocessors and SoCs. Nonvolatile caches and registers have a great impact on highly efficient PG so-called nonvolatile PG (NV-PG) [1,2]. Recently, we proposed a NV-SRAM cell [3,4] and a nonvolatile flip-flop (NV-FF) [2,5,6] based on pseudo-spin-transistor architecture. These are suitable for NV-PG logic systems. The NV-SRAM cell is configured with a standard 6T-SRAM cell and PS-MOSFETs that consist of an STT-MTJ and an ordinary MOSFET [3]. For the stable operations of the cell, sufficient SNMs for the normal SRAM operations are required as well as ordinary 6T-SRAM cells. However, SNMs have not been systematically analyzed for our proposed NV-SRAM cell and other recently-proposed NV-SRAM cells using STT-MTJs [7-9].

BET is an important index for PG applications of the NV-SRAM cells, which determines a minimum shutdown period of PG [4-6]. For our proposed cell, the BET is mainly governed by the write operation to the STT-MTJs owing to the relatively high write current (I_{MTJ}) for current-induced magnetization switching (CIMS) of the STT-MTJs. Static leakage currents in the cell also affects the BET. For the NV-FF based on the PS-MOSFET architecture, the BET can be reduced by optimizing the size of its constituent transistors [6]. On the other hand, this technique is not suitable for the NV-SRAM cell owing to the restriction in its cell design (the transistor size cannot be freely changed in SRAM design). Thus, microarchitecture for the NV-SRAM cell based on bias control for reducing the BET is highly important.

In this paper, SNMs of our proposed NV-SRAM cell using PS-MOSFETs and other NV-SRAM cells using STT-MTJs were computationally analyzed. NV-PG performance of the NV-SRAM using PS-MOSFETs was also evaluated.

Simulation method

Figure 1(a) shows the circuit configuration of the NV-SRAM cell using PS-MOSFETs [3,4]. The circuit operations were analyzed by the HSPICE program with our developed STT-MTJ model [2] and the standard 65 nm MOSFET technology model [10]. The device parameters used in this paper are listed in Table 1, which were determined by reference to an optimized 6T-SRAM cell [11] and a recently-developed perpendicular-magnetization STT-MTJ [12]. Other NV-SRAM cells [7-9] were also examined using these device parameters unless otherwise noted.

SNM analysis

Solid curves in Figs. 1(b), (c), and (d) show butterfly curves for the read, write, and hold operations of the NV-SRAM cell, respectively. Dashed curves in the figure show those of a standard 6T-SRAM cell as a reference. The shapes of the butterfly curves of the NV-SRAM cell are completely consistent with those of the 6T-SRAM cell. Thus, the SNMs of the NV-SRAM cell are the same as those of the 6T-SRAM cell. This feature comes from the electrical separation of the normal SRAM operation and the NV-PG operation, which is achieved by the PS-MOSFETs in the cell. On the other hand, the SNMs of other NV-SRAM cells [7-9] are severely deteriorated by the effect of the STT-MTJs connected to the inverter loops, as shown in Figs. 2-4. For all the cells, the read operation is the worst case in the SNMs. Figure 5 shows the comparison of the SNMs for the read operation.

Break-even time power-efficiency analyses

Figure 6 schematically shows time evolution of the total static leakage current of our proposed NV-SRAM cell with the currents required for the PG operations, in which the definition of BET and all the parameters used in the BET analysis are also shown. The sequence includes the normal SRAM operation, store, shutdown, restore, and sleep modes. The BET can be divided into two components, i.e., BET_{SR} (the 1st term of Eq. (1)) for the PG operations and BET_L (the 2nd and 3rd terms of Eq. (1)) for the total static leakage current. BET_{SR} can be successfully reduced by decreasing V_{SR} and V_{CTRL}, as shown in Fig. 7(a), although BET_{SR} depends on the leakage current I_L^{SD} during the shutdown mode. BET_{SR} can also be reduced by shortening the write pulse width (τ_{wpw}) for the STT-MTJs [4]. Note that the reduction of τ_{wpw} requires increase in the write current margin I_{MTJ}/I_C.

The leakage current (I_L^{NV}) of the NV-SRAM cell during the normal SRAM operation mode is higher than that of the 6T-SRAM cell owing to the addition of the PS-MOSFETs. However, I_L^{NV} can be sufficiently reduced by applying V_{CTRL}, as shown in Fig. 7(b). Furthermore, this effect is also effective in the sleep mode, as shown in the figure. Thus BET_L can also be minimized by the leakage control. Figure 8 summarizes these bias controls for the BET reduction. The write bias control and the leakage control can effectively reduce BET_{SR} and BET_L, respectively. When the NV-SRAM cell is applied to caches, the shutdown operation without the write operation to the STT-MTJs, i.e., store-free shutdown, is available. This is because there frequently exists the situation that data already-stored in the STT-MTJs of the NV-SRAM cells before shutdown are required after the PG operation. In this case, it is not necessary to rewrite data to the STT-MTJs in the shutdown operation. BET_{SR}^0 is given by substituting E_{store}^0 (which represents the energy required for the store-free shutdown) with E_{store} in Eq. (1). The energy dissipation of this shutdown operation can be highly saved, resulting in very low BET_{SR}^0, as shown in Fig. 8.

The average power P_{ave} of the static leakage power and the power for the PG operations during τ_{cyc} can be reduced by increasing the duty ratio $r_{SD} (= \tau_{SD}/(\tau_{SD}+\tau_{exe}))$ of the shutdown mode, as shown in Fig. 9. The reduction of P_{ave} is enhanced by increasing $r_{sleep} (= \tau_{sleep}/\tau_{exe})$ of the sleep mode. I_L^{SD} affects the P_{ave} reduction rate, as shown in Fig. 9. I_L^{SD} should be suppressed within $\sim 0.2 \times I_L^{NV}$ by optimizing power domain size and power switch design.

References

[1] S. Sugahara, and J.Nitta, *Proc. IEEE*, 98, 2124 (2010). [2] S. Yamamoto, and S. Sugahara, *JJAP*, 48, 043001 (2009). [3] Y. Shuto *et al.*, *JAP*, 105, 07C933 (2009). [4] Y. Shuto *et al.*, *JJAP*, 51, 040212 (2012). [5] S. Yamamoto, and S. Sugahara, *JJAP*, 49, 090204 (2010). [6] S. Yamamoto *et al.*, *IET Electronics Letters*, 47, 1027 (2011). [7] K. Abe, *et al.*, *SSDM 2010*, paper F-9-3, (2010). [8] W. Zhao, et al., *IEEE Trans. Magn.*, 45, 3784 (2009). [9] T. Ohsawa, *et al.*, *SSDM 2011*, paper F-1-2, (2011). [10] Predictive Technology Model (PTM), http://ptm.asu.edu/ [11] K. Nii, *et al.*, *IEEE Solid-State Circuits* 44, 977 (2009). [12] H. Yoda, *et al.*, *Current Appl. Phys.*, 10, e87 (2010).

978-1-4673-0996-7/12 $31.00 © 2012 IEEE

Table 1: Device parameters

MOSFET					STT-MTJ						
L(nm)/W(nm)				$\|V_{th}\|$ (V)	R_P (kΩ)	R_{AP} (kΩ)	TMR ratio	V_{half} (V)	J_C (A/cm²)	I_C (μA)	Junction area (nm²)
Load	Driver	Access	PS-MOSFET								
65/90	65/130	65/90	65/90	0.17	10.2	20.4	100%	0.5	5×10^5	9.8	$\pi\times25^2$

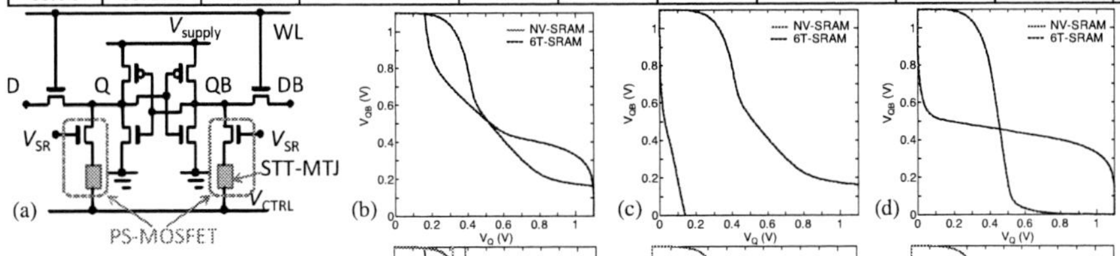

Fig. 1: (a) Circuit configuration of our proposed NV-SRAM cell and its butterfly curves of (b) the read, (c) write, and (d) hold operations. Resistance R_P of the STT-MTJs in parallel magnetization: 5kΩ, 10kΩ, and 20kΩ.

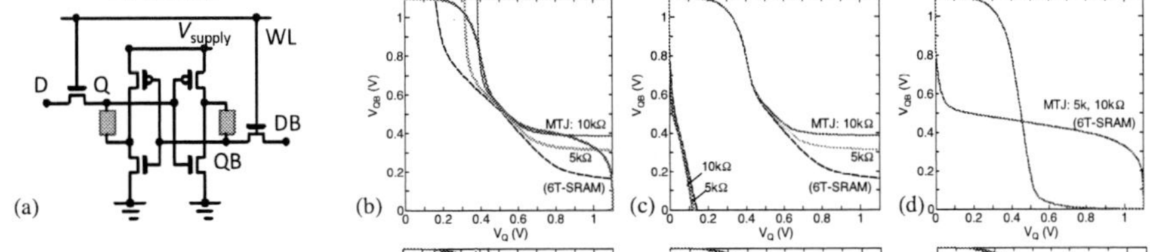

Fig. 2: (a) Circuit configuration of a reference NV-SRAM cell [7] and its butterfly curves of (b) the read, (c) write, and (d) hold operations. R_P: 10kΩ and 5kΩ.

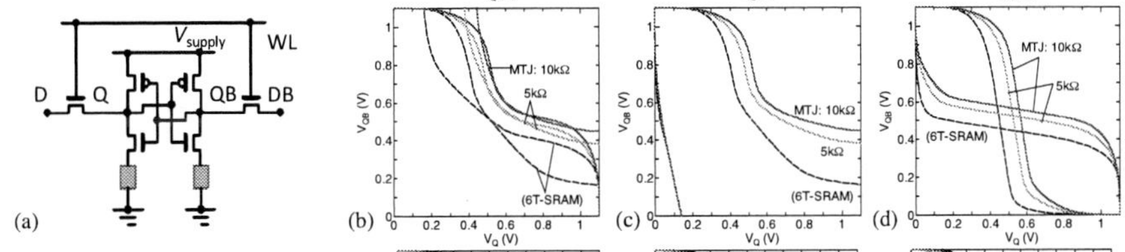

Fig. 3: (a) Circuit configuration of a reference NV-SRAM cell [8] and its butterfly curves of (b) the read, (c) write, and (d) hold operations. R_P: 10kΩ and 5kΩ.

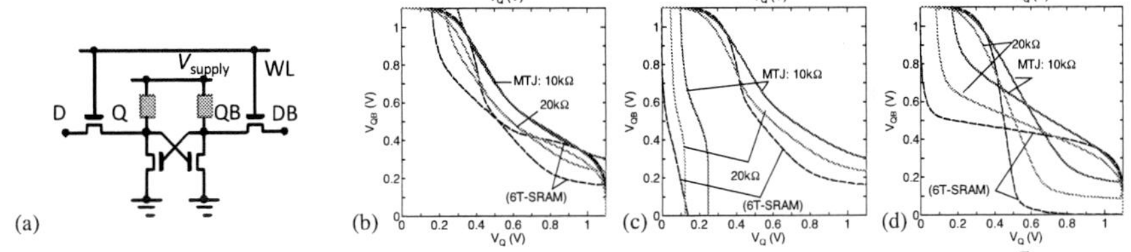

Fig. 4: (a) Circuit configuration of a reference NV-SRAM cell [9] and its butterfly curves of (b) the read, (c) write, and (d) hold operations. R_P: 10kΩ and 20kΩ.

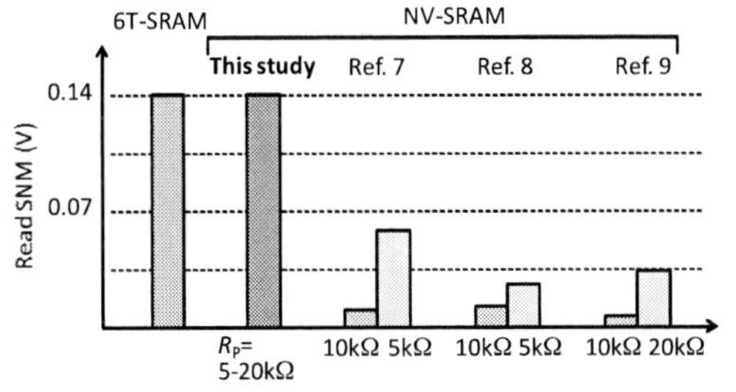

Fig. 5: Comparison of the SNMs for the read operation of the 6T-SRAM and various NV-SRAM cells. R_P is the resistance of the STT-MTJs in parallel magnetization

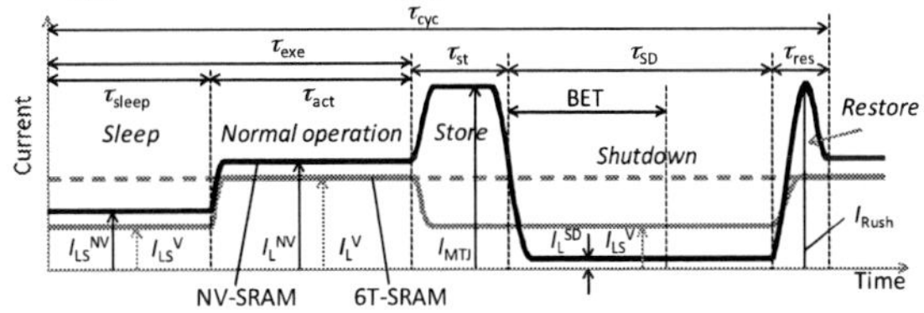

$$BET = BET_{SR} + \eta_{LS}^{NV} r_{sleep} \tau_{exe} + \eta_L^{NV} \frac{V_{DD}}{V_{sleep}} (1 - r_{sleep}) \tau_{exe} \quad \cdots\cdots(1)$$

$$BET_{SR} = \frac{(1-r_{SF})E_{store} + r_{SF}E_{store}^0 + E_{restore}}{(I_{LS}^V - I_L^{SD})V_{sleep}}, \quad \eta_{LS}^{NV} = \frac{I_{LS}^{NV} - I_{LS}^V}{I_{LS}^V - I_L^{SD}}, \quad \eta_L^{NV} = \frac{I_L^{NV} - I_L^V}{I_{LS}^V - I_L^{SD}}$$

Fig. 6: Time evolution of the static leakage current and the current required for PG of the NV-SRAM cell. In the store operation mode, data on the inverter loop are stored into the STT-MTJs of the cell by CIMS, and in the restore operation mode, the stored data are restored to the inverter loop by pull-up of V_{supply}[3].

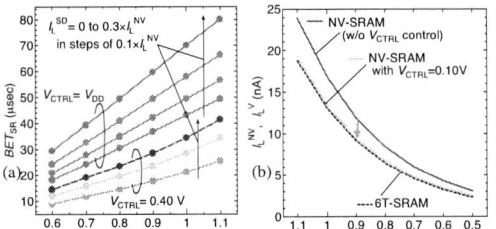

Fig. 7: (a) BET_{SR} as a function of V_{SR} for $V_{CTRL} = V_{DD}$ and 0.4V. I_L^{SD} is assumed to be varied from 0 to $0.3I_L^{NV}$ in steps of $0.1I_L^{NV}$ (b) I_L^{NV} as a function of V_{supply} with or without V_{CTRL} control. I_L^V of the 6T-SRAM cell is also shown in the figure.

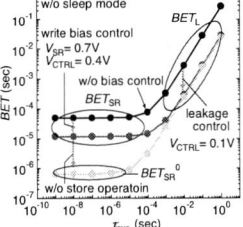

Fig. 8: BET as a function of τ_{exe} without the sleep mode, in which various BET reduction techniques are employed.

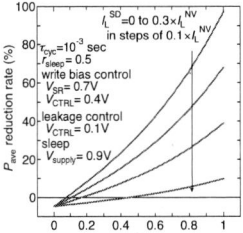

Fig. 9: Reduction rate of P_{ave} as a function of r_{SD} for $r_{sleep} = 0.5$, in which P_{ave} is compared to that of the 6T-SRAM cell with the sleep mode.

978-1-4673-0996-7/12 $31.00 © 2012 IEEE

Recent Progress of Resistive Switching Random Access Memory (RRAM)

Yi Wu, Shimeng Yu, Ximeng Guan, H.-S. Philip Wong,
Department of Electrical Engineering, Stanford University, Stanford, CA 94305, USA

Abstract

This paper gives an overview of recent works on metal oxide resistive switching memory (RRAM). We explored the stochastic nature of resistive switching in metal oxide RRAM and a 2-D analytical solver was established to explain the switching parameter variations in HfOx-based RRAM. As an example of application beyond digital memory/storage, AlOx-based RRAM was explored for neuromorphic computing.

Introduction

Metal oxide RRAM has attracted considerable attention in recent years. It is one of the most promising candidates for future non-volatile memory application [1-2]. Suitable materials include complex metal oxides such as the pervoskite oxides of SrTiO3[3], SrZrO3 [4], ferromagnetic materials such as (Pr, Ca)MnO3[5] and the binary metal oxides such as NiO [6], TiO2 [7], HfOx [8], AlOx [9]. Among these resistive switching materials, binary metal oxides are extensively studied because of the simplicity of the material and compatibility with silicon CMOS BEOL fabrication process. Recent advances in binary metal oxide RRAM reported in literature are summarized in **Table 1**. Besides the metal oxide material itself, the materials for the top and the bottom electrodes are also critical in determining the memory device characteristics [29]. It is found that even for the same oxide material but with different electrode materials, the switching mode can be drastically different. The early RRAM had large device areas, large switching currents (~mA), slow programming speeds (~µs), low endurance cycles (<10^3 cycles), and required a 1T1R. Today, many of these shortcomings have been addressed. RRAM device sizes now aggressively scale down to 10nm × 10nm; programming currents are now in the order of µA range; programming and read speed are on the order of ns; endurance cycles are up to 10^{12}; retention time is near 3000 hours at 150°C; the forming process can be eliminated. Chip-scale memory array with Mb size were demonstrated. Demonstrations of multi-bit operation have been made, and rudimentary demonstration of integrated 1D1R type device for 3D integration looks promising even though they do not meet all the requirements at this point.

However, further understanding of the underlying physics of RRAM, especially the switching parameter variations, is required for industry large-scale manufacturing. Such understanding is crucial for establishing the scalability of RRAM memory technology and making reliability (such as retention and endurance) projections for product qualification. In a recent work [30-31], we have established a Monte Carlo 2D numerical simulator to quantify the electron conduction and the stochastic generation, recombination, and migration process of oxygen vacancies (Vo) / ions during the switching. This model empirically reproduced the experimentally observed abrupt set process, the gradual reset process, the current fluctuations and the switching parameter variations. It provided insights into the origin of the tail bits of high resistance states (HRS) distribution.

While RRAM has the potential as a non-volatile memory technology, another emerging application is the use of RRAM as electronic synapse element for hardware implementation of neuromorphic computing [29]. Due to RRAM's multilevel storage capability and low power consumption, it can behave like an analog memory emulating the function of plastic synapses [32] in a neural network. Recently, TiOx [33], WOx [34], and HfOx [35] based synapses have shown spike timing-dependent plasticity (STDP) behavior at the device level. In an earlier work, we have demonstrated multi-level resistance states controllability of AlOx-based RRAM for low power memory device application [36]. We further explore the feasibility of using AlOx-based RRAM as electronic synapse element [37]. Gradual resistance modulation with less than 1.7% resistance change per switching was demonstrated. STDP-like curve was shown which suggests this AlOx-based has the potential for use as a neuromphic computation device.

Stochastic Nature of Resistive Switching in RRAM

One of the commonly recognized bipolar switching mechanisms can be interpreted in this way: the set process (high resistance state (HRS) to low resistance state (LRS)) is attributed to the formation of conduction filaments (CFs) by generation of oxygen vacancy (Vo), while the reset process (LRS to HRS) is attributed to the rupture of CFs by recombination of the Vo with the oxygen ions that migrate from the oxygen reservoir at the electrode/oxide interface [38].

TiN/HfOx/Pt memory stack was fabricated and electrical measurement results were reported in [39]. The insensitivity of the measured current to the temperature shown in Fig. 2 suggests that trap-assisted-tunneling (TAT) current is the dominate conduction mechanism [40]. Low frequency noise (LFN) measurement was conducted and the $1/f^{\alpha}$-like DC noise power spectral density (PSD) was reported in Fig. 3 [41]. $\alpha \approx 1$ for LRS and $\alpha \approx 2$ for HRS. At HRS, the index α ranges from 1 to 2 at a certain cut-off frequency which indicates that there is a tunneling gap between electrodes and the residual of the CFs in the HRS. AC conductance measurement was performed and f^{β}-like ($\beta \approx 2$). AC conductance is reported in Fig. 4 [42]. The index β is independent of resistance values of different HRS, suggesting that the AC conductance is due the electron hopping between the nearest neighbor traps within the CFs.

Based on the TAT model, we built a numerical solver to calculate the electron occupation probability of all the traps which accounts for all the possible tunneling paths between the traps, and between the traps and the electrodes. Fig. 5 shows the simulated switching I-V curves of forming/reset/set and the corresponding Vo configuration. Experimental and simulated DC I-V and reset transient current curves of the HfOx-based RRAM device are compared in Fig. 6 and Fig. 7, respectively. For both DC and pulse programming, the reset process is gradual and current fluctuations are observed. By tracking the evolution of Vo in the simulation, it is found that a current jump is due to the generation of a new Vo in the gap region. In Fig. 8, the simulated HRS distribution after 1000-times pulse cycling correlated with their Vo configuration is shown. It is revealed that the log-normal spread of HRS is due to the variation of the average gap distances, while the tail bits of HRS are because of the Vo generation near the electrode at the end of a programming pulse. To reduce the tail bits of HRS, the reset-verify technique [16] can be used but this may introduce over-reset bits in the HRS. We propose using an

978-1-4673-0996-7/12 $31.00 © 2012 IEEE

additional buffer oxide layer which has a larger oxygen ion migration barrier to confine the switching within the active oxide to prevent over-reset.

AlOx-based RRAM for Neuromorphic Computing

A typical neuromorphic computation system is based on neuron circuits and synapses. The synapses are the connection bridges between the neuron circuits and each neuron may have more than 1000 synapse connections with other neurons. So challenges for hardware implementation of neuromorphic computation system are to develop the electronic synapse that is ultra-high integration density and has very low power consumption.

TiN/Ti/AlOx/TiN memory stack with active area of $0.48\mu m \times 0.48\mu m$ was fabricated on Ti/SiO$_2$/Si substrate. A sketch of the device structure was given in Fig. 9(a) along with a transmission electron microscopy (TEM) image of the cross-section area in Fig. 9(b). All tests were conducted under the condition that the bottom electrode TiN side was grounded and the signals were applied to the top electrode.

A typical current-voltage (I-V) characteristic of AlOx-based RRAM was given in Fig. 10 with inset of the forming curve. During the DC set process, compliance current (CC) is set by the semiconductor parameter analyzer to prevent excessive current overshooting. It is known that the LRS resistance of RRAM can be controlled by CC levels and the HRS can be modulated as function of the V$_{STOP}$ voltages [36]. Five discrete resistance levels are obtained either by changing the CCs or the V$_{STOP}$ magnitudes (Fig. 11). The resistance modulation windows using the two techniques are almost identical but the underlying physics is different. Increasing the CC during set widens the dominant filament. Enlarging the V$_{STOP}$ could extend the tunneling gap distance between electrode and residual CF at HRS. Next, we consider using this AlOx-based RRAM as electronic synaptic element. For use as a synaptic element, very fine control of the RRAM resistance, i.e. close to 1% change per synaptic activity is required [43]. Fig. 12 shows that one can gradually increase the conductance by changing the CC levels from $50\mu A$ to $900\mu A$ using $500\mu s$ positive SET pulses at 1.5V. Similarly the conductance can gradually change by modifying the magnitudes of RESET negative pulses from -1V to -1.6V. An order of magnitude conductance change was obtained through 85 steps for SET and 60 steps for RESET transitions: average 1.2% and 1.7% resistance change per switching for SET and RESET respectively.

Spike-timing dependent plasticity (STPD), for which synaptic weight changes depend on relative spike timings of pre- and post-synaptic neurons, has been discovered in several biological systems [44-45]. For an electronic synapse to emulate the behavior of the biological synapse, it would be necessary that the conductance of the electronic synapse could be continuously modified by the relative arrival timing of the input impulses. A scheme to translate the input pulse arrival timing differences into RRAM programming pulses was proposed (Fig.13). Using the scheme above, a STDP-like curve could be reproduced for using the RRAM fabricated above (Fig. 14). The relative conductance changes were determined from data shown in Fig. 11. The timing has arbitrary unit because it depends on the length of the timeslot. This result illustrates that the AlOx-based RRAM has the potential to be served as electronic synapse element in a neuromphic computing system.

Conclusion

In this paper, we review recent progress of metal oxide RRAM focusing on device modeling and new application in neuromorphic system. The metal oxide RRAM has been developing rapidly in the past several years from down-scaling of device sizes to chip-scale memory array demonstration. However, challenges still remain before RRAM is ready for manufacturing: (1) better uniformity in device characteristics for using the RRAM in large memory array; (2) more complete understanding of the conduction and resistive switching mechanism. Meanwhile, it would be interesting to further investigate the potential using nano-scale RRAM as artificial synapse element in hardware neuromorphic computing at the system level.

Acknowledgement

This work is supported in part by DARPA, the National Science Foundation, the Nanoelectronics Research Initiative of the Semiconductor Research Corporation, and the Stanford Non-Volatile Memory Technology Research Initiative. Y.W. is additionally supported by a Fellowship from the O.G. Villard Engineering Fund at Stanford and S. Yu is supported by the Stanford Graduate Fellowship and IEEE Electron Devices Society Masters Student Fellowship. The authors thank our collaborators at Industrial Technology Research Institution, Hsinchu, Taiwan for providing the AlOx-based RRAM, discussions, and summer internship for Y.W.

References

[1] R. Waser, et al., Adv. Mater., vol. 21, pp. 2632-2663, 2009; [2] M. J. Lee, et al., Nano Lett., vol. 9, pp. 1476-1481, 2009; [3] Y. Watanabe, et al., Appl. Phys. Lett., vol. 78, pp. 3738-3740, Jun 4 2001.; [4] A. Asamitsu, et al., Nature, vol. 388, pp. 50-52, Jul 1997; [5] C. C. Lin, et al., IEEE Trans. on Electron Devices, vol. 54, pp. 3146-3151, 2007; [6] B. Lee, et al., VLSI, pp. 28-29, Kyoto, Japan, 2009; [7] S. Kim, et al., IEEE Trans. on Electron Devices, vol. 56, pp. 3049-3054, 2009; [8] H. Y. Lee, et al., IEDM, pp. 297-300, 2008; [9] Y. Wu, et al., IEEE Electron Device Lett., vol. 31, pp. 1449-1451, 2010; [10] I. G. Baek, et al., IEDM, 2004, pp. 587-590; [11] A. Chen, et al., IEDM, 2005, pp. 746-749; [12] D. Lee, et. al., IEDM, pp. 30.8.1-30.8.4, 2006; [13] K. Tsunoda, et al., IEDM, pp. 767-770, 2007.; [14] Z. Wei, et al., IEDM, pp. 721-724, 2011; [15] B. Lee, et. al., VLSI, pp. 28-29, 2009; [16] Y. S. Chen, et al., IEDM, pp. 95-98, 2009; [17] H. Y. Lee, et. al., IEDM, pp. 19.7.1-19.7.4, 2010; [18] Y. H. Tseng, et. al., IEDM, pp. 99-102, 2009; [19] Y. Sakotsubo, et. al., VLSI, pp. 87-88, 2010; [20] C. Ho, et. al., IEDM, pp. 436-439, 2010; [21] W. C. Chien, et al., IEDM, pp.440-443, 2010; [22] C. H. Cheng, et al., IEDM, pp. 448-451, 2010; [23] J. Lee, et al., IEDM, pp. 19.5.1 - 19.5.4, 2010; [24] K. Wanki, et al., VLSI, pp. 22-23, 2011; [25] X. A. Tran, et. al., VLSI, pp. 44-45, 2011; [26] J. Yi, et. al., VLSI, pp. 48-49, 2011; [27] Y.-B. Kim, et al., VLSI, pp. 52-53, 2011; [28] B. Govoreanu, et al., IEDM, pp. 729-732, 2011; [29] H.-S. P. Wong, et. al., Proc. IEEE, invited review, 2011; [30] X. Guan, et. al., IEEE Trans. Electron Devices, vol. 59, pp. 1172-1182, 2012; [31] S. Yu, et. al., IEEE Trans. Electron Devices, vol. 59, pp. 1183-1189, 2012; [32] G. S. Snider, NANOARCH, pp. 85-9292, 2008; [33] K. Seo, et al., Nanotechnology, vol. 22, pp. 254023, Jun 24 2011; [34] T. Chang, et al., ACS Nano, vol. 5, pp. 7669-7676, Sep 2011; [35] S. Yu, et al., IEEE Trans. on Electron Devices, vol. 58, pp. 2729-2737, 2011; [36] Y. Wu, et. al., J. Appl. Phys., pp.110-114, 2011; [37] Y. Wu, et. al., accepted, IMW, 2012; [38] N. Xu, et al., Symp. VLSI Technol., 101, 2008; [39] S. Yu, et al., VLSI-TSA, 2011; [40] S. Yu, et al., Appl. Phys. Lett., vol. 99, 063507, 2011; [41] S. Yu, et al., Appl. Phys. Lett., vol. 99, 232105, 2011; [42] S. Yu, et al., Phys. Rev. B, vol. 85, 045324, 2012; [43] D. Kuzum, et al., Nano Lett., June 14, 2011; [44] L. I. Zhang, et al., M. Nature, 395, 1998; [45] H. Markram, et al., Science, 275, 1997.

978-1-4673-0996-7/12 $31.00 © 2012 IEEE

	NiO IEDM 2004	Cu₂O IEDM 2005	Cu: MoOₓ IEDM 2006	Ti:NiO IEDM 2007	TaOₓ IEDM 2008	HfOₓ IEDM 2008	NiO VLSI 2009	HfOₓ IEDM 2009 &2010	TiON IEDM 2009	Ta₂O₅/ TiO₂ VLSI 2010	WOₓ IEDM 2010	WOₓ IEDM 2010	GeO/Hf ON IEDM 2010	ZrO₂/ HfOₓ IEDM 2010	N:AlO VLSI 2011	HfOₓ/ AlOₓ VLSI 2011	TiOₓ/ AlₓOᵧ VLSI 2011	TaOₓ/ Ta₂O₅ VLSI 2011	Hf/HfOₓ IEDM 2011
Switching type	unipolar	bipolar	bipolar	unipolar	bipolar	bipolar	unipolar	bipolar	unipolar	unipolar	bipolar	bipolar	bipolar	bipolar	bipolar	unipolar	bipolar	bipolar	bipolar
Structure	1T-1R	1T-1R	1R	1T-1R	1T-1R	1T-1R	1R	1T-1R	1T-1R	1T-1R	1R	1T-1R	1R	1R	1T-1R	1R	1T-1R	1R	1T-1R
Cell Area (μm²)	~0.2	~0.03	~25	~0.49	~0.25	~0.1	0.0023 (48nm)	0.0009 (30nm)	0.19	~3	8.1E-5 (9nm)	0.0036 (60nm)	11300	0.0025 (50nm)	1	~6000	0.0029 (54nm)	~9000	0.0001 (10nm)
Speed	~5us	~50ns	~10ns	~5ns	~10ns	~5ns	180ns	~300ps	~10us	~1us	~1us	~50ns	~20ns	~40ns	N/A	~30ns	~10ns	~10ns	~10ns
DC Peak Voltage	<3V	<3V	<2V	<3V	<2V	<1.5V	<2V	<2.5V	<4V	<3V	<4V	<3V	<3V	<2V	<2V	<2.5V	<3V	<2.5V	<1.5V
DC Peak Current	~2mA	~45μA	~0.5mA	~100μA	~17μA	~25μA	~90μA	~200 μA	~150 μA	~200 μA	~1μA	~1mA	~100nA	~50μA	~50nA	~1mA	~20μA	~30μA	~50 μA
HRS/LRS Ratio	>10	>10	>10	>90	>10	>1,000	>25	>1000	>20	>100	>10	>10	>700	>10	>100	>1000	>10	>100	>10
Endurance	10⁶	600	10⁶	100	10⁹	10⁵	7x10³	10¹⁰	10⁶	10⁶	200	10⁶	10⁵	10⁵	10⁵	10⁵	N/A	10¹²	5x10⁷
Retention	300h@ 150°C	30h@ 90°C	28h@ 85°C	1000h@ 150°C	3000h@ 150°C	10h@ 200°C	N/A	28h@ 150°C	1000h@ 150°C	120h@ 100°C	280h temp. N/A	2000h@ 150°C	3h@ 125°C	28h@12 5°C	28h@ 125°C	28h@ 150°C	100h@ 150°C	3h@ 200°C	30h@ 250°C

Table 1 A representative list of binary metal oxide RRAM device characteristics. Data are collected from Baek 04 [10], Chen 05 [11], Lee 06 [12], Tsunoda 07 [13], Wei 08 [14], Lee 08 [8], Lee 09 [15], Chen 09 [16], Lee 10 [17], Tseng 09 [18], Sakotsubo 10 [19], Ho 10 [20], Chien 10 [21], Cheng 10 [22], Lee 10 [23], Kim 11 [24], Tran 11 [25], Yi 11 [26], Kim 11 [27], Govoreanu 11 [28].

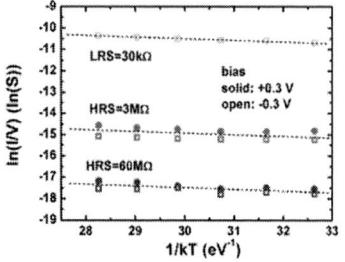

Fig. 2 ln(I/V) vs. 1/kT for multilevel resistance states of TiN/HfOx/Pt RRAM at both positive and negative biases. Small activation energy obtained and this insensitivity of measured current to temperature suggests that trap-assisted-tunneling (TAT) current is the dominate conduction mechanism.

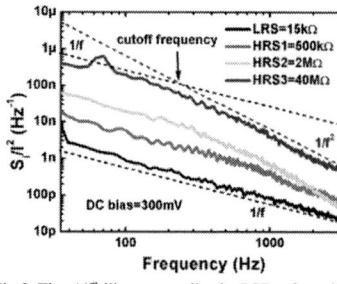

Fig.3 The $1/f^{\alpha}$-like normalized PSD for different resistance states. The slope index $\alpha \approx 1$ for LRS because multiple transition with various relaxation times is allowed, and α changes to 2 for HRS at the cutoff frequency because the shortest tunneling path cause a minimum relaxation time. From the LFN characterization, it is suggested that the switching between LRS and HRS is due to a formation of tunneling gap by partially rupturing the CFs.

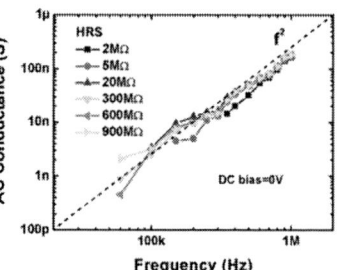

Fig. 4 f^{β}-like ($\beta \sim 2$) AC conductance was observed in HfOx based resistive switching memory. The rise of AC conductance above a certain corner frequency is attributed to the electron hopping between the nearest neighbor traps in the CFs.

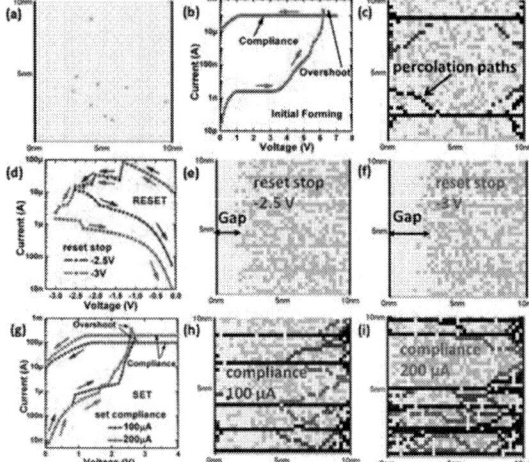

Fig. 5 Cross-section view of the simulated cell (left electrode: positive bias for forming/set and negative bias for reset, pink points are Vo), simulated cell (10 nm10 nm) corresponds to the weak spot region of a RRAM cell, e.g. the grain boundary. (a) initially randomly distributed Vo in as-fabricated cell; (b) forming process; (c) percolation paths after forming; (d) reset process with different stop voltages; (e) & (f) smaller/larger gap due to smaller/ larger reset stop voltage; (g) set process with different compliance current; the current overshoot during forming/set is also shown; (h) & (i) fewer/more percolation paths due to smaller/larger compliance current. Percolation paths are found by the Dijkstra algorithm and color-coded in grey scale according to the conducting strength (darker means stronger).

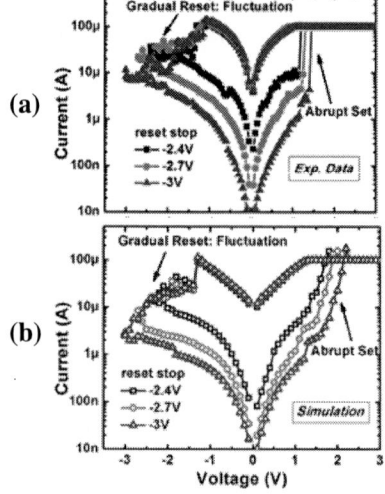

Fig. 6 (a) Experimental and (b) simulated I-V characteristics of HfOx memory for different reset stop voltages. Abrupt set and gradual reset are reproduced.

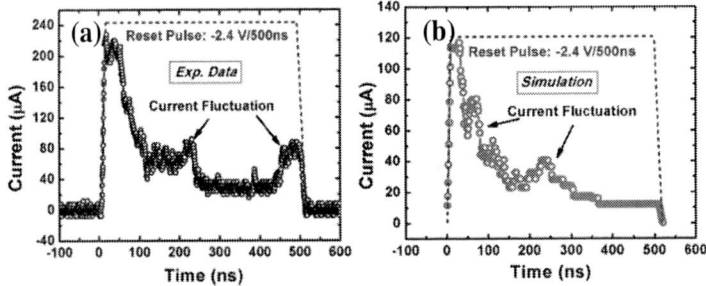

Fig. 7 (a) Experimental and (b) simulated pulse transient current in the reset process. Current fluctuation is observed, and before the end of pulse, current jumps due to new Vo generation.

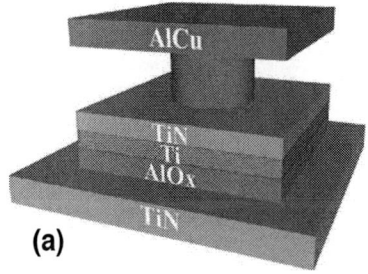

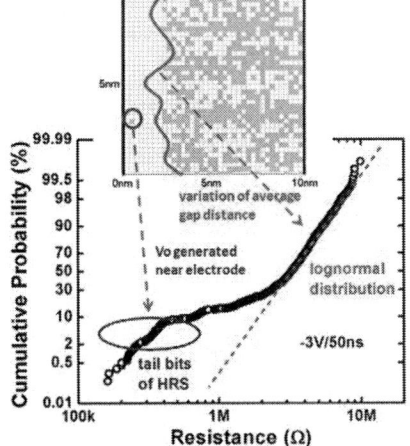

Fig. 8 Simulated HRS distribution after 1000-times pulse cycling. The lognormal distribution is due to the Gaussian distribution of the average gap distances. The tail bits are due to new Vo near the electrode generated at the end of the pulse near the electrode at the end of a programming pulse. To reduce the tail bits of HRS, the reset-verify technique [16] can be used but this may introduce more over-reset bits of HRS.

Fig.9 (a) A schematic of the memory structure; (b) TEM picture of the Ti/AlOx/TiN memory device.

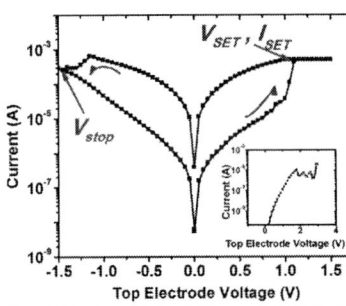

Fig. 10 I-V characteristic of AlOx-based RRAM with inset of forming curve.

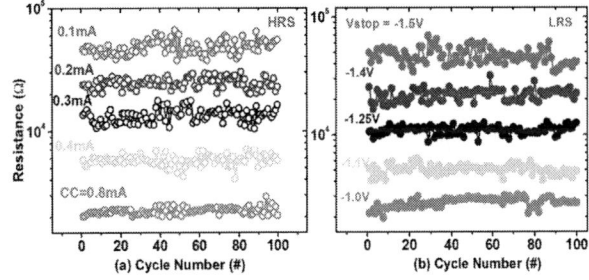

Fig. 11 (a) DC endurance property by changing the compliance current during SET; (b) by changing the increasing the stopping voltage amplitude during RESET.

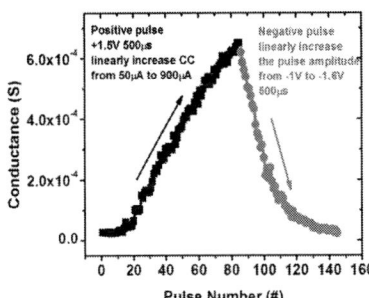

Fig. 12 Gradual SET of the memory device resistance is implemented by using pulses with increasing compliance current setup; gradual RESET is by increasing the negative applied voltage amplitudes. An order of magnitude conductance change in AlOx-based RRAM was obtained through 85 steps for SET and 60 steps for RESET transitions: average 1.2% and 1.7% resistance change per switching for SET and RESET respectively

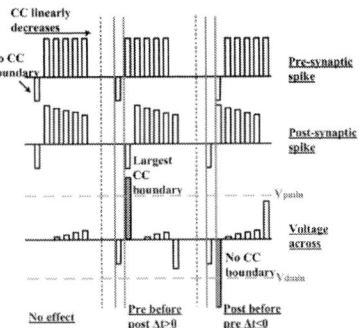

Fig. 13 Schematic of the spike-timing-dependent plasticity (STDP) realization scheme. Vpmin and Vdmin are the minimum voltage amplitudes which induce potentiation and depression, respectively. Only the blue-shaded pulses beyond Vpmin or Vdmin could actually change the resistance of memory device.

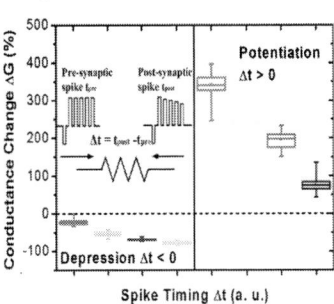

Fig. 14 STDP-like curve using with the resistance modulation data in Fig. 11. . Potentiation based on data in Fig.4 (a) using compliance current levels control resistance and depression based on data in Fig.4 (b) using RESET stop voltages modulate conductance.

978-1-4673-0996-7/12 $31.00 © 2012 IEEE

Bidirectional Selection Device Characteristics of Ultra-Thin (<3nm) TiO$_2$ layer for 3D Vertically Stackable ReRAM Application

Jiyong Woo[1], Jubong Park[2], Jungho Shin[2], Godeuni Choi[1], Seonghyun Kim[2], Wootae Lee[2],
Sangsu Park[3], Daeseok Lee[1], Euijun Cha[3], and Hyunsang Hwang[1]

[1]Department of Materials Science and Engineering, Pohang University of Science and Technology (POSTECH)
Pohang, Republic of Korea, Phone: +82-54-279-5123, Fax: +82-54-279-2399, E-mail: jiyongis@postech.ac.kr
[2]School of Materials Science and Engineering, [3]Department of Nanobio Materials and Electronics
Gwangju Institute of Science and Technology (GIST), Gwangju, Republic of Korea

Abstract

We propose the feasibility of bidirectional selection device characteristics in ultrathin (<3nm) TiO$_2$ layer. We utilized the localized conducting path as virtual electrode to investigate device property at extremely scaled area. By using electrical method such as "forming" and "reset" processes in oxide, virtual electrode/sub-3nm-thick TiO$_2$/virtual electrode structure was achieved. The measured current-voltage characteristics of fabricated device exhibited uniform bidirectional selection behavior with a high selectivity ($\sim10^5$) and showed the feasibility of high current density ($>10^6$A/cm^2).

Introduction

Cross-point array architectures have been considered as promising candidates for next-generation nonvolatile memory architectures as their scalability and simplicity can help resolve the physical scaling limitation of conventional flash memory structures [1]. However, to prevent the undesired sneak path current, a selection device having non-linear characteristics is necessary, as shown in Fig. 1. Several types of selection devices, such as metal-insulator transition (MIT) [2], ovonic threshold switching (OTS) [3], mixed ionic-electronic conductors (MIEC) [4], and metal-oxide-based metal-insulator-metal (MIM) devices [5] have been proposed.

In this paper, we fabricated and investigated a selection property in ultrathin (<3nm) TiO$_2$ layer with virtual electrode/sub-3nm-thick TiO$_2$/virtual electrode structure by using electrical methods. The characteristics of the uniform bidirectional selection device were observed, for example, such as high selectivity and high current density.

Experiment

For the selection behavior layer, a 3-nm-thick TiO$_2$ film was deposited at 150°C by an atomic layer deposition (ALD) system using tetraisopropoxide as the precursor and H$_2$O as the oxidizer. Then, a Pt top electrode (TE) was deposited using an RF sputtering system.

Results and discussion

Fig. 2 show cross-sectional TEM image of 3nm-thick TiO$_2$ layer in 250nm via-hole structure. To investigate the device property at extremely scaled area, here we fabricated device with ultrathin TiO$_2$ layer by utilizing localized conducting path as virtual electrode. Fig. 3 shows the fabrication steps to form ultrathin TiO$_2$ layer by using electrical methods such as "forming" and "reset" processes. Fig. 4 shows the final structure of ultrathin TiO$_2$ layer.

Fig. 5 (a) shows the initial I-V characteristics of the Pt/TiO$_2$/Pt device. When a 3 V bias was applied to the TE, the device showed forming process. To explain the forming process in the TiO$_2$ layer, we performed a conductive AFM (C-AFM) analysis on the as-deposited state of the TiO$_2$ layer. Several conducting paths were observed in the TiO$_2$ layer and area of conducting paths were estimated to be about 5~10nm, as shown in Fig. 5 (b). It is believed that the oxygen deficient path or metallic Ti channel was formed by migrating oxygen ions and oxygen vacancies in the TiO$_2$ layer under an applied bias. Consequently, the localized conducting path was formed by forming process in the TiO$_2$ layer.

The electrical analysis of the forming and reset processes in the Pt/TiO$_2$/Pt device is shown in Fig. 6. By utilizing filament, which was observed by C-AFM, as virtual electrode, we ruptured filament by reset process. As shown in Fig. 6, the reset process could be achieved under either positive or negative voltage during the second sweep. That is, the reset process did not depend on the bias polarity, and Joule heating is considered to be the dominant reset mechanism. In addition, the hottest spot is located in the middle of the filament, as was determined from a simulation result for the temperature distribution in the reset process. As a result, the remaining filament after the reset process can act as the virtual TE (BE) instead of the Pt TE (BE), and the symmetric device structure of the virtual electrode/sub-3nm-thick TiO$_2$/virtual electrode can be achieved.

Fig. 7 (a) shows the bidirectional non-linear I-V characteristic of ultrathin TiO$_2$ layer. The high selectivity (on the order of 10^5), which was calculated by the ratio of the currents at 2 V and at 4 V, is observed. In a non-linear region, the conduction behavior is dominated by Fowler-Nordheim (F-N) tunneling across the barrier, as shown in Fig. 7 (b). Fig. 7 (c) shows the on and off current distribution at 2V and 4V for 100 switching cycles. Furthermore, highly scaled device offers to the high current density. As a results, a current density as high as 10^6 A/cm^2, which is sufficient for operating the ReRAM device, is obtained by assuming the effective area of the virtual electrode is about 10 nm, which was estimated by C-AFM. Stable selection behavior in pulse endurance characteristics over 10^4 repeated cycles are confirmed, as shown in Fig. 7 (d).

Summary

The feasibility of bidirectional selection behavior in ultrathin TiO$_2$ layer with reduced active area was investigated. The ultrathin TiO$_2$ layer was achieved by utilizing localized conducting path as virtual electrode formed by electrical methods. The measured device characteristics at extremely scaled area showed uniform bidirectional selection behavior with high selectivity ($\sim10^5$) and high current density ($>10^6$A/cm^2).

Acknowledgements

This work was supported by the National Research Foundation of Korea (NRF) grant funded by the Korea government (MEST) (No. 2011-0018646) and by R&D Program of the Ministry of Knowledge Economy.

References

[1] J. Borghetti, et al., Nature. Vol. 464, pp. 873-876, April (2010)
[2] M. –J. Lee, et al., Adv. Mater., 19, pp. 73 (2007)
[3] D. Kau, et al., IEDM tech. Dig., pp. 617 (2009)
[4] K. Gopalakrishnan, et al., Symp. on VLSI Technology, pp. 205-206 (2010)
[5] J. –J. Huang, et al., IEEE Electron Device Letters, 32, pp.1427-1429 (2009)

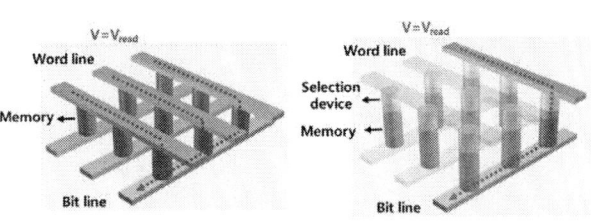

Fig. 1 (a) The interference in the arrays from neighboring cells in the form of sneak path currents leads to a misreading problem during read operation. (b) Rectified read operation by suppressing the sneak path currents.

Fig. 2 Cross-sectional TEM image of Pt/TiO₂/Pt device in 250nm via-hole structure.

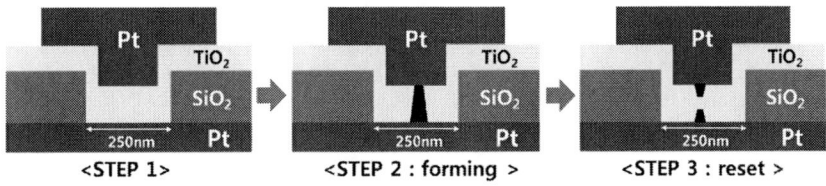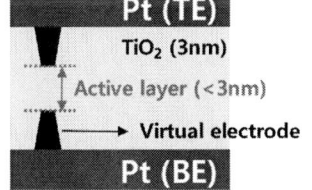

Fig. 3 Illustration of steps to fabricate ultrathin TiO₂ layer. Deposition of 3nm TiO₂ layer by ALD at step 1. Formation of localized conducting path (filament) by forming process at step 2. Rupturing of filament by reset process at step 3.

Fig. 4 The active layer is sub-3nm and area of device is between virtual electrodes.

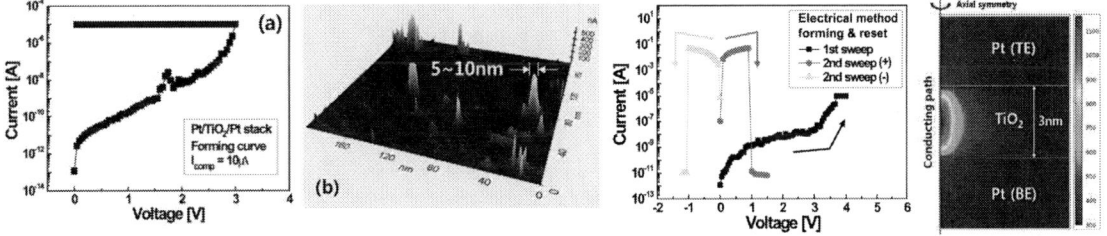

Fig. 5 (a) Forming curve of Pt/TiO₂/Pt device. (b) C-AFM image of 3nm-thick TiO₂/Pt layer.

Fig. 6 The electrical analysis (left) of electrical methods such as forming and reset process and thermal simulation result (right) at reset process.

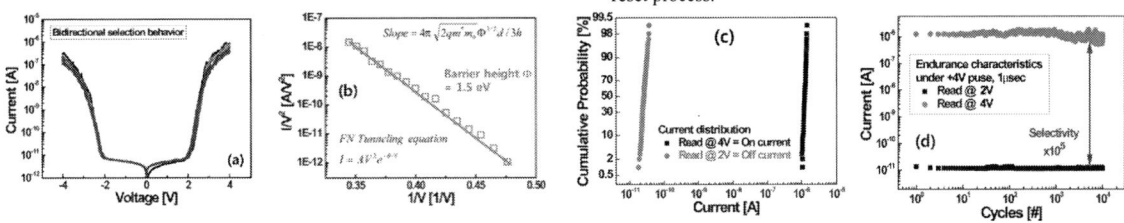

Fig. 7 (a) The I-V characteristics of bidirectional selection behavior and (b) ln (I/V²) versus 1/V curves for fitting of Fowler-Nordheim tunneling mechanism. (c) Cumulative probability of current distribution and (d) pulse endurance characteristics of ultrathin selection device.

Co-existed Unipolar and Bipolar Resistive Switching Effect of HfO$_x$-Based RRAM

B. Chen[1,2], B. Gao[2], Y.H. Fu[1,2], R. Liu[2], L. Ma[1,2], P. Huang[2], F.F. Zhang[2], L.F. Liu[2], X.Y. Liu[2], *J.F. Kang[2], G.J. Lian[3]; *E-mail: kangjf@pku.edu.cn

[1]Peking University Shenzhen Graduate School, Shenzhen 518055, China [2]Institute of Microelectronics, Peking University, Beijing 100871, China, [3]School of Physics, Peking Uinversity, Beijing 100871, China

Abstract

Both unipolar and bipolar resistive switching behaviors are demonstrated and investigated in the TaTiN/HfOx/Pt structured RRAM devices. A physical model based on the recombination among the electron-depleted oxygen vacancies (V_O^{2+}) and the oxygen ions (O^{2-}) released from the TaTiN electrode is proposed to clarify the co-existed bipolar and unipolar resistive switching effect. In the proposed physical model, Joule heating controlled O^{2-} decomposition and electric-field controlled O^{2-} drift dominate the unipolar and bipolar resistive switching behaviors, respectively.

Index Terms: HfOx based RRAM, Co-existed bipolar and unipolar resistive switching

Introduction

Transition metal oxide-based resistive-switching random access memory (RRAM) has emerged as one of the most promising candidates for future nonvolatile memory application [1-3]. Various resistive switching phenomena such as unipolar and bipolar switching have been extensively studied but the correlated physical mechanisms are still in debated [4-9]. In this paper, the Pt/HfOx/TaTiN structured HfOx based RRAM devices were fabricated. The co-existed bipolar and unipolar switching effects were measured and investigated in the fabricated HfOx based RRAM. A physical model based on the electric-field controlled O^{2-} drift and Joule-heating controlled O^{2-} decomposition [8,9] is proposed to clarify the physical origin of the co-existed bipolar and unipolar resistive switching effect.

Experiment

Fig. 1 (a) & (b) schematically show the structure of HfO$_x$ based RRAM devices and the fabrication process flow, respectively. Electrical measurements were performed using Agilent4156C and Keithley4200 analyzers.

Result and Discussion

After a current sweep forming process [7], the HfOx based RRAM devices were switched to high resistance states (HRS). Fig.2 shows the typical I-V curves of the HfOx based RRAM during forming and first RESET cycle. Then the devices can be switched by using bipolar and unipolar modes. The typical I-V curves of both bipolar and unipolar switching are shown in Fig. 3. For the bipolar switching case, the RESET process from LRS to HRS is gradually change with the reset voltage, and correlated with current-compliance during set process. Otherwise, for the unipolar switching case, the RESET process is sharply transition from LRS to HRS with larger programming current.

Fig. 4 (a) compares the measured resistance and reset current distributions at bipolar and unipolar switching modes. The voltage sweep mode with 1mA set current-compliance and -1.5V reset voltage is used for the measurements. Larger reset current and resistance variation were observed under unipolar switching mode.

The co-existed unipolar and bipolar resistive switching effect mainly occurs in the resistive switching layer of HfOx near TaTiN electrode, which could be attributed to recombination effect among the charged oxygen vacancies (V_O^{2+}) in the conducting filaments of HfOx layer and the oxygen ion (O^{2-}) released from the TiTaN electrode [8], as shown in Fig.5. The SET process for unipolar switching is similar to but the RESET process is different from the bipolar switching. In the SET process, O^{2-} is generated from oxygen vacancies in the HfO$_x$ resistive switching layer then drifted to the TaTiN TE under the SET voltage or electric-field. The left V_O^{2+} can be occupied by electrons (to change into neutral V_O) and construct the conducting filaments. In the RESET process, the neutral V_O can change into the charged V_O^{2+} due to electron-depletion occupied in V_O when the applied bias or electric-field reaches a critical value [8]. The O^{2-} stored in TaTiN TE can be released either by the RESET electric-field or by the Joule-heating of the RESET current then drifted or diffused into HfOx layer to recombine the charged effect V_O^{2+} [8, 9]. Generally, the RESET occurs only when both the electron-depletion in V_O and O^{2-} release take place same time. Based on the proposed model, we can predict that electron-depletion dominates the RESET process when the filament is thick with low LRS resistance but the O^{2-} release dominates the RESET when the filament is thin with high LRS resistance.

In order to verify the proposed physical model, the dependence of the RESET characteristics on the LRS resistance (determined by the SET current compliance) is measured, as shown in Figs.6-8. The sharp transition from LRS to HRS @ high current compliance (corresponding to low LRS resistance) is observed during bipolar SET process as shown in Fig.6, consistent with the model prediction. The measured dependence of R_{HRS} and RESET current on the last LRS resistance, as shown in Figs.7&8, is also consistent with the predictions of the proposed physical model.

Conclusion

The co-existed unipolar and bipolar resistive switching behaviors are demonstrated in the TaTiN/HfOx/Pt structured RRAM devices and explained by a proposed physical model. This helps to deeply understand the physical origin of resistive switching in the oxide-based RRAM.

Acknowledgement

We thank IME@A*STAR for the device fabrication. This work is partly supported by the 973 and NSFC Programs (2011CBA00600, and 60906040).

Reference

[1] M. J. Kim et al, IEDM2010 p444 [2] H.Y. Lee et al, IEDM2010 p460; [3] W. C. Chien et al, IEDM2010 p440; [4] G. Bersuker et al, IEDM2010, p456; [5] L. Goux et al, VLSI2011, p24; [6] R. Waser et al, Adv. Mater. 2009, 21, p2632; [7] B.Chen et al, EDL, 32, 282 (2011); [8] B. Gao et al, IEDM 2011, p417; [9] Y.S. Chen et al, J. Phys. D, 45, 065303 (2012)

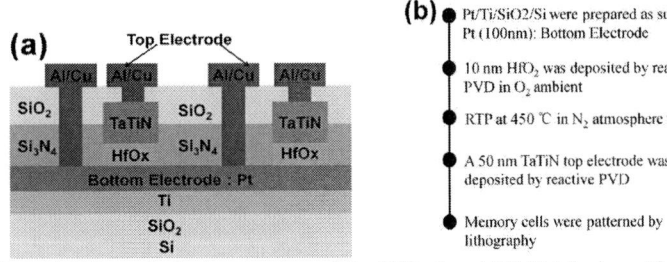

Fig.1 (a) Schematic device structure of the HfOx based RRAM devices. The HfO$_X$ resistive switching layer and the TaTiN top electrode (TE) are deposited by reactive PVD, respectively. (b) Process flow of the device fabrication.

Fig.2 Typically I-V characteristics of forming process measured using current sweep and the first reset process after forming.

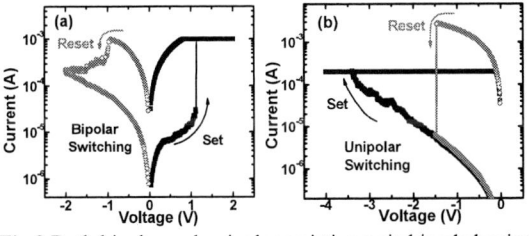
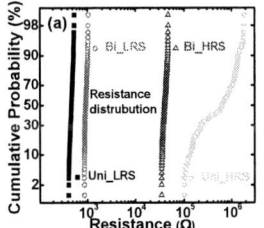

Fig.3 Both bipolar and unipolar resistive switching behaviors were measured in the HfOx based RRAM. **(a)** Measured I-V curves for bipolar switching. **(b)** Measured I-V curves for unipolar switching.

Fig.4 Measured distributions of bipolar and unipolar switching parameters for 100 cycles. **(a)** for HRS and LRS resistance; **(b)** for reset current. Bipolar switching shows the better uniformity.

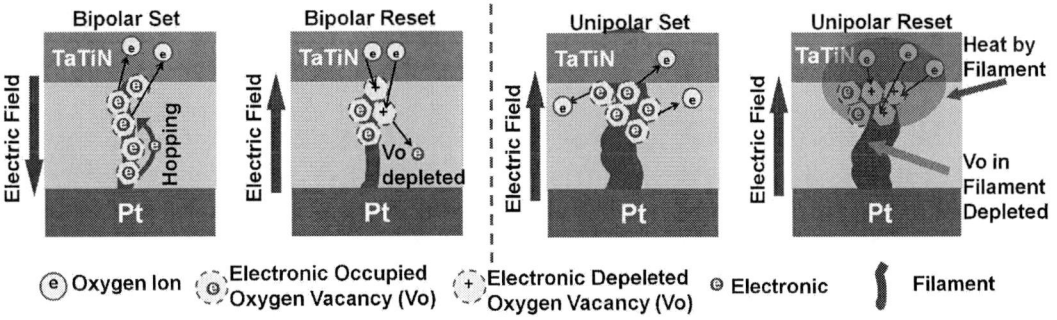

Fig.5 Schematic view of the proposed physical model to clarify the coexisted bipolar and unipolar switching behaviors in the HfOx-based RRAM. The electric-field controlled oxygen ions drift and Joule-heating controlled oxygen ions thermal-decomposition dominate the bipolar- and unipolar switching behaviors.

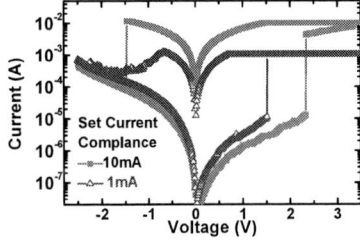
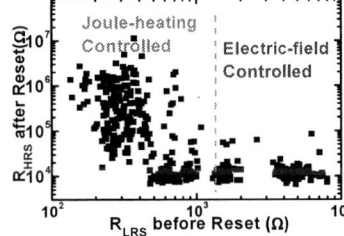
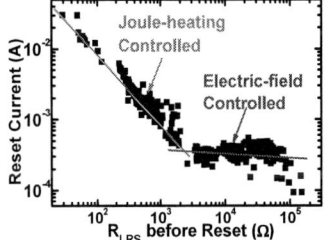

Fig.6 Current compliance effect of SET process on the switching behaviors. The sharp transition from LRS to HRS @ high RESET current is observed, indicating that Joule-heating effect dominates the resistive switching.

Fig.7 Measured dependence of R$_{HRS}$ on the last LRS resistance for bipolar switching. Joule-heating controlled switching at low R$_{LRS}$ shows significant variation compare with the electric-field controlled switching at high R$_{LRS}$.

Fig.8 Measured dependence of reset current on the last LRS resistance for bipolar switching. Significantly increased RESET current with R$_{LRS}$ is observed in Joule-heating controlled switching process.

978-1-4673-0996-7/12 $31.00 © 2012 IEEE

4kb nonvolatile nanogap memory (NGpM) with 1 ns programming capability

T.Takahashi, S.Furuta, Y.Masuda, S.Kumaragurubaran, T.Sumiya, M.Ono, Y.Hayashi[*],
T.Shimizu[**], H.Suga[**],M.Horikawa[**] and Y. Naitoh[**]

Funai Electric Advanced Applied Technology Research Institute (FEAT), TCI-A37, 2-1-6 Sengen, Tsukuba 305-0047, Japan
E-mail: takahashi.t@funai-atri.co.jp, Phone:+81-29-886-6500, FAX:81-29-886-6511
*Tsukuba Device Solution Center, 2-3-10 Umezono, Tsukuba, 305-0045 Japan
** Nanosystem Research Institute, National Institute of Advanced Industrial Science and Technology (AIST), 1-1-1 Higashi
Tsukuba 305-8562, Japan, E-mail: ys-naitou@aist.go.jp

Abstract

A 4k bits nonvolatile high-speed nanogap memory device
was fabricated with a newly developed vertical nanogap
structure and its memory characteristics were evaluated. The
newly developed vertical nanogap structures realized
controllable electrode gap and higher yield compared to the
initial phase lateral type nanogap structure. The structures
were integrated on a CMOS chip. The specially embedded
measurement circuit revealed programming speed from a low
resistance state to a high resistance state (from on to off state)
to be 1 ns.

Introduction

One of the authors found that thin film metal electrodes with
a gap less than 10 nm on an insulating substrate showed
nonvolatile memory effect in vacuum [1]. Similar nanogap
resistance change were widely observed for metals, such as
Au, Pd, Pt, Ta [2], and even for Si [3] and carbon nanotubes
[4] not only in vacuum but also in inert gases [10]. The
current between the electrodes is due to electron tunneling.
And the resistance corresponding to the gap distance was
changed and controlled by applying voltage to the electrodes
[5]. We have studied this phenomenon and applied for
nonvolatile nanogap memory [8], because of its superior
properties, i.e. high speed resistance switching, operation in a
wide temperature range up to 200 C[1] and intrinsic high bit
density. In this paper, results of integration of the nanogap
memory (NGpM) on a CMOS LSI structure and evaluation of
high speed switching capability using a specially embedded
measurement circuit are reported.

Fabrication

NGpM is initially configured laterally on the insulating
substrate as shown in Fig.1a, which is called as lateral type. It
has been used to investigate the property and the mechanism
of nanogap resistance change [1],[6], [7], however it is not
suitable for memory array. We have developed various
vertical type nanogap elements and finally adopted a trench
type and a hole-in-line type each of which is shown in Fig.1b
and 1c respectivly. [8] Using these types of nanogap element,
a 4kb NGpM is integrated on a CMOS LSI in which
transistors for each nanogap element selection, control
circuits and the specially designed embedded measurement
circuits were furnished. The technology of the CMOS LSI is
0.35μ m high-voltage process of a foundry. Fabrication
process on the chip is summarized in table 1. In this vertical
type the gap distance between 1st and 2nd metal is
determined by the spacer thickness, which is 10nm or less.
The device size factor F is 40nm ϕ for hole-in-line type and
60nm for trench type. In the vertical type, the nanogap
element is located at the cross point of upper and lower
electrode, then $4F^2$ minimum size needed for a memory
element. A nanogap element finer than 40nm also operates
properly. [7]

Programming

In our previous work, it was very difficult to investigate the
intrinsic programming speed of the nanogap elements,
because of limitation of commercially available pulse
generators with high-voltage high-speed drivability (10V,
10ns) and also an influence of the parasitic capacitance of
lead pad, test fixtures and coaxial cable.

In this paper, we embedded a variable pulse generator and a
constant current source with measurement circuits into 4kb
memory chip to confirm the suitable programming operation
method and to evaluate the intrinsic program speed. Variable
programming conditions are illustrated in Table 2.

Results and discussion

A microphotograph of an experimental 4kb memory chip
with the block assignment is shown in Fig.2. Chip size is 3.4x
3.4mm, and with 43 pads for lead. At first a DC voltage
sweep is applied between lower and upper electrode of each
nanogap element, which is called "Forming". The cycling of
program and read is performed. The resistance distributions
during 100 cycles are depicted in Fig.3. In the figure,
program pulse condition is 8V, 64ns to off state and 2uA,
100us to on state. To evaluate intrinsic programming speed,
the embedded variable pulse generator and measurement
circuit embedded in TEG area were used. Program to set off
state is possible by 1ns pulse as illustrated in Fig.4. Program
to set on state is shown in Fig 5 as about 8ns which still
includes a circuit delay of probably 1 or 2ns. Those results
are approaching the intrinsic value far from our previous
works. ([5],[8],[9])

Although the number of bits or nanogap elements measured
is small, superiority of nanogap memory element has been
confirmed in the programming speed, the element size and
integrity to a conventional CMOS technology. Endurance
cycles more than10^5,data retention, wide range operation
under -80~150C[11] were already reported.

Summary

We have developed 4kb nonvolatile nanogap memory
(NGpM) and evaluated endurance under a high speed
programming and confirmed 1nsec operation (to the off state).
This is the first time demonstrational application of nanogap
element to memory device. We suggest that NGpM is one of
the ultra-high speed nonvolatile memories for the next
generation.

Acknowledgements

This research was supported by a grant from the New
Energy and Industrial Technology Development Organization
(NEDO) as a part of Innovation Research Project on
Nanoelectronics Materials and Structures. The device
fabrication for this work was partially conducted in
Kyoto-Advanced Nanotechnology Network and the AIST
Nano-Processing Facility, supported by "Nanotechnology
Network"

[1] Y. Naitoh, M. Horikawa, H. Abe, and T. Shimizu, Nanotechnology **17**, 5669 (2006).
[2] S. Furuta, T. Takahashi, Y. Naitoh, M. Horikawa, T. Shimizu, and M. Ono, Jpn. J. Appl. Phys., **47**, 1806 (2008).
[3] Y. Naitoh, Y. Morita, M. Horikawa, H. Suga, and T. Shimizu Appl. Phys. Express 1., 2008, 103001
[4] Y. Naitoh, K. Yanagi, H. Suga, M. Horikawa, T. Tanaka, H. Kataura, and T. Shimizu, Appl. Phys. Express, **2**, 035008 (2009).
[5] Y. Masuda, T. Takahashi, S. Furuta, M. Ono, T. Shimizu, and Y. Naitoh, App. Surf. Sci., **256**, 1028 (2009).
[6] Y. Naitoh, H. Suga, and M. Horikawa, Jpn. J. Appl. Phys., 50, 06GF10 (2011).

[7] T. Mizukami, Y. Miyato, K. Kobayashi, K. Matsushige, and H. Yamada: Appl. Phys. Lett. 98, 2011, p.083120
[8] S. Furuta, Y. Masuda, T. Takahashi, S. Kumaragurubaran, M. Ono, H. Suga, Y. Naitoh, and T. Shimizu, Nonvolatile Memory Technology Symposium NVMTS-2011, P-R-02
[9] S. Kumaragurubaran, T. Takahashi, Y. Masuda, S. Furuta, T. Sumiya, M. Ono, T. Shimizu, H. Suga, M. Horikawa, and Y. Naitoh, Appl. Phys. Lett. 99, 263503 (2011)
[10] Y. Naitoh, M. Horikawa, and T. Shimizu: Jpn. J. Appl. Phys. 49, 2010, p.01AH08
[11] Handout paper 2010CEATEC, 20110SEMICON Japan..

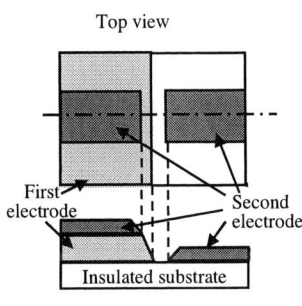

Fig.1a Lateral type

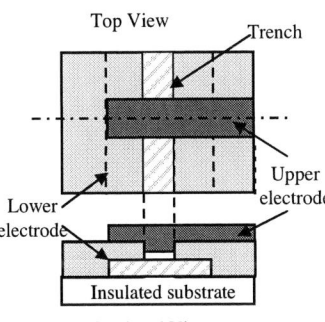

Fig.1b Trench type

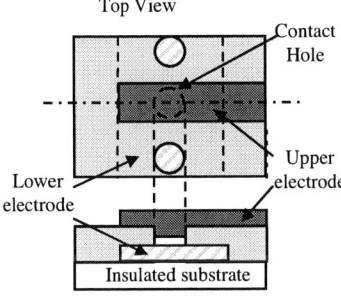

Fig.1c Hole-in-line type

Table1. 4kb chip fabrication process flow

	Lith.Type
1. Lower Metal : lith., dep. and lift off.	Photo
2. Spacer & Inter layer: dep.	
3. Pad-window for Lower Mt : lith. and etch.	Photo
4.Upper metal Lead wire lith., dep. and lift off.	Photo
5. Trench or Hole on Inter layer dry etch	EB
6. Upper metal electrode : lith., dep. and lift off.	EB
7.Removal of spacer	

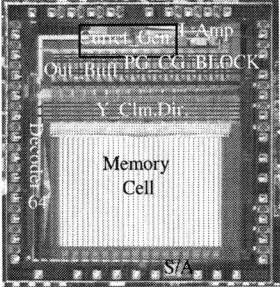

Fig.2 4k Memory outlook

Table 2. 4kb programming mode

To Off (L→H)	To On (H→L)
Voltage Pulse 0.5～64ns (8Steps), DC From outer lead	Constant Current (2 step possible) 0.1～100 μ s (variable clock by FPGA) ,2steps i=0.5,1,2,3, 4, 6,8,12 μ A

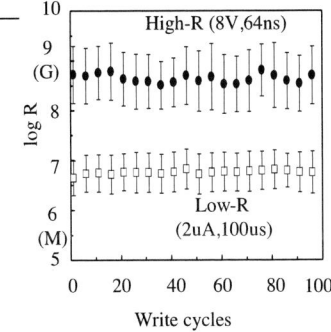

Fig.3 Resistance distribution per Programming cycles : Pont is mean and error bar is standard deviation

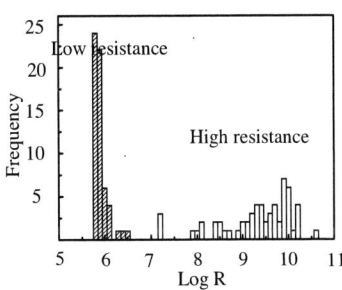

Fig.4 Histogram of resistance after1ns write pulse applied. Programming conditions:
{ to Off state :12V 1ns
{ to On state : 2uA, 100 us

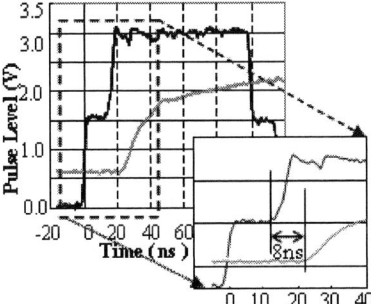

Fig5. Set on transition timing by measurement of TEG-B Black is in-put waveform which is overlapped incident and refraction waves. Red is out-put waveform.

978-1-4673-0996-7/12 $31.00 © 2012 IEEE

Characteristics of Metal/Ferroelectric(PVDF-TrFE)/Graphene (MFG) Device

[1]H.J. Hwang, [2]E.J. Paek, [1]J.H. Yang, [1]C.G. Kang, [1,2]B.H. Lee

[1]School of Materials Science and Engineering, [2]Department of Nanobio Materials and Electronics
Gwangju Institute of Science and Technology, Oryong-dong 1, Buk-gu, Gwangju, Korea
Phone: +82-62-715-2308, Fax: +82-62-715-2304, e-mail: bhl@gist.ac.kr

Abstract

Characteristics of new reconfigurable graphene device with Metal/ Ferroelectric (PVDF-TrFE)/ Graphene (MFG) stack is presented. Key features include programming speed< 100nsec, retention up to 1000sec, endurance upto 1000 cycles and more than 775% on/off ratio. While memory like functionalities are primarily presented in this paper, MFG device has many versatile applications such as reconfigurable interconnect resistor or logic device, pressure sensitive touch sensor and so on.

Introduction

Practical applications of graphene are still limited to a few such as transparent electrode, low noise amplifier and sensors due to its drawbacks such as ambipolar conductivity and zero/small band- gap [1-3]. However, a remarkable progress in large area graphene process up to 32inch lowered a barrier for the implementation of graphene in other electronic applications. Thus, research to find proper applications of graphene attracted a lot attention recently. Among several approaches, a hybridization of graphene with other functional materials has been suggested as a way to detour the technical limits of graphene [4-6]. For example, metal/ferroelectric/ graphene (MFG) stack or metal/ piezoelectric/graphene(MPG) stack can provide interesting functionalities [7-9]. Especially, MFG device has conductivity modulation functions like a metal/ ferroelectric/ insulator/silicon (MFIS) device, but with a flexibility of implementation in both FEOL and BEOL structure.

In this work, functionalities of single and array MFG devices are successfully demonstrated using Metal /Ferroelectric (PVDF- TrFE)/Graphene (MFG) stack.

Experiment

Detail fabrication process for MFG device is summarized in Fig.1. PVDF-TrFE(75:25) was spin-coated on monolayer CVD graphene/SiO_2 substrate with Au contacts. Then, the PVDF-TrFE layer was patterned using O_2 plasma etch to open source and drain contacts.[10] Fig. 2 shows the SEM images of MFG device from in the middle of fabrication steps. The graphene channel was clearly patterned and top gate electrodes on the PVDF-TrFE were overlapped with source and drain electrode to minimize the series resistance. The quality of graphene after the device process was confirmed by Raman spectrum (Fig. 3). After the curing process, the PVDF-TrFE shows good roughness and morphology (Fig. 4). β phase peaks indicating the presence of permanent polarization were maintained before and after O_2 plasma etching(Fig.5). Also, ~8$\mu C/cm^2$ of polarization was comparable with the values reported in the literature (Fig. 6).

Results and Discussion

Operation principle of MFG device is schematically shown in Fig.7 and actual device I_d-V_{bg} curves are shown in Fig.8. Before poling PVDF-TrFE, charge neutrality point was located near 0V. Under a positive bias, the charge neutrality point was shifted to negative direction due to polarized dipoles inducing negative charges in the graphene. Vice versa, the negative charge was appeared near the graphene channel when the negative bias was applied and the charge neutral point was shifted to right side. The maximum on/off ratio of MFG was record high, 778% at V_{bg}= -10V, compared to prior works [4-6]. The changes in field effect mobility were not significant before and after poling, indicating the polarization of PVDF-TrFE primarily affected the Fermi level of graphene without causing charge doping effects.

The reconfigurability of MFG device was maintained upto 1,000 cycles even at ±30V poling cycle (Fig. 9). In each cycle, the polarization was reversed for 0.5sec and I_d was measured at V_d=1V without any external electric field. Around 100 cycles, the abrupt on/off ratio increase due to concurrent charge trapping was observed but eventually saturated. The range of conductance modulation was strongly dependent on the poling bias. Poling effect was activated ~20V at 0.5sec poling time and rapidly increased up to 35V. At higher polling bias, the on/off ratio increased and maintained the state until 1,000 sec (Fig.10).

Polling speed was tested at 15V and 20V by applying a short gate pulses. In both biases, the switching action was observed down to 100nsec. Theoretically, <1nsec switching is possible, but the test structure was too big to test such high frequency switching (Fig.11). While both polarity switching was feasible, a positive poling bias was more effective in controlling the distribution because the PVDF-TrFE already has a small initial polarization before the poling (Fig. 12).

Conclusion

Novel hybrid device combining a graphene with a ferroelectric material (PVDF-trFE) has been successfully demonstrated with a single device. Since this structure can overcome the depolarization problem of MFIS device, there are many useful applications such as memory, reconfigurable interconnect with a very high switching speed << 100nsec (<1nsec theoretically).

Acknowledgement: This work was supported by basic science research program and WCU program through NRF grant funded by MEST (No.R31 -10026, and 2011-0019159) and Inter-ER Cooperation Projects of MKE/KIAT.

Fig. 1 Schematic flow of MFG device fabrication. Starting substrate was a monolayer CVD graphene on 100nm SiO$_2$/Si substrate.

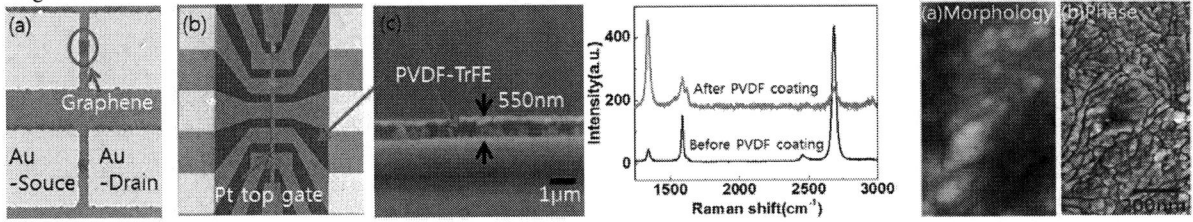

Fig. 2 SEM images of device. (a) After graphene channel patterning (b) final devices with top gates and (c) X-sectional SEM showing the thickness of PVDF-TrFE layer.

Fig. 3 Raman spectrum of CVD graphene before and after PVDF-TrFE coating.

Fig. 4 AFM images of PVDF-TrFE layer.

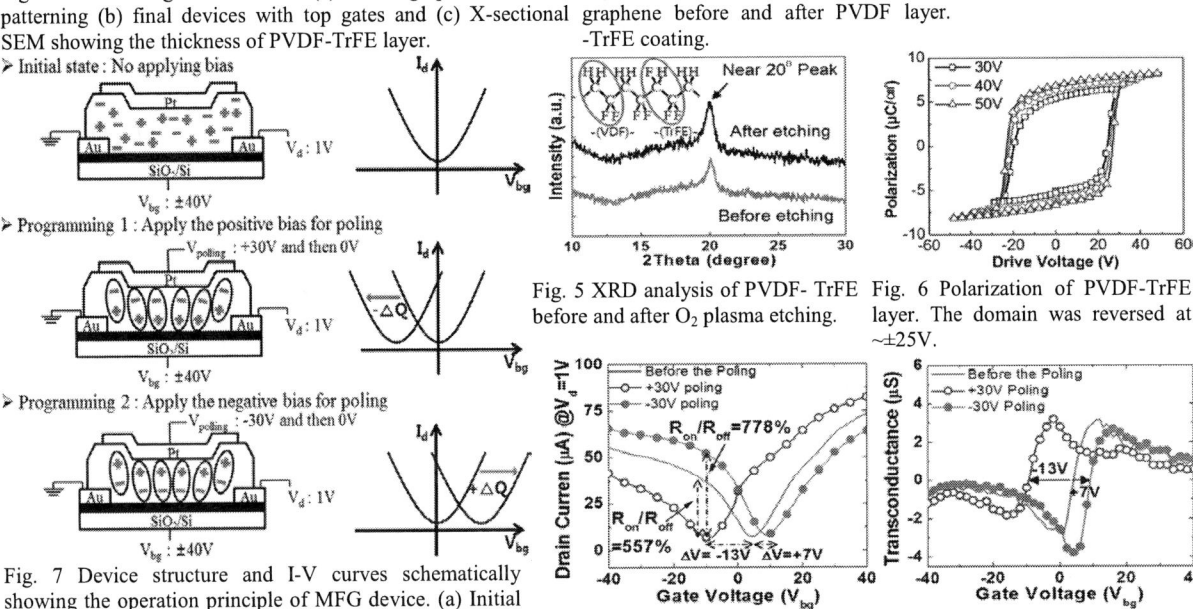

Fig. 7 Device structure and I-V curves schematically showing the operation principle of MFG device. (a) Initial state of MFG device: the domain is randomly arranged. (b) After applying +V$_{poling}$ s and (c) after applying -V$_{poling}$.

Fig. 5 XRD analysis of PVDF- TrFE before and after O$_2$ plasma etching.

Fig. 6 Polarization of PVDF-TrFE layer. The domain was reversed at ~±25V.

Fig. 8 I$_d$-V$_{bg}$ curves of MFG devices. Maximum on/off ratio =778%, no degradation in electron g$_{m,max}$ but g$_{m,max}$ of hole was reduced at +V$_{poling}$, and increased at -V$_{poling}$

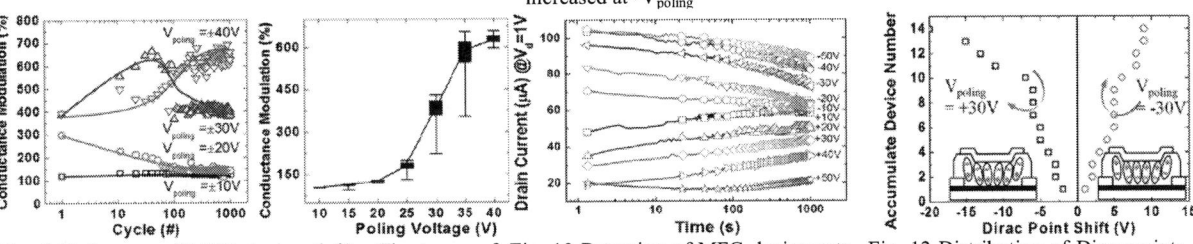

Fig. 9 Endurance of MFG device (left), The range of conductance modulation as a function of V$_{poling}$. (right)

Fig. 10 Retention of MFG device upto 1,000 times at different V$_{poling}$.

Fig. 12 Distribution of Dirac points after negative and positive poling.

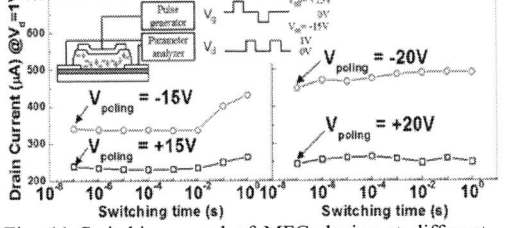

Fig. 11 Switching speed of MFG device at different poling voltages. Rising and falling time of pulse were100nsec.

References:

[1] X. Li et al., Nano Lett. 9(12), p.4359, 2009.
[2] S. Bae et al., Nat. Nanotechnol. 5, p.574, 2010.
[3] K.S. Kim et al., Nature, 568, p.706, 2009.
[4] R.S. Dahiya et al., Appl.Phys.Lett. 95(3), p. 034105, 2009.
[5] Y. Zheng et al., Phy.Rev.Lett. 105, p.166602, 2010.
[6] X. Hong et al., Phy.Rev.Lett. 102, p.136808, 2009.
[7] E.J. Paek et al., Ext. Abstract of SSDM,p.1331, 2011.
[8] E.J. Paek et al., Korean patent no. 2011-0030182, 2011.
[9] H.J. Hwang et al., Proc. of Nano Korea, 2011
[10] W.Y. Kim, et al., Micro. Eng. 88, p.1576, 2011.

Silicon Single-Electron Transfer Devices: Ultimate Control of Electric Charge

Akira Fujiwara, Gento Yamahata, Katsuhiko Nishiguchi, Gabriel P. Lansbergen, and Yukinori Ono[*]

NTT Basic Research Laboratories, NTT Corporation, 3-1 Morinosato Wakamiya, Atsugi, Kanagawa 243-0198, Japan
Phone: +81-046-240-2643   Fax : +81-046-240-4317   E-mail: fujiwara.akira@lab.ntt.co.jp

Downscaling of electronic devices using state-of-the-art nanotechnology provides us with the opportunity to control electric charges at the level of a single electron (SE). So called SE devices [1] have been studied since a few decades ago in a variety of fields including future low-power LSI, high-sensitivity sensors, and solid-state quantum computing. Among the categories of SE devices shown in Fig. 1, the single-electron transistor (SET) has been the most widely investigated because it is a field-effect transistor that is likely to be suitable for a wide range of applications.

An SE transfer device is another category that enables ultimate control of electric charges based on a clocked transfer. Such a function is promising for device concepts including current standards in metrology [2] and on-demand electron sources for a circuit using an SE as a bit of information [3]. SE transfer is also ultimate in terms of low power because there is no static energy dissipation. In principle the energy dissipation can be reduced to a level that is related to the transfer of only an SE.

The SE current standard is a potential candidate for achieving a new SI base unit ampere, which has very recently been proposed by a committee of the International Committee for Weights and Measures (CIPM) [4]. The new ampere is set based on a fixed value of the elementary charge instead of the force between two parallel conductors. The quantum metorogy triangle [5] shown in Fig. 2 is an experiment to test the consistency of three electrical quantum standards: the SE current standard, the Josephson voltage standard, and the quantum Hall resistance standard. Two technical challenges for this experiment are (i) high current $I = nef$ (n: integer, f: clock frequency) at a nanoampere level and (ii) high transfer accuracy with an error rate below 10^{-8}. Semiconductor based devices are beneficial to high frequency operation since electrically tunable tunnel barriers can be utilized.

In this paper we describe our recent efforts to develop SE transfer devices based on Si nanotechnology.

The device comprises gate array Si nanowire MOSFETs on a silicon-on-insulator wafer. Figure 3(a) shows the top-view scanning-electron-microscope (SEM) image. Si nanowire is covered by fine poly-Si gates to form tunnel barriers and a charge island with a charging energy of approximately 10 meV. A wide upper poly-Si gate (UG) is then formed as an island gate as well as an implantation mask to define the n-type source and drain. Figures 3(b) and 3(c) shown the operating principle of the SE turnstile [6,7] and ratchet [8-10], respectively. In the turnstile two clock signals are applied in turn to transfer electrons from the source to the island, and then from the island to the drain, while the ratchet needs only one AC signal with a relatively larger amplitude to transfer electrons using a cross coupling of the barrier gate to the island. The number of transferred electrons per cycle is controlled using the upper gate voltage, V_{UG}. Figures 4(a) and 4(b) show the current by the SE transfer at 20 K for the turnstile ($f = 10$ MHz) and the ratchet ($f = 2.3$ GHz), respectively. Clear current plateaus, $I = nef$, are observed and nanoampere current is demonstrated by $3ef$.

To achieve precise and direct evaluation of the transfer accuracy, a technique called shuttle error measurement [2] must be employed in which the DC charge sensor can detect the error in the high-frequency transfer. Figure 5(a) shows the SEM image of the device for such a measurement [11]. The error in the shuttle transfer between the node and the source results in a change in the time-averaged number of electrons in the node, which is monitored by the charge sensor. Two operating points used for the shuttle transfer are depicted in Fig. 5(b), which shows usual current measurement data of the SE turnstile using the three gates. So far we have carried out single-shot measurement of the SE transfer as shown in Fig. 5(c), where one error event is observed during the repetitive capture/ejection of the SE to/from the node. The best error rate obtained was 2.1×10^{-3} at 17 K, which was the first demonstration of the accuracy evaluation in a semiconductor-based SE transfer device [11]. According to the theory of thermal error, we can expect an error rate as low as 10^{-8} below 4 K. Lower temperature measurement is now in progress.

Recently, we reported on dopant-based SE transfer [12] that could be a different approach for higher transfer current. The concept is to transfer electrons via multiple dopants, each of which can contain an SE, thereby giving $I = N_D ef$ (N_D: number of dopants). By implanting multiple arsenic donors into a charge island the current plateaus of SE transfer via dopants are clearly observed up to $N_D = 6$ in a low gate voltage region as shown in Fig. 6.

In conclusion, Si nanowire MOSFETs represent a promising candidate for achieving high-speed, high-current, and highly-accurate SE transfer devices.

This work was partly supported by the Funding Program for Next Generation World-Leading Researchers of JSPS (GR103) and KAKENHI 20241036.

[*]Present address: Toyama University, Japan

978-1-4673-0996-7/12 $31.00 © 2012 IEEE

References

[1] K. Likharev, Proc. IEEE **87** (1999) 606.
[2] M. Keller et al., Appl. Phys. Lett. **69** (1996) 1804.
[3] K. Nishiguchi et al., Appl. Phys. Lett. **88** (2006) 183101.
[4] http://www.bipm.org/utils/en/pdf/24_CGPM_Resolution_1.pdf
[5] M. W. Keller , Metrologia **45** (2008) 102.
[6] L. P. Kouwenhoven et al., Phys. Rev. Lett. **67**, (1991) 1626.
[7] A. Fujiwara et al., Appl. Phys. Lett. **84** (2004) 1323.
[8] A. Fujiwara et al., Appl. Phys. Lett. **92** (2008) 042102.
[9] B. Kaestner et al., Phys. Rev. B **77** (2008) 153301.
[10] S. P. Giblin et al., New J. Phys. **12**(2010) 073013.
[11] G. Yamahata et al., Appl. Phys. Lett. **98** (2011) 222104.
[12] G. P. Lansbergen et al., Nano Lett. **12** (2012) 763.

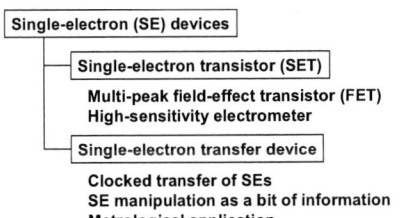

Fig. 1. Single-electron device categories.

"Quantum metrology triangle"

Josephson
$V = f/K_J$
$K_J = 2e/h$

"Current standard"

$I = nef$

Clocked transfer
of single electrons

Quantum hall
$V = R_K / I$
$R_K = h/e^2$

Fig. 2. Quantum meteorology triangle.

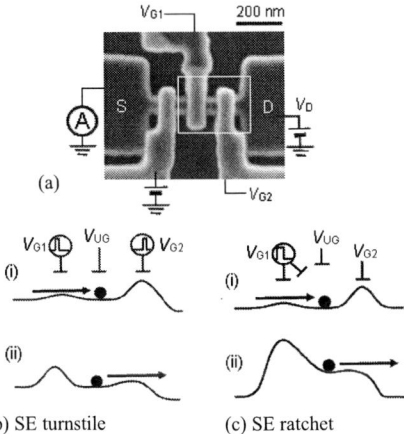

(b) SE turnstile   (c) SE ratchet

Fig. 3. (a) Top-view scanning electron micrograph of Si nanowire MOSFETs before upper gate formation. (b) and (c) Schematic potential diagram of tunable-barrier SE transfer.

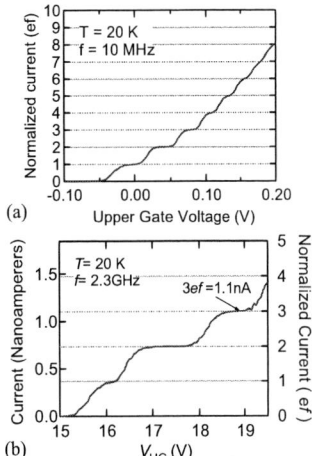

Fig. 4. (a) Normalized current as a function of the upper gate voltage for an SE turnstile. (b) Nanoampere SE transfer at $f = 2.3$ GHz using an SE ratchet [8].

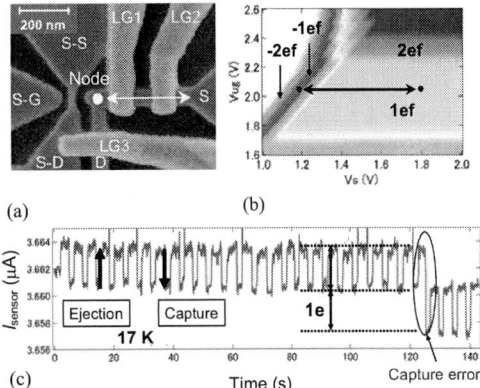

Fig. 5. (a) Scanning electron micrograph of the device for SE shuttle transfer. (b) Contour plot of current by SE turnstile using three gates (LG1, LG2, and LG3) as a function of V_S and V_{UG}. Arrow indicates operation points used for shuttle transfer between the node and the source. (c) Single-shot measurement of the shuttle transfer [11].

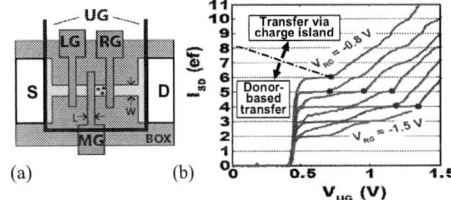

Fig. 6. (a) Top view of donor-based SE transfer device. (b) Current as a function of V_{UG} at 36 K and $f = 2.5$ MHz with changing V_{RG}, showing SE transfer with 6 donors [12].

978-1-4673-0996-7/12 $31.00 © 2012 IEEE

Reinvestigation of Dot Formation Mechanisms in Silicon Nanowire Channel Single-Electron/Hole Transistors Operating at Room Temperature

Ryota Suzuki, Motoki Nozue, Takuya Saraya, and Toshiro Hiramoto

Institute of Industrial Science, University of Tokyo, 4-6-1 Komaba, Meguro-ku, Tokyo 153-8505, Japan

Phone: +81-3-5452-6264, Fax: +81-3-5452-6265, E-mail: r-suzuki@nano.iis.u-tokyo.ac.jp

Abstract: Dot formation mechanisms of single-electron transistors (SETs) and single-hole transistors (SHTs) are reinvestigated. "Shared channel" SET/SHTs in form of nanowire (NW) channel FETs are fabricated and characterized. It is suggested that, in addition to quantum confinement effect (QCE), the positive charges create parasitic dots in SHT channels resulting in multiple-dot SHTs. It is concluded that a <110> SET is the best structure to obtain room temperature (RT) operating single-dot device with high yield.

1. Introduction

A SET/SHT is an emerging electron device for future ultimately-high-density LSIs. However, its circuit application, especially at RT, has made little progress because of difficulty in fabrication. Although a single-dot SET is favorable due to their suppressed variability compared with a multi-dot one, controlled fabrication techniques of a SET/SHT with an extremely small single-dot have not been established yet.

To investigate dot formation mechanisms in Si-based SETs/SHTs, a "shared channel" device have been fabricated, which has a channel shared by both n- and p-type source and drain (S/D) electrodes [1-3] (Fig. 1). Since both a SET and a SHT which share the channel exhibited Coulomb oscillation, it has been thought that QCE is the dominant formation mechanism of tunnel barriers and dots [1] (Fig. 2). It was also found that clearer Coulomb oscillations were observed in SHTs than SETs [2,3] although clear explanations were not given in terms of carrier polarity dependence and channel direction dependence of QCE.

In this study, we reinvestigate dot formation mechanisms in RT operating Si NW channel SETs/SHTs for yield improvement from viewpoint of influences of channel direction and carrier polarity on dot formation.

2. Device Structure and Fabrication Process

As in the previous works, shared NW channel SETs/SHTs were fabricated on a (100) SOI wafer by a similar fabrication process to gate-all-around NW FETs. NW channels in <110> and <100> directions with width of around 10 nm were defined by electron-beam lithography (EBL) (Fig. 3). The slight thermal oxidation method [4] was applied to obtain sub-3-nm-wide NWs with shape fluctuations, which is necessary for dot formation. When no dot is formed in the channel, the device behaves just as a normal NW FET. NW FETs are also useful to evaluate carrier and channel direction dependence of QCE through V_{th} modulation arising from channel width variation [3].

3. Results and Discussion

Fig. 4 shows I_d-V_g characteristics of the shared channel NW FETs which do not exhibit clear Coulomb oscillation. V_{th} correlations between nFETs and pFETs are plotted in Fig. 5. In <110> NW pFETs the V_{th} increase by width scaling is smaller than in <110> NW nFETs, while in <100> NW FETs magnitudes of V_{th} modulation are almost the same for nFETs and pFETs. These results obtained from large number of devices support the results in the previous work [3]. However, in terms of bulk effective mass, V_{th} modulation in <100>

pFETs is expected to be larger because NW channels are surrounded by (100) surfaces. Actually, larger V_{th} increase in (100) ultra-thin-body SOI pMOSFETs has been reported [5], indicating stronger QCE in pFETs. On the other hand, a theoretical work predicts comparable modulation of conduction and valence band edges in sub-4nm-wide <100> silicon NWs [6], supporting the experimental data in Fig. 5.

These V_{th} correlations well explain characteristics of <110> shared channel SET/SHTs, which is shown in Fig. 6, where only SETs exhibit clear Coulomb oscillations because sufficiently high tunnel barriers are formed only in conduction band by carrier dependent QCE.

However, characteristics of <100> shared channel SET/SETs in Fig. 7 are difficult to be explained by comparable QCE in both bands which is suggested by V_{th} modulation. It should be noted that some devices show single-dot like behavior only in one of SET/SETs. Moreover, complex and larger oscillations due to stochastic Coulomb blockade in multi-dot systems appear more often in SHTs than SETs. These results suggest that NW channels of such devices have significant difference in potential structure between both bands and more dots tend to be formed in valence bands. Clearly, these results in Fig. 7 could not be explained by formation of tunnel barriers and dots only by QCE.

To explain these results, an additional dot formation mechanism is proposed (Fig. 8), where neighboring charge (especially positive) raises existing insufficiently-high tunnel barriers due to QCE or even creates a new tunnel barriers, result in formation of a parasitic dots only for SHTs. Although such positive charge contributes large Coulomb oscillations of SHTs, the tunnel barrier position is unlikely to be controlled and resulting dots may be affected by background-charge effect. On the other hand, dot formation by QCE can be controlled if nanofabrication technique is established. Therefore, <110> SETs are more suitable for controlled fabrication of single-dot devices since formation of parasitic dots by positive charge can be avoided.

4. Conclusions

Dot formation mechanisms of NW channel SET/SHTs have been reinvestigated using shared NW channel devices. Large oscillations that are often observed in <100> SHTs are likely to originate from multi-dot systems. To obtain single-dot devices operating at RT with high yield, <110> SET is the best structure.

Acknowledgment

This work was partly supported by Special Coordination Funds for Promoting Science and Technology and Grant-in-Aid for Scientific Research from MEXT. The authors would like to thank Prof. K. Uchida and Dr. M. Saitoh for helpful discussion.

References

[1] H. Ishikuro and T. Hiramoto, *Appl. Phys. Lett.*, 1999.
[2] M. Saitoh and T. Hiramoto, *Jpn. J. Appl. Phys.* 2003.
[3] M. Kobayashi and T. Hiramoto, *J. Appl. Phys.*, 2008.
[4] M. Saitoh *et al.*, *Jpn. J. Appl. Phys.* 2005.
[5] K. Uchida *et al.*, *IEDM*, 2002.
[6] J.-A. Yan *et al.*, *Phys. Rev. B.*, 2007.

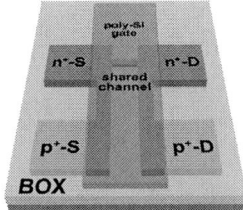

Fig. 1: Schematic of a shared NW channel device. When n-type S/D electrodes are used, the device operates as a nFET or a SET. In the same way, p-type S/D are used for a pFET or a SHT.

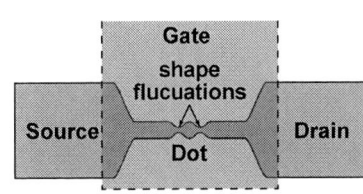

Fig. 2: Mechanism of dot formation by QCE in silicon NW channel SETs and SHTs. Shape fluctuations in extremely narrow (< 3 nm) NW locally enhance lowest subbands both in conduction and valence bands, and create tunnel barriers. If two sufficiently-high barriers are adjacently formed, a device works both as a SET and a SHT.

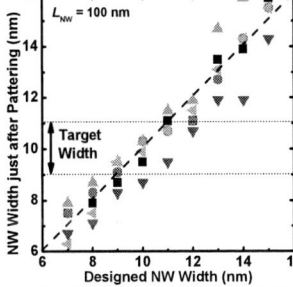

Fig. 3: Initial widths of Si NWs defined by EBL. Final width becomes less than 3 nm by wet etching and thermal oxidation if initial width is around 10 nm.

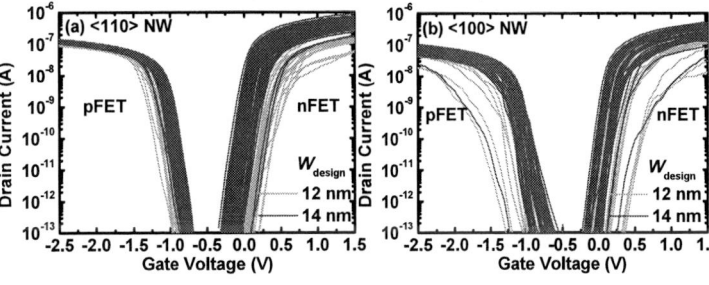

Fig. 4: I_d-V_{gs} characteristics of shared channel NW FETs. Significant V_{th} variability is observed since actual finial widths of the NW channels vary as shown in Fig. 3 even if they have the same designed width. The width is converted into V_{th} modulation by QCE, and sensitivity depends on the NW channel direction and carrier polarity.

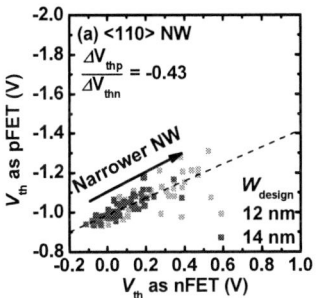

Fig. 5: V_{th} of shared channel NW FETs. A data point represents V_{th}'s of a nFET and a pFET which share the NW channel. A Slope of the line indicates the ratio between the strength of QCE of electrons and holes.

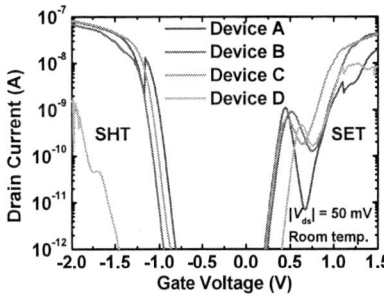

Fig. 6: Characteristics of <110> channel SET/SHTs. Clear single-dot like Coulomb oscillations are observed only in SETs due to absence of sufficiently-high tunnel barriers in the valence bands of <110> NWs.

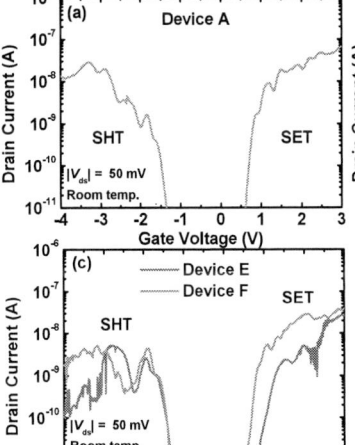

Fig. 7: Characteristics of <100> channel SET/SHTs. (a) A device exhibits similar single-dot-like oscillations both in SET and SHT. (b) Devices with single-dot-like oscillations observed only in one of SET/SHTs. (c) Multi-dot devices with significantly asymmetric oscillations.

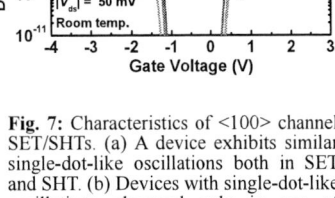

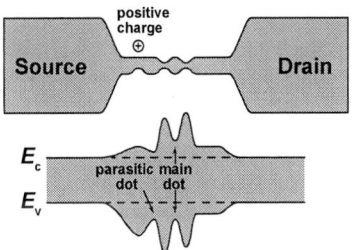

Fig. 8: Schematic of dot formation mechanism. In addition to dots formed by QCE, a positive charge such as a fixed charge in oxide and an interface-trapped hole raises or create a tunnel barrier to form a parasitic dot only for SHTs.

978-1-4673-0996-7/12 $31.00 © 2012 IEEE

Quantum Transport Property in FETs with Deterministically Implanted Single-Arsenic Ions Using Single-ion Implantation

M. Hori[1], T. Shinada[2*], F. Guagliardo[3], G. Ferrari[3], and E. Prati[4**]

[1]School of Science and Engineering, Waseda University, 3-4-1 Ohkubo, Shinjuku, Tokyo 169-8555, Japan
[2] Waseda Institute for Advanced Study, Waseda University, 1-6-1 Nishiwaseda, Shinjuku, Tokyo 169-8050, Japan
[3]Politecnico di Milano, Piazza Leonardo da Vinci, 32, I-20133, Milan, Italy
[4] Laboratorio MDM, CNR-IMM, Via Olivetti 2, I-20864 Agrate Brianza, Italy
* Tel.: +81-3-5286-2110, Fax: +81-3-5272-2470, Email: shinada@aoni.waseda.jp
** Tel.: +39-039-6035556, Fax: +39-039-6881175, Email: enrico.prati@cnr.it

1. Introduction

With the gate length of MOSFETs approaching 10 nm, the channel region contains only one or a few dopant atoms. The random dopant distribution will cause fatal fluctuation of the device performances, such as threshold voltage or transconductance [1-4]. On the other hand, one positive outcome of ultimately scaled-down doped devices has been demonstrated [5-8]. In previous work, we have fabricated transistors with deterministically implanted a few phosphorus (P) atoms and have revealed the quantum transport properties at low temperature for the first time [9]. Here we report the fabrication of single dopant devices by controlling both the position and number of arsenic (As) atoms with the single-ion implantation (SII) method [10], and investigation of the electron transport around the "critical density" which causes the transition from the localized electron states to the Hubbard bands formation. The results of electrical measurement highlight the value of deterministic doping towards the doped-channel device limits.

2. Experimental

Figure 1 illustrates a structure of the transistor for evaluating the quantum transport at low temperature. Figure 2 shows an experimental process flow. Transistors were fabricated on n-type (100) silicon-on-insulator (SOI) substrates patterned using standard photolithography. The channel width is 100 nm. The length and thickness of the channel are 200, 90 nm, respectively. A 15 nm silicon dioxide layer covers the channel region to prevent any contaminations. The silicon transistors have been electrically contacted with the highly phosphorus doped n-type source and drain regions. The drain current (I_d) is controlled by the gate bias (V_g) from the substrate thorough the 125-nm-thick buried oxide. The device shows an accumulation-mode n-type transistor operation. Single arsenic (As) ion was implanted at 60 keV using SII method. Two types of samples with different dopant arrays (single and three dots along the channel) were prepared as shown in fig. 3 and fig.4. Two As ions were placed at each dot. To electrically activate the implanted ions the samples were then lamped-annealed at a temperature of 900°C for 1 min in N_2. The quantum transport in the sub-threshold region was evaluated by measuring I_d-V_g characteristics at 4-20 K. The results were obtained under the condition of a 1 or 6 mV drain voltage (V_d).

3. Results and Discussions

Figure 3 shows the electron transport property of a transistor with the single dot (2 arsenic ions). The figure shows four current peaks originating from the Coulomb blockade and single-electron tunneling. Two pairs of Coulomb blockade peaks are attributed to the D^0 and D^- states of the two implanted donors. Thus, the number of the peaks coincides with double of number of dopants. No other peaks above the fourth one have been observed, in agreement with the experimental evidence that no electrostatic unintentional dots are formed in the channel by the gate bias. This indicates the high reliability of the single-ion doping method that enables us to control the ion number precisely.

Figure 4 shows the electron transport property of a transistor with three dots (6 arsenic ions). The three-dot device has a higher donor concentration than the single dot device. The upper and lower Hubbard bands are formed out of impurity-state wave functions when the concentration of impurities in the material is sufficient to produce a critical overlap of the wave functions from adjacent impurity atoms [11]. This result indicates that the electron transport in single-dopant device is sensitive to the variation of the number and position of arsenic ions.

4. Conclusions

We fabricated silicon transistors containing two and six arsenic ions implanted in one dimensional array along the channel by single-ion implantation method. The quantum transport was measured through the D^0 and D^- states of the arsenic ions at low temperature. We observed two different quantum transport regimes from the individual donor regime to the intermediate doping regime in which Hubbard bands are formed in agreement with the theoretical models. These results indicate that our deterministic single-ion doping method is more effective and reliable for single-dopant transistor development and pave the way towards single atom electronics for extended CMOS applications [12].

978-1-4673-0996-7/12 $31.00 © 2012 IEEE

Acknowledgments

This work was supported in part by the Semiconductor Research Corporation (SRC) No. 1676.001, Grants-in-Aid for Scientific Research Nos 22681020, 23226009 and 20241036 from MEXT, Japan, Grant-in-Aid for JSPS Fellows, Japan, and the Short Term Mobility Program from Consiglio Nazionale delle Ricerche, Italy.

References

[1] T. Shinada, et al., *Nature* **437**, 1128 (2005).
[2] K. Takeuchi, et al., *IEDM Tech. Dig.* p467 (2007).
[3] T. Tsunomura, et al., *VLSI Symp. Tech. Dig.* p97 (2010).
[4] M. Hori, et al., *Appl. Phys. Lett.* **99**, 1128 (2010).
[5] H. Sellier, et al., *Phys. Rev. Lett.* **97**, 206805 (2006).
[6] Y. Ono, et al., *Appl. Phys. Lett.* **90**, 102106 (2007).
[7] E. Prati, et al., *Phys. Rev. B.* **80**, 165331 (2009).
[8] M. Tabe, et al., *Phys. Rev. Lett.* **105**, 016803 (2010).
[9] T. Shinada, et al., *IEDM Tech. Dig.* p697 (2011).
[10] I. Ohdomari, et al., *J. Phys. D:Appl. Phys.* **41**, 043001 (2008).
[11] P. Norton, *Phy. Rev. Lett.* **37**, 164 (1976).
[12] ITRS 2011 Edition, Emerging Research Material.

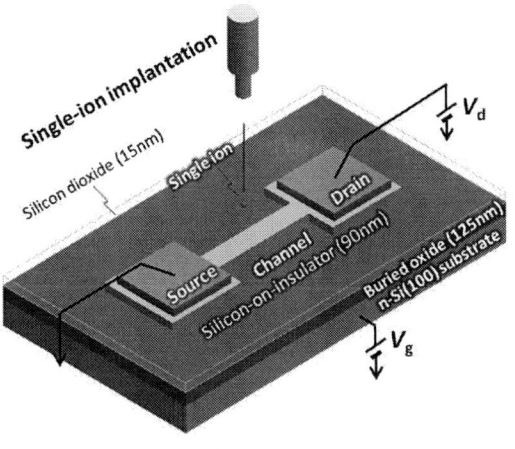

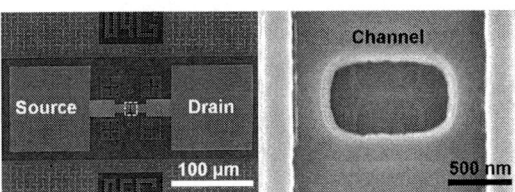

Fig.1. Illustration of device structure and the scanning electron microscopy (SEM) images. The drain current (I_d) is controlled by the gate bias (V_g) from the substrate thorough the 125-nm-thick buried oxide. Arsenic ions are implanted one-by-one using single ion implanter. The channel length is 200 nm and the width is 100 nm.

Fabrication of sample transistors
P-doped n-type (100) silicon-on-insulator (SOI) substrate
Initial channel doping concentration: $<1 \times 10^{15}$ cm^{-3}
(Average number of dopants: <1 in (100nm)^3)
L_g = 200 nm, W = 100 nm, $tSOI$ = 90 nm
Highly P-doped n-type source/drain
Accumulation-mode n-type transistor operation

Single-ion implantation
Arsenic, 60keV
Number of dopants: 2 per dot

Rapid thermal anneal
at 900°C for 1min in N$_2$

V_g-I_d measurement at 1 mV at 4-20 K

Fig.2. Fabrication process flow.

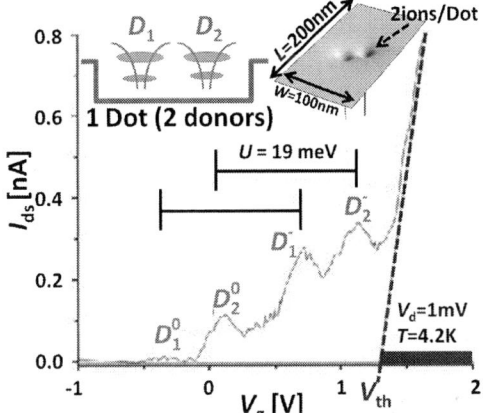

Fig.3. *I-V* characteristic for transistor with single dot (2 donor ions). The four current peaks of the two donors are visible for the sample with the isolated donors. Number of the peaks coincides with double of number of dopants.

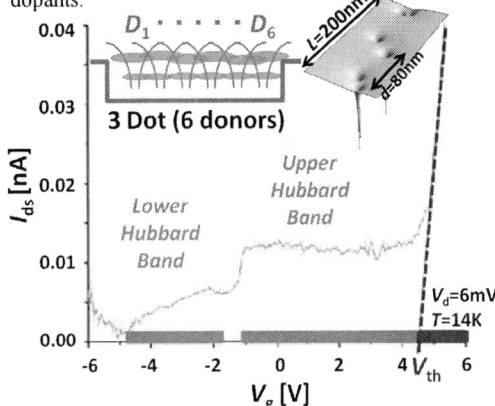

Fig.4. *I-V* characteristic for transistor with three dot (6 donor ions). The broad characteristic in which Coulomb blockade peaks merged was observed. This was attributed to the high overlap of the donor wavefunction between two contiguous sites.

High-frequency properties of Si single-electron transistor

Hiroto Takenaka, Michito Shinohara, Takafumi Uchida , Masashi Arita, Akira Fujiwara[†], Yukinori Ono[†],
Katsuhiko Nishiguchi[†], Hiroshi Inokawa[††], and Yasuo Takahashi

Department of Information Science and Technology, Hokkaido University, Sapporo 060-0814 Japan
Tel: +81-11-706-6794, Fax: +81-11-706-6457, E-mail: h-takenaka@ist.hokudai.ac.jp
[†]NTT Basic Research Laboratories, NTT Corporation, 3-1 Morinosato Wakamiya, Atsugi, 243-0198, Japan
[††]Research Institute of Electronics, Shizuoka Univ., 3-5-1, Johoku, Hamamatsu, 432-8011, Japan

Abstract—High-frequency limit of Si single-electron transistor (SET) is investigated. Since the SETs inevitably have tunnel barriers, the operation speed is thought to be low. To measure the high frequency properties of SETs, we employed their special rectification characteristics, which occurred due to the asymmetry of Coulomb diamond when alternating current voltage was applied to the drain terminal. By the use of the effect, we evaluated the high-frequency properties of Si SETs.

I. INTRODUCTION

The small electron devices operated with high speed and low-power are desired. Although single-electron devices have attracted as a low power device, their high speed operation are not expected because they have to have tunnel barriers. Here we investigated how fast SETs operate.

One of the critical issues to measure the high-frequency transfer characteristics of SETs is that the voltage applied to the drain terminal was limited to be small so as to keep the Coulomb blockade condition. This makes it difficult to measure the high-frequency characteristics of SETs.

Here we propose a new method to measure the high-frequency characteristics by the use of special rectifying effect of SETs. It is well known that the rectification occurs according as the gate voltage due to the asymmetry of Coulomb diamond when alternating current (AC) voltage is applied to drain or source terminal of SET [1]. Using this phenomenon, we evaluated the cut-off frequency of SETs.

II. DEVICE FABRICATION AND MEASUREMENT

The SETs used for the measurement were Si SETs made by pattern-dependent oxidation (PADOX) method [2]. The device was formed in the top Si layer of a SOI wafer as shown in Fig. 1. The SET island was formed in the center of the Si nanowire, and was covered by a top gate (phosphorus-doped poly-Si). Notice that parasitic MOSFETs are inevitably formed both side of the SET.

We evaluated the high-frequency properties of Si SETs by applying high-frequency voltage of 100 Hz ~ 50 MHz to the drain terminal and by measuring the rectifying current at 8 K.

III. RESULTS AND DISCUSSION

Since the usable high-frequency range is limited up to 50 MHz, we have to measure SETs with high tunnel resistance so as to detect the cutoff in the measurable frequency range. In general for Si SETs fabricated by PADOX, higher tunnel resistance will be achieved in the lower gate voltage region close to the first electron appearance. Measured rectifying characteristics of a SET are shown in Fig. 2. Pairs of minus and plus peaks produced by the rectifying effect are shown at the gate voltage (V_g) of the initial SET oscillation peak shown in the inset. When the frequency of applied AC voltage increased, the drain current (I_d) peak heights decreased. This phenomenon occurred in the lower frequency at lower V_g, and moved to higher V_g direction as the frequency increased. Fig. 3 shows the cutoff characteristics of peak-to-peak value of I_d evaluated to the rectifying characteristics shown in Fig. 2. It is clear that the cutoff frequency f_C increased as V_g increased. Although the tunnel resistance of the fourth peak is larger than that of third peak, f_C of the fourth peak is lager than that of third one. The higher f_C is achieved by the higher V_g rather than by lower tunnel resistance. These results are attributed to the idea that the f_C is determined by f_C of the parasitic MOSFET.

We estimated f_C of the parasitic MOSFET from the gate capacitance and conductance of the MOSFET fabricated simultaneously on the same wafer as shown in the solid line of Fig. 4. f_C measured for four SETs were also plotted on the same figure. It is clearly shown that the plotted data fit well to the estimated f_C of the parasitic MOSFET. These results suggest that f_C is limited by the MOSFET in the lower V_g region. In order to evaluate f_C of SETs in the present available frequency range, i.e., up to 50 MHz, the V_g must be set at or larger than 1.2 V, where f_C of the MOSFET should be larger than 50 MHz.

We found out a SET which has relatively high tunnel resistance (low oscillation current peak) at the V_g larger than 1.2 V. I_d-V_g characteristics of the SET are shown in Fig. 5. The tunnel resistance (R_t) of the first peak is about 2.5 GΩ and the gate capacitance (C_g) is evaluated to be 0.47 aF from the oscillation period. The cutoff frequency of the first peak is estimated to be 135 MHz according to the equation $f_c = 1/2\pi R_t C_g$.

Fig.6 shows the rectifying I_d-V_g characteristics measured at around the V_g of the first peak at various frequencies. Since no reduction of $I_{d(p-p)}$ was observed, f_C of the SET is larger than 50MHz, which is consistent with the f_C of 135 MHz estimated above.

The lowest tunnel resistance R_t of about 65 kΩ can be achieved in the same SET at the V_g of 5.7 V as shown in Fig. 5. From this R_t, the cutoff frequency of SETs is expected to be larger than 1 THz.

IV. CONCLUSION

We evaluated the high-frequency characteristics of SETs by the use of special rectifying effect of SETs. First of all, we measured the cutoff frequency of the parasitic MOSFET

978-1-4673-0996-7/12 $31.00 © 2012 IEEE

formed both side of the SET and found out that the applied V_g should be higher than 1.2 V to minimize their influence.

Even though we use the SET with high tunnel resistance, we have not been measured the cutoff frequency due to the usable frequency limit of measuring system. Therefore, we can estimate the cutoff frequency is larger than 50 MHz.

Since we can find a SET with much lower tunnel resistance of about 65kΩ, we can conclude that it is possible to achieve SETs which transfer a frequency higher than 1 THz.

ACKNOWLEDGEMENT

This work was partly supported by Grants-in-Aid from the Japan Society for the Promotion of Science (JSPS KAKENHI 2156081, 22240022).

REFERENCES

[1] J.Weis et al.,Semicond, Sci. Technol. **10**, 877 (1995)
[2] Y. Takahashi et al., Electron. Lett., **31**, 136 (1995)

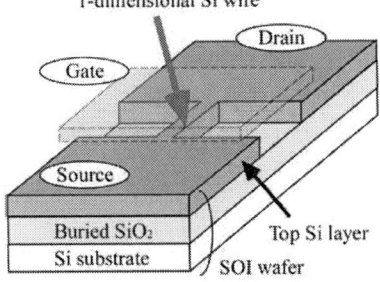

Fig. 1. Initial device structure of the SET before PADOX.

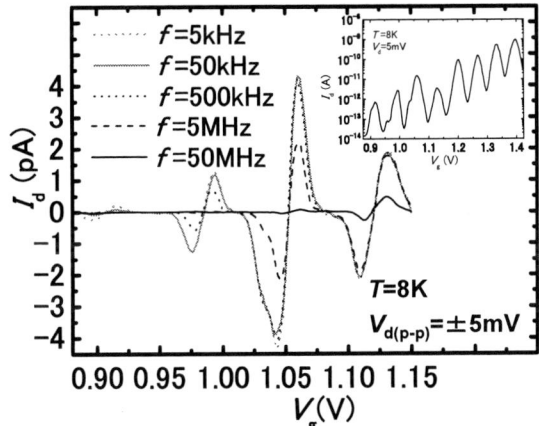

Fig. 2. Rectifying I_d-V_g characteristics of a SET at several frequencies measured at $V_{d(p-p)} = \pm 5$ mV and 8 K. The inset is normal I_d-V_g characteristics of the SET measured.

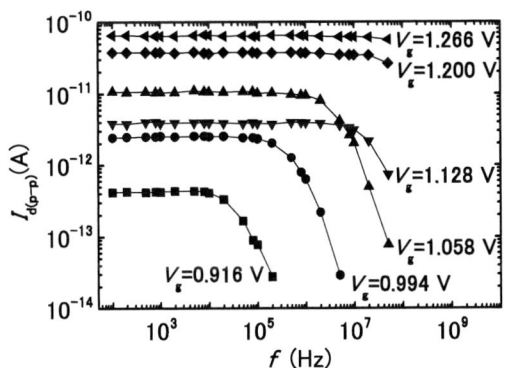

Fig. 3. Cutoff characteristics of a SET with a parasitic MOSFET measured at 8 K. Parameters are the gate voltage used for measuring rectifying characteristics.

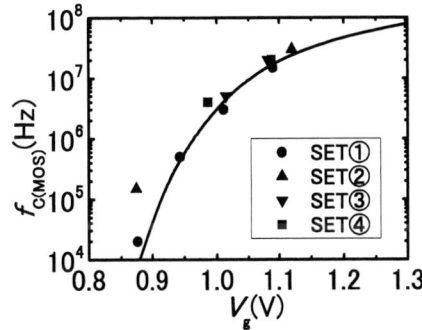

Fig. 4. Measured f_C (points) and expected f_C of the parasitic MOSFET (line).

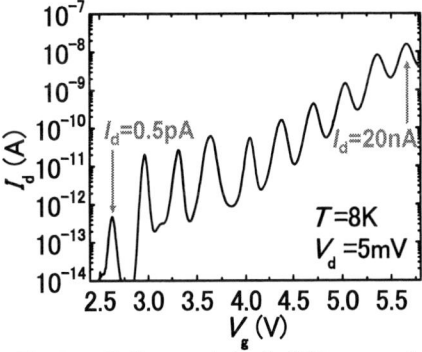

Fig. 5. I_d-V_g Characteristic of a SET measured

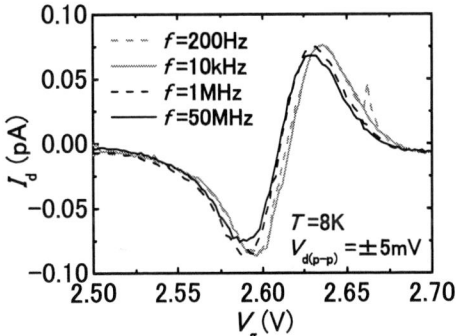

Fig. 6. Rectifying I_d-V_g characteristics of a SET at several frequencies measured at $V_{d(p-p)} = \pm 5$ mV.

Negative Differential Resistance Devices with Ultra-High Peak-to-Valley Current Ratio Based on Silicon Nanowire Structure

Sunhae Shin, Min Woo Ryu, and Kyung Rok Kim*

School of Electrical and Computer Engineering, Ulsan National Institute of Science and Technology
Ulsan, 689-798 Korea, Tel: +82-52-217-2122, Fax: +82-52-217-2109, *E-mail: krkim@unist.ac.kr

Abstract — **Negative differential resistance (NDR) devices are proposed with ultra-high peak-to-valley current ratio (PVCR) over 10^4 based on silicon nanowire structure.**

I. INTRODUCTION

The negative differential resistance (NDR) devices have a promising device operation principle for multi-functionality with a reduced complexity. The typical NDR device based on silicon (Si) materials is an Esaki tunnel diode with a heavily-doped pn junction and its key performance metric is the peak-to-valley current ratio (PVCR). Most of experimentally reported PVCR based on band-to-band tunneling (BTBT) mechanism, however, is limited below 10 by the trap-assisted tunneling current through forbidden band-gap [1, 2].

In this paper, we propose novel NDR device with ultra-high PVCR based on simple *pn* junction and Si nanowire (NW) structure. Device operation principle and various NDR characteristics according to the carrier injection mechanism will be presented based on the three-terminal structure with the controllable PVCR characteristics.

II. DEVICE OPERATION PRINCIPLE

Figure 1(a) shows the proposed NDR device structure based on the combination of diode and NW transistors with 10 nm radius and 3.5 nm dielectric thickness. As shown in the circuit schematics of Fig. 1(b) and (c), we proposed two types of NDR devices combined with *p*-type and *n*-type Si NW transistors, respectively. Basic operation principle of these three-terminal NDR devices is that carriers injected from or to the diode region can be controlled by the gate voltage creating NDR region with a completely suppressed valley current. Figure 2 (a) shows the conventional 2-terminal Esaki tunnel diode simulation results based on our numerical BTBT model [3, 4], which reproduce NDR with the realistic valley current owing to trap-assisted tunneling (TAT) with PVCR below 10.

In this work, however, since Si NW transistors without junctions can control the charge flow by the channel depletion from the gate bias and workfunction control as in Fig. 2(b) [5], completely suppressed valley current and ultra-high PVCR can be obtained by the proper biasing combinations in the combined structure of the junction diode and NW transistors.

Figure 3 illustrates the operation principle of the combined diode with *p*-type NW transistor (Fig. 1(b)) by the energy band diagrams. Both of the moderately doped junction and heavily doped tunnel junction can inject the electrons and drain the holes at low gate bias. If we increase the gate bias faster than source bias, the potential barrier for holes increases and thus, NDR can be obtained with ultra low valley current. In case of tunnel junctions, the higher peak current can be expected due to the BTBT at lower source bias (Fig. 3(c)).

III. DEVICE CHARACTERISTICS

Figure 4 shows the NDR characteristics of the *p*-type NW transistors connected to the junction diode with various doping concentrations. As expected from the device operation principle, peak current has been reduced by the increase of built-in potential as doping increases. After the tunnel junction is obtained by degenerate doping, however, both of the increased peak current and lower voltage operation can be achieved due to the additional carrier injection from the tunnel junction at lower bias condition. The proposed NDR devices show the PVCR over 10^4 even though TAT is allowed with a realistic level in Fig. 2(a), since valley current can be completely suppressed due to the depletion under the higher gate bias. In addition, we can control the peak current by designing junction area as shown in Fig. 5. More interestingly, the multiple peak and valley can be observed in the smaller tunnel junction area, which dominates the current flow with higher resistance. The first peak and valley correspond to the inherent NDR from tunnel diode at lower bias and the second peak and valley come from the competing between thermal carrier injection and current suppression by higher gate bias of NW transistors

IV. COMPARISON OF CARRIER INJECTION MECHANISMS

Figure 6 illustrates the operation principle of combined diode with *n*-type NW transistors. In this structure, the floating potential of *p*-region can be affected by both the drain and gate bias simultaneously. In case of junction diode with low doping, the injected carriers from *n*-type NW can easily flow to the drain when drain bias increases (Fig. 6(a)). If the gate voltage decreases from on-state to off-state while drain bias increases, carrier injection is inhibited by the increased potential barrier due to the depletion under gate region (Fig. 6(b)). As shown in Fig. 7, doping-dependent NDR characteristics indicate that the peak current increases and shift to lower voltage due to the reduction of potential barrier in the floating *p*-region when doping concentration of junction diode decreases as an opposite behavior from the previous *p*-type NW case of Fig. 1(b). One of the advantages in this mechanism is that we can accommodate the same steps of gate and drain voltage shift onto each current-voltage data point for more compact NDR circuit configuration.

V. CONCLUSION

We proposed the novel three-terminal NDR devices with an ultra-high PVCR based on Si NW structure. The controllable NDR characteristics with PVCR over 10^4 have been demonstrated by the successful suppression of valley current through the depletion of Si nanowire. Comparison study between various carrier injection mechanisms on the proposed NDR device has been performed by the careful analysis of the current components.

978-1-4673-0996-7/12 $31.00 © 2012 IEEE

ACKNOWLEDGEMENT: This work was supported by the National Research Foundation of Korea (NRF) funded by the Ministry of Education, Science and Technology (2011-0005342).

REFERENCES

[1] K. R. Kim et. al., IEEE Trans. Nanotechnology 4(3), p.317 (2005)
[2] A. Ramesh et. al., IEEE Trans. Electron Dev., 59(3) p.602 (2012)
[3] K. R. Kim, et. al., *SISPAD* (2005) p.159
[4] Sentaurus TCAD Manual v. D-2010.03 (Synopsys, 2010)
[5] J.-P. Colinge et. al., Appl. Phys. Lett. 96, 073510 (2010)

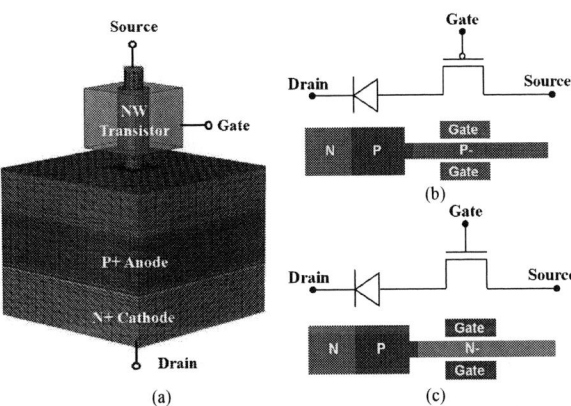

Fig. 1. Device structure and schematics of the proposed three-terminal NDR devices: (a) 3-dimensional schematic view, (b) circuit symbol and 2D schematic of combined diode with p-type nanowire transistor, and (c) circuit symbol and 2D schematic of combined diode with n-type nanowire transistor.

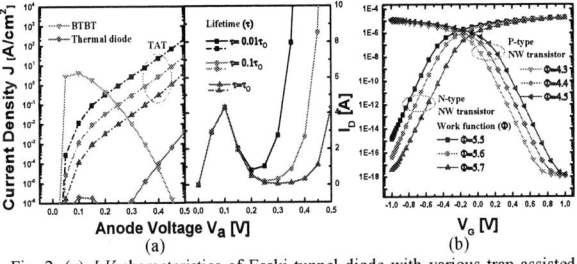

Fig. 2. (a) J-V characteristics of Esaki tunnel diode with various trap-assisted tunneling (TAT) currents (b) I_D-V_G curves of n-type and p-type NW transistors with various workfunctions of gate material.

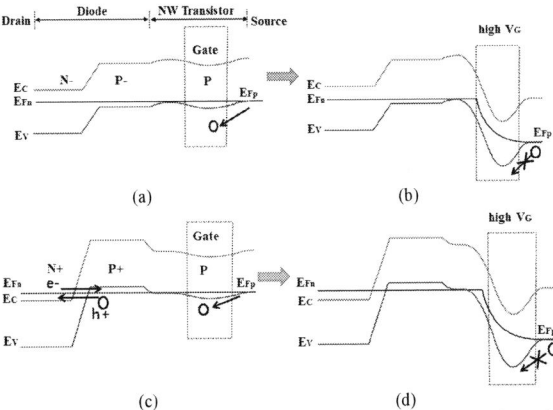

Fig. 3. Energy band diagram of NDR device with p-type NW transistor: (a) Initial band diagram and (b) final band diagram at V_G= 1.6 V and V_S= 1.0 V with low doping junction diode. (c) Initial band diagram and (d) final band diagram at V_G= 1.6 V and V_S= 1.0 V with high doping tunnel junction diode.

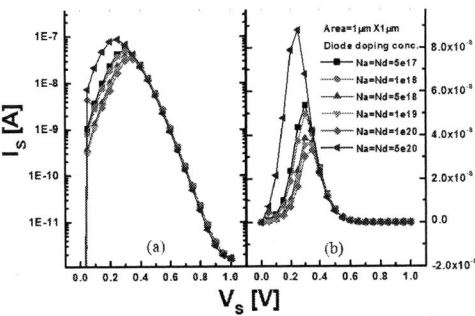

Fig. 4. The I_S-V_S characteristic of NDR device with p-type NW transistor according to the various doping concentration of *pn* junction diode: (a) log scale (b) linear scale. Gate voltage V_G has been applied to each data point with larger increasing step of 80 mV than V_S increasing step of 50 mV in this *I-V* plot.

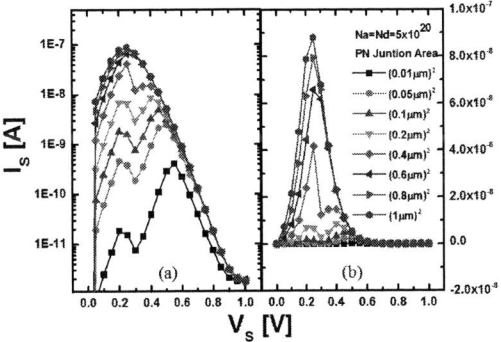

Fig. 5. The I_S-V_S characteristic of NDR device with p-type NW transistor according to the various area of junction diode: (a) log scale (b) linear scale.

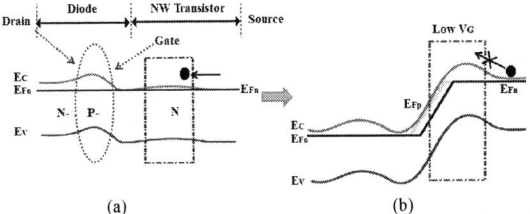

Fig. 6. Energy band diagram of NDR device with *n*-type NW transistor: (a) Initial band diagram at increasing current with high V_G and (b) final band diagram at valley current suppression with low V_G.

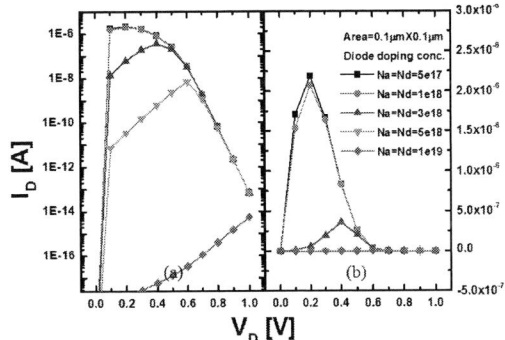

Fig. 7. The I-V characteristic of NDR device with n-type NW transistor according to the various doping concentration of diode: (a) log scale (b) linear scale. Gate voltage V_G has been applied to each data point decreasing step of 50 mV which is same with V_D increasing step in this *I-V* plot.

Mapping of single donors in nano-scale MOSFETs at low temperature

J. Verduijn[1,2], G.C. Tettamanzi[1], R. Wacquez[3], B. Roche[3], B. Voisin[3], X. Jehl[3], M. Sanquer[3], S. Rogge[1,2]

1) Centre for Quantum Computation and Communication Technology, School of Physics, University of New South Wales, Sydney, New South Wales 2052, Australia, 2) Kavli Institute of Nanoscience, Delft University of Technology, Lorentzweg 1, 2628 CJ Delft, The Netherlands, 3) CEA-LETI Minatec, Grenoble, F-38054, France

Introduction

Variability of device characteristics is a major problem for future CMOS technology nodes [1,2]. In particular, a number of studies have shown that large variations of the threshold voltage and degraded sub-threshold slope can be induced by dopants, see for example refs. [1,2,3]. Even for devices with no intentional doping in the channel the variation in threshold can be enormous due to dopants from the contact that diffuse into the channel [3,4]. This effect has been investigated in detail at low temperature in ultra-scaled MOSFETs [3,4,5]. By utilizing the SOI substrate as an additional gate we demonstrate how only a few dopants (~ 5) can deteriorate device characteristics of nano-scale MOSFETs. This technique allows for the mapping of dopants in the channel and provides new insights in their role in ultra-scaled device. In the context of dopant-based devices this may provide useful isights [6].

Device structure and experimental set-up

Nano-wire MOSFETs have been fabricated from a SOI wafer with a gate length and channel width down to 20 nm (Fig. 1). The SOI film chosen to be either undoped or doped with phosphorus at concentrations 10^{17} cm^{-3} to set the channel doping level of the finished device. The thickness of the body was either 20 or 12 nm. To reduce charge noise and gate leakage as much a possible, a polycrystalline gate in combination with a 5 nm thick SiO_2 gate dielectric was used. A 15 nm silicon nitride spacer prevents dopants to end up in the channel during the formation of the source/drain contacts. Further, the handling wafer, below a 145 nm buried oxide, was doped to allow using it as an additional gate. For the electrical measurements, the device is mounted on a stage that can be submerged in liquid helium at 4.2 K. The source, V_{sd}, top gate, V_{tg}, and back gate, V_{bg}, voltages are set with respect to the grounded drain, while the drain current I_d is recorded. All measurements are performed in the linear regime, that is V_{sd} = 1 mV for the room temperature measurements and 100 μV for the low temperature measurements.

Results and discussion

Measurements on two devices (Fig. 2), with and without channel doping, were performed at room temperature. The undoped device has a threshold of -145 mV whereas the other device, with 10^{17} cm^{-3} phosphorus channel doping (~5 in the channel region), has a threshold of -280 mV (Fig. 2) and a sub-threshold slope of 107 and 153 mV/decade respectively. The sub-threshold slope of the doped device is significantly worse than the one of the undoped device (note that the theoretical limit is 60 mV/decade). Furthermore, when we measure the same devices as low temperature we find a threshold, i.e. the location of the first resonance, to be at 14 mV for the undoped device and as -11 mV for the doped device. The similarities between these values indicate that, even though the low temperature threshold is essentially that of an undoped device, the few dopants present in the channel mediate thermally smeared transport at room temperature. As a result of the presence of only a small number of dopants, the threshold voltage is lowered significantly and the the sub-threshold slope is degraded [3,4]. And, since the exact positions of the dopants is random, this results in large device-to-device threshold variations. To confirm this idea we used the back gate at low temperature to scan the channel potential landscape.

While maintaining the same small source/drain bias, sweeps in V_{tg} at different V_{bg} are recorded. This is again done for an undoped- and a 10^{17} cm^{-3} doped channel device (Fig. 3). At low temperature the undoped device shows a very regular, nearly linear threshold shift in top gate as a function of the back gate voltage. This is what is expected for a MOSFET where the back gate is just changing the channel potential in a monotonous way. The slope for positive (negative) V_{bg} is slightly different from the negative threshold case because the current flow either at the top (bottom) of the channel [6]. Comparing this result with the same measurement on the doped device, the influence of the channel doping becomes apparent; it induces disordered resonances in the plane of the V_{tg} and V_{bg}, because donors are located at random positions is the channel. Further, a pronounced kink at V_{bg} ~ 4 V can be attributed to screening of the gates by ionized donors [7]. Looking in more detail at one of the resonances, it turns out it allows us to study how a single donor couples to the gates and the source/drain contacts.

As mentioned before the device which is doped at a concentration of 10^{17} cm^{-3} has on average only ~5 dopants in the channel region. The resonances generated by confined quantum states just above the band edge at the channel/gate dielectric interface are indicated with a red dashed line in the left panel of Fig. 3. Below these resonances a resonance, with a different capacitive gate coupling is observed. Its sub-threshold position together with the anomalous capacitive coupling leads us to identify this resonance with a single phosphorus donor between the source and drain contacts in the body of the channel. In addition we observe that the tunnel current at the resonance varies over more than two orders of magnitude which shows that the tunnel coupling to the contacts strongly is modified by the back gate. This is an important result in as tuneability of tunnel coupling is often desired for new donor based devices [6].

978-1-4673-0996-7/12 $31.00 © 2012 IEEE

Summary and conclusions

Using low temperature measurements we have been able to identify the influence of only about five donors in the channel the channel of an ultra-scaled MOSFET as the source of an anomalously low room temperature threshold voltage and large sub-threshold slope. Further we observe the influence of these dopants on the low temperature threshold voltage shift as a function of applied back gate voltage. The understanding of this behavior allows us to identify resonant tunneling mediated by a single donor in the channel of a doped channel device and we show that the back gate strongly modifies the tunnel coupling. These results give new insights in dopant transport in ultra-scaled MOSFETs, which is relevant for conventional device characteristics as well as for new dopant-based device architectures [6].

Acknowledgements

This work was supported by the EC FP7 FET-proactive NanoICT project AFSiD (214989).

Corresponding author: J. Verduijn, address: CQC2T, UNSW Sydney, 2052 NSW, Australia, email: a.verduijn@unsw.edu.au, telefone: +61 (0) 938 55 655

References

1) C. L. Alexander, et al., IEEE Trans. Nanotechnology 4, 339 (2005)
2) S. Markov, et al., IEEE EDL 33, 315 (2012)
3) M. Pierre, et al., Nature Nanotechnology 5, 133 (2010)
4) R. Wacquez, et al., IEEE VLSI Technology Symp., 193 (2010)
5) H. Sellier, et al., APL 90, 073502 (2007)
6) L. C. L. Hollenberg, et al., PRB 69, 113301 (2004)
7) Y. Ono, et al., PRB 74, 235317 (2006)

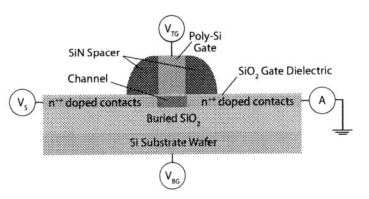 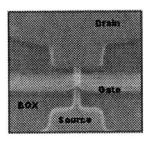

Fig. 1: (left panel) Schematic cross-section of the investigated devices with the location of the used voltage biases indicated. The current is measured on the drain side, which is kept grounded. (right panel) On the left a TEM cross-sectional image of a typical device fabricated in this batch on the right a SEM image shows how the gate is wrapped around the etched nano-wire.

Fig. 2: The top (bottom) two panels show room temperature and 4 K current-voltage characteristics of a device with a channel width of 40 nm (40 nm), a length of 60 nm (40 nm) and height of 20 nm (12 nm). The top device has received a 10^{17} cm^{-3} phosphorus channel doping, whereas the bottom device has no intentional doping in the channel. The extracted threshold voltages are indicated by a (red) dashed line in all graphs. The low temperature threshold is take as the position of the first resonance.

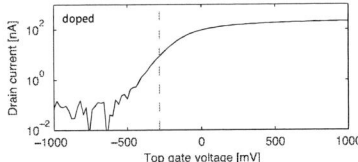
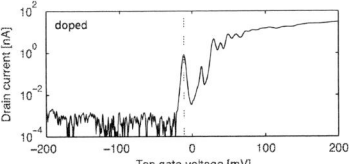
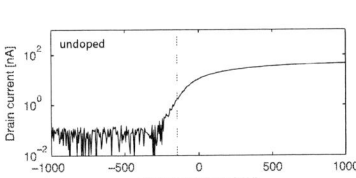
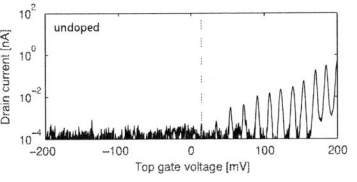

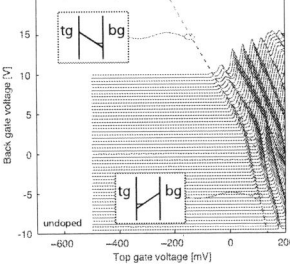
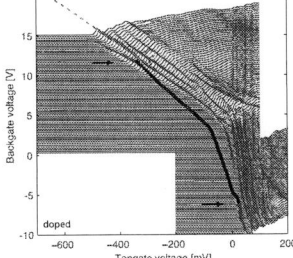

Fig. 3: (left panel) The drain current as a function of V_{tg} and V_{bg} for a undoped channel device with a channel width/length of 60/40 nm and a body thickness of 12 nm is measured. The data in the right panel is from the same doped device as the top panels of Fig 2. For clarity the scales are the same for both graphs. A clear difference in threshold voltage between the undoped channel device on the left and the phosphorus doped device is shown. The first resonance in between the arrows on the right panel, marked by black circles, is due to a single phosphorus dopant. These two arrows mark the points where the donor state aligns with the states at the interface. Since the first resonance slope is different from the states at the interface indicating it is located somewhere in the body of the channel.

978-1-4673-0996-7/12 $31.00 © 2012 IEEE

A Single Atom Transistor

M.Y. Simmons

Centre of Excellence for Quantum Computation and Communication Technology,
University of New South Wales, Sydney, NSW 2052, Australia

Over the past decade we have developed a radical new strategy for the fabrication of atomic-scale devices in silicon [1]. Using this process we have demonstrated few electron, single crystal quantum dots [2], conducting nanoscale wires with widths down to ~1.5nm [3] and most recently a single atom transistor [4]. We will present atomic-scale images and electronic characteristics of these atomically precise devices and demonstrate the impact of strong vertical and lateral confinement on electron transport. We will also discuss the opportunities ahead for atomic-scale quantum computing architectures and some of the challenges to achieving truly atomically precise devices in all three spatial dimensions.

[1] F.J. Rueß et al., Nano Letters 4, 1969 (2004).

[2] M. Fuchsle et al., Nature Nanotechnology 5, 502 (2010).

[3] B. Weber et al., Science 335, 64 (2012).

[4] M. Fuchsle et al., Nature Nanotechnology 7, 242 (2012).

Statistical Variability Study of a 10nm Gate Length SOI FinFET Device

Binjie Cheng[1], Andrew R. Brown[2], Xingsheng Wang[1] and Asen Asenov[1,2]

[1]School of Engineering, University of Glasgow, Glasgow, U.K.   [2]Gold Standard Simulations, Glasgow G12 8LT, U.K.

Email: Binjie.Cheng@glasgow.ac.uk Tel: +44 141 3304792

Introduction

Traditional bulk transistor scaling is rapidly approaching its limit due to the random dopant fluctuation [1]. Attributed to the much-improved electrostatic integrity introduced by side gates, FinFET devices can tolerate a very low channel doping concentration, resulting in a much-reduced statistical variability (SV) level compared with its bulk counterpart. FinFET devices can be fabricated on either bulk or SOI substrates. However, a recent study published by the SOI Industry Consortium indicates that SOI FinFETs can offer long-term benefits with regard to process variability [2]. In this work, we present a comprehensive simulation study on the SV aspect of a 10nm gate length SOI FinFET template device.

Device Design and Characteristics

The device design is based on a generic SOI FinFET structure that has been implemented directly in the GSS 'atomistic' simulator GARAND [3]. Fig. 1 shows a schematic picture of the device structure. Physical and electrical specifications of the device have been taken from the ITRS 2010 update [4] for a 10nm gate-length high-performance multi-gate device. Dual metal with gate-last process is assumed in order to eliminate metal gate granularity and the associated work function variability. The channel is effectively un-doped with a very light doping of 1×10^{15} cm^{-3}. The basic geometrical and electrical parameters are summarized in Fig.1 with the device Id-Vg characteristics presented in Fig.2, demonstrating the excellent electrostatic integrity of the template device.

Statistical Variability in the 10nm FinFET

Although FinFET devices can tolerate very low channel doping, the random discrete dopants (RDD) in the source/drain region can still play an active role in variation of device characteristics. Due to the 3D nature of the FinFET device, line edge roughness (LER) in FinFETs not only introduces traditional gate edge roughness (GER) variation, but also introduces fin edge roughness (FER) variation. Consequently, the sources of variability considered in this study are RDD, GER and FER, as illustrated in Fig. 3. The impact of individual SV sources on device figures of merit: threshold voltage (V_{TH}), on-current (I_{ON}) and DIBL are presented in Fig. 4. The impact of RDD and FER on device characteristics is operation-region dependent; while FER directly manipulates the quantum confinement in the fin, presenting a consistent impact on all of the figures of merit. The impact of GER and FER on device characteristics with respect to the magnitude of the edge roughness is presented in Fig.5. Due to the dominant influence of FER on device characteristics, the reduction of edge roughness can significantly improve the FinFET variability performance. In principal, the inter-correlation strengths between figures of merits can ultimately determine the size of the statistical parameter set used to capture the impact of SV in compact models. As demonstrated in Fig.6, there are strong correlations between subthreshold figures of merits for individual SV sources, and these correlations are maintained in the combined-sources scenario. However, there are degrees of disentangling between sub-threshold and on-current regions, which is manifested in the combined-sources scenario, as

demonstrated by the inset of Fig.7. The FER configuration determines the best-case DIBL performance, as demonstrate by device *159* in Fig.7; while GER plays an important role in determining the worst-case DIBL behaviour, as illustrated by the S/D configuration in device *718*.

On top of combined static SV sources, statistical reliability simulation were carried out at three different trapped-charge sheet density N_T levels: 1×10^{11}cm^{-2}, 5×10^{11}cm^{-2} and 1×10^{12}cm^{-2}, corresponding to the early, intermediate and later degradation stages. The μV_{TH} is increased from 140mV for the fresh state to 164mV at the later degradation stage, with σV_{TH} increased from 31mV to 36mV. The worst-case V_{TH} increase with degradation is 85mV with device *926*. As illustrated in Fig.8, among total traps of 7 in this device, 6 traps are located near the source side of the channel region, and distributed at both sides of fin, such configuration can effectively cut the current path under saturation V_{TH} bias condition, introducing a large V_{TH} shift. Device short-channel behaviour can be significantly manipulated by NBTI/PBTI degradation. In this particular ensemble of 1000 devices, device *323* has the worst-case DIBL degradation with value shifted from 29mV/V to 68mV/V ($\Delta V_{Tsat} < \Delta V_{Tlin}$); while device *443* has the best-case DIBL improvement with value shifted from 65mV/V to 44mV/V ($\Delta V_{Tsat} > \Delta V_{Tlin}$). The corresponding configurations of fin and trap locations of these two extreme devices are presented in Fig.9. Fig.10 demonstrates that number of traps is not the only factor determined the magnitude of device performance degradation, and furthermore, V_{TH} and I_{ON} shift are not perfectly correlated.

Similar results are obtained for pFinFET device as well. All device variability information is transferred into BSIM CMG model by a two-stage statistical compact model parameter extraction strategy [5]. A 6T SRAM cell with minimum size configuration is used as a benchmark circuit in this study. The device work function is tuned for the SRAM application, and the SNM simulation result is presented in Fig.11. The μSNM is 154mV and σSNM is 13mV, which indicates that it can pass the "μ-6σ" yield criterion [6] with considerable margin.

Conclusions

A comprehensive statistical variability simulation study of a 10nm gate length FinFET device is presented. The FER-induced quantum confinement variation has a consistent impact on all device operation regions; while the RDD induced S/D resistance variation has little impact on the sub-threshold, but has relatively strong impact on the on-current, which is in contrast with the impact of GER on device characteristics. The statistical reliability simulation results indicate that the impact of NBTI/PBTI on individual device is the combined results of trap and fin configurations. Both statistical variability and reliability simulations demonstrate some degree of disentangling between sub-threshold and on-current behaviour. The advantage of FinFET technology is demonstrated by the result of statistical SRAM cell simulation.

Acknowledgement

This work was supported in part by the European Union through the FP7 Integrated Project Trams.

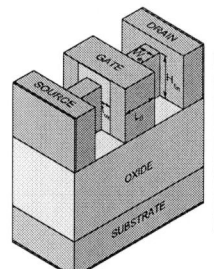

	ITRS	TCAD
L_G, nm	9.7	10
EOT, nm	0.57	0.585
H_{fin}, nm	-	12.5
W_{fin}, nm	4.8	5
I_{DSAT}, mA/μm	2.0	1.9
I_{OFF}, nA/μm	100	71
SS, mV/dec	-	71
DIBL, mV/V	-	53

Fig.1 Schematic of the 3D FinFET, and the structure and electrical parameters of 10nm FinFET device

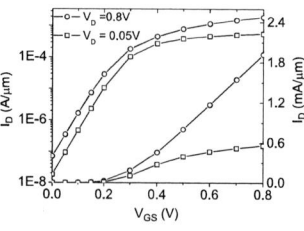

Fig.2 Gate transfer characteristics of the 10nm FinFET device.

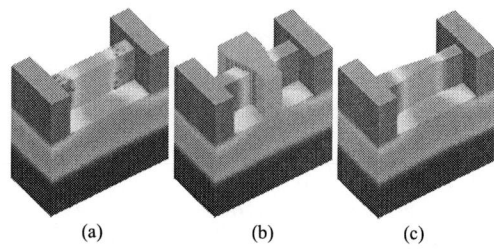

Fig.3 The effect on the structure and electrostatic potential of the different SV sources: (a) RDD, (b) GER and (c) FER.

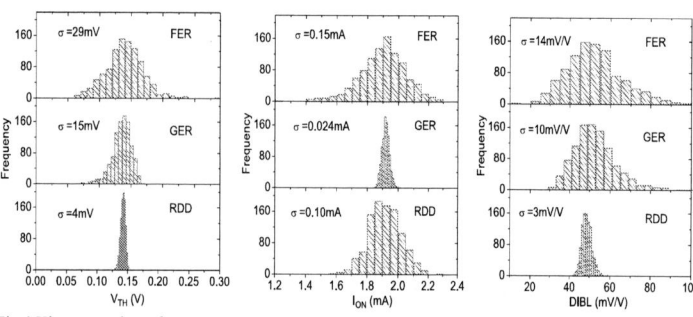

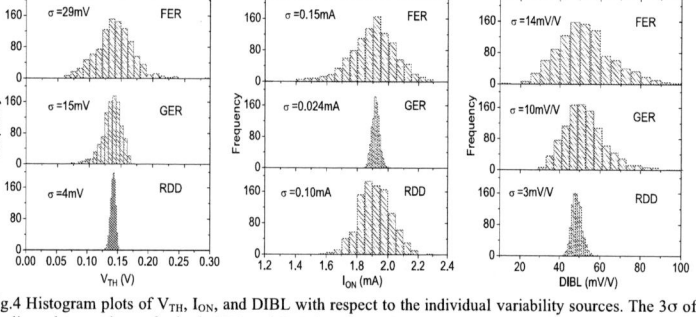

Fig.4 Histogram plots of V_{TH}, I_{ON}, and DIBL with respect to the individual variability sources. The 3σ of the line edge roughness for both GER and FER is 2nm, and the correlation length is 30nm. RDD have a mild influence on the subthreshold region, but a relatively large contribution in the on-current region; while for GER the opposite trend holds.

Fig.5 Impact of GER, FER on V_{TH}, I_{ON}, and DIBL with respect to magnitude of roughness, RDD is presented as a reference.

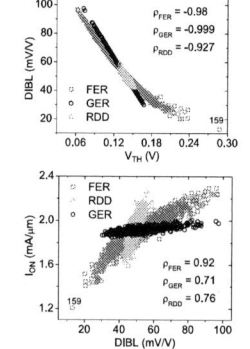

Fig.6 Correlations among device figures of merits

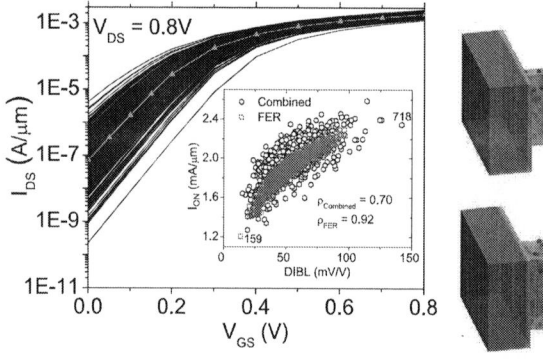

Fig.7 Id-Vg characteristics for an ensemble of 1000 devices with combined SV sources; the 3σ of the line edge roughness for both GER and FER is 2nm, symbol is the uniform device; inset: the correlation between on-current and DIBL under the influence of FER and combined SV sources; right: devices with the best and worst DIBL, showing the fin and discrete S/D dopant configurations.

Fig.8 Scatter plot between V_{TH} of fresh devices and later degradation stage devices; bottom-right inset: Q-Q plot of V_{TH} at different degradation stages; top-left inset: device with the worst V_{TH} degradation, showing the fin, discrete dopant and trap locations, and current iso-surface.

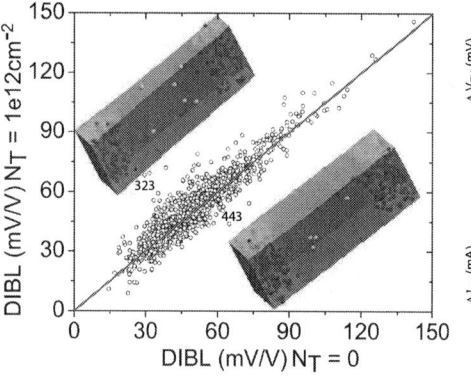

Fig.9 Scatter plot between DIBL of fresh devices and later degradation stage devices; bottom-right inset: device with the best DIBL improvement; top-left inset: device with the worst DIBL degradation, showing the fin, discrete dopant and trap locations.

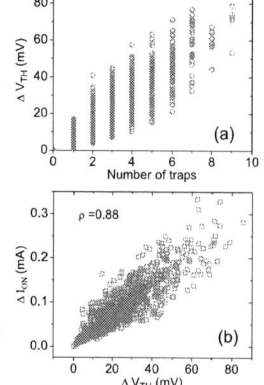

Fig.10 (a) Correlation between V_{TH} shift and number of traps, (b) Correlation between V_{TH} shift and I_{ON} shift, with N_T of 1e12cm^{-2}.

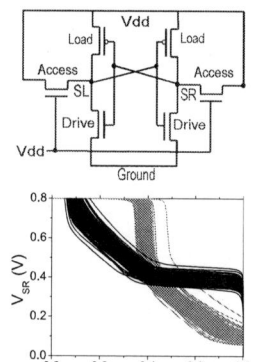

Fig.11 Bias configuration for SNM of SARM cell, and the static transfer characteristics of SRAM cell with cell ratio of 1 in fresh state.

Reference:

[1] A. Cathignol, B. Cheng, et al., IEEE Electron Device Letters, Vol. 29, pp. 609-611, 2008

[2] http://www.soiconsortium.org

[3] www.GoldStandardSimulations.com

[4] www.itrs.net

[5] B. Cheng, S. Roy, et al., Solid-State Electron., vol 49, pp. 740 (2005)

[6] P. A. Stolk, H. P. Tuinhout, et al., Tech. Digest of IEDM, pp. 215-218, 2001.

Reduced Drain Current Variability in Fully Depleted Silicon-on-Thin-BOX (SOTB) MOSFETs

T. Mizutani[1], Y. Yamamoto[2], H. Makiyama[2], T. Tsunomura[2], T. Iwamatsu[2], H. Oda[2], N. Sugii[2], and T. Hiramoto[1]

[1]Institute of Industrial Science, The University of Tokyo, [2]Low-power Electronics Association & Project (LEAP)

4-6-1 Komaba, Meguro-ku, Tokyo 153-8505, Japan, Phone：+81-3-5452-6264, E-mail: mizutani@nano.iis.u-tokyo.ac.jp

Abstract

Drain current variability in silicon-on-thin-BOX (SOTB) MOSFETs are analyzed by decomposing into current variability components and compared with conventional bulk MOSFETs. It is found that drain current variability in SOTB MOSFETs is largely suppressed thanks to not only reduced V_{TH} variability but also reduced current-onset voltage (COV) variability due to intrinsic channel.

Introduction

Random variability in scaled transistors is one of the most serious issues for further device scaling and supply voltage reduction [1-4]. Although V_{TH} variability and its origins have been intensively studied, drain current variability is also a serious concern for logic and memory circuit applications. Tsunomura et al. [5] have investigated the origins of drain current variability in conventional bulk MOSFETs and found that, besides V_{TH} and Gm components, the effects of "current-onset voltage" (COV) variability [6,7] caused by channel potential fluctuations largely contribute to the current variability.

On the other hand, fully depleted (FD) SOI MOSFETs are attractive for low power applications [8,9]. It has been shown that FD SOI MOSFETs with intrinsic channel has much smaller V_{TH} variability than bulk MOSFETs due to the absence of random dopant fluctuation (RDF) [9,10]. Since COV variability is also suppressed in FD SOI [10], it is expected that drain current variability is greatly suppressed in FD SOI MOSFETs.

In this study, random drain current variability in FD SOI MOSFETs is intensively measured using device-matrix array (DMA) TEG and compared with that in bulk MOSFETs. It is found by decomposing drain current variability components that very small drain variability is caused by the reduction of V_{TH} variability as well as COV variability.

Measurements

Fully depleted NMOSFETs with intrinsic channel were fabricated on silicon-on-thin-BOX (SOTB) substrate by 65nm technology [11]. SOI thickness is 13nm, BOX thickness is 10nm, and T_{INV} is 2.8nm. The minimum gate length (L) is 46nm. For reference, conventional bulk NMOSFETs were also fabricated. The variability of 1k transistors was measured using DMA TEG, and the variability of bulk and SOTB transistors were compared.

Fig. 1 compares I-V characteristics of 1k bulk and 1k SOTB MOSFETs with L=56nm. V_{TH} variability as well as drain current variability is suppressed in SOTB MOSFETs. Fig. 2 compares cumulative plots of drain current (I_{on}). Small drain current variability (σ=2.5% in linear region and σ=2.8% in saturation region) is attained in SOTB. Figs. 3 and 4 compare distribution of V_{THC} and V_{THEX}, respectively, where V_{THC} is threshold voltage defined by subthreshold constant current and V_{THEX} is threshold voltage defined by extrapolation (V_{gs} intercept of tangent line with largest slope in I_{ds}-V_{gs}). Smaller V_{TH} variability in SOTB MOSFETs is validated.

Fig. 5 compares distribution of COV, where COV is defined by the difference between V_{THEX} and V_{THC} (COV = V_{THEX} − V_{THC}) [5-7]. It has been clarified that COV variability is caused by large channel potential fluctuations due to RDF. From Fig. 5, COV variability is also suppressed in SOTB MOSFETs because of intrinsic channel. Fig. 6 compares distribution of maximum Gm (G_{mmax}), which also shows smaller Gm variability in SOTB.

Drain Current Variability Components

In order to examine the origins of drain current variability, the drain current variability was decomposed into three components (V_{THC}, COV, and Gm) [5]. Fig. 7 shows the correlation coefficients among these three factors, confirming that three factors are almost independent both in bulk and SOTB.

Figs. 8 and 9 compare contributions of three drain current variability components in linear region and saturation region, respectively. It is found that COV component is very small, particularly in linear region, in SOTB MOSFETs. This is because the channel of a SOTB transistor is intrinsic. On the other hand, Gm component is the largest among three components in SOTB MOSFETs. In this decomposition method, the effect of drain/source resistance is included in Gm components. Large Gm component, particularly in linear region, is caused by large parasitic resistance variability due to very thin SOI.

Benchmarking

The drain current variability obtained in this study is compared with the literature [5,12-15]. The drain current variability in SOTB transistors is smaller than bulk transistors [5,12,13] and FD SOI MOSFETs [15] and comparable with nanowire transistors [14].

Conclusion

Random drain current variability in fully depleted SOTB MOSFETs is intensively investigated and it is s clarified that the drain current variability is small thanks to the reduction of Vth component and current-onset voltage (COV) component due to intrinsic channel. Fully depleted SOTB transistors are promising for ultra low voltage applications.

Acknowledgement

This work is supported by the Ministry of Economy, Trade and Industry (METI) and New Energy and Industrial Technology Development Organization (NEDO).

References

[1] M. J. M. Pelgrom et al., IEEE JSSC, vol. 24, p. 1433, 1989.
[2] K. Takeuchi et al., IEDM, p.467, 2007.
[3] K. J. Kuhn, IEDM, p. 471, 2007.
[4] T. Tsunomura et al., IEEE TED, vol. 56, p. 2073, 2009.
[5] T. Tsunomura et al., VLSI Tech. Symp., p. 97, 2010.
[6] A. Kumar et al., Silicon Nanoelectronics Workshop, p. 7, 2010.
[7] T. Mizutani et al., Silicon Nanoelectronics Workshop, p. 81, 2010.
[8] R. Tsuchiya et al., IEDM, p. 631, 2004.
[9] O. Weber et al., IEDM, p. 58,, 2010.
[10] T. Hiramoto et al., IEEE SOI Conference, p. 170, 2010.
[11] Y. Yamamoto et al., to be presented in VLSI Tech. Symp., 2012.
[12] H. Masuda et al., CICC, p. 593, 2005.
[13] M. Saitoh et al., VLSI Tech. Symp., p. 114, 2009.
[14] M. Saitoh et al., VLSI Tech. Symp., p. 132, 2011.
[15] J. Mazurier, IEDM, p. 575, 2011.

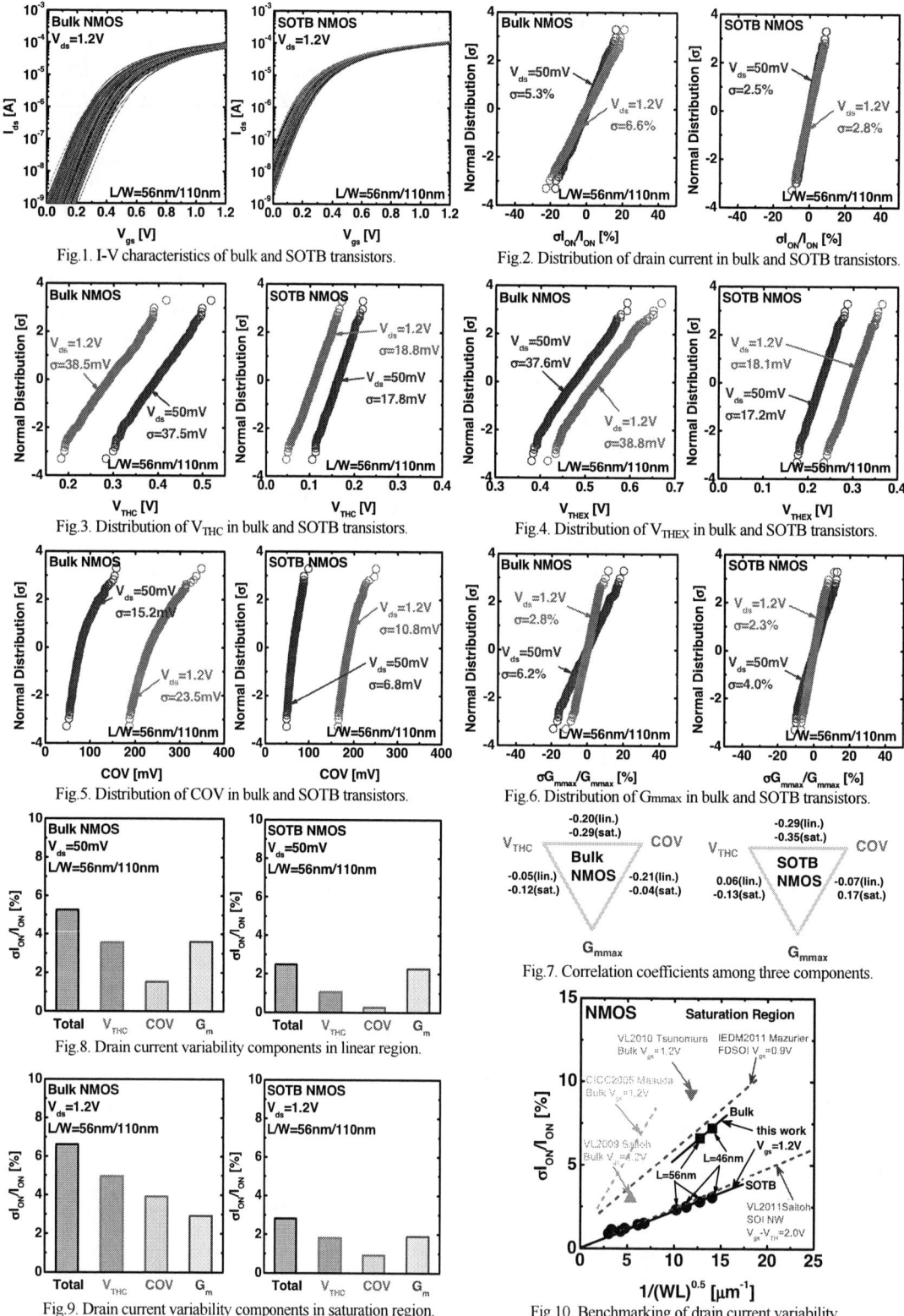

Fig.1. I-V characteristics of bulk and SOTB transistors.

Fig.2. Distribution of drain current in bulk and SOTB transistors.

Fig.3. Distribution of V_{THC} in bulk and SOTB transistors.

Fig.4. Distribution of V_{THEX} in bulk and SOTB transistors.

Fig.5. Distribution of COV in bulk and SOTB transistors.

Fig.6. Distribution of G_{mmax} in bulk and SOTB transistors.

Fig.7. Correlation coefficients among three components.

Fig.8. Drain current variability components in linear region.

Fig.9. Drain current variability components in saturation region.

Fig.10. Benchmarking of drain current variability.

978-1-4673-0996-7/12 $31.00 © 2012 IEEE

The Impact of the Carrier Transport on the Random Dopant Induced Drain Current Variation in the Saturation Regime of Advanced Strained-Silicon CMOS Devices

E. R. Hsieh[1], Steve S. Chung[1], C. H. Tsai[2], R. M. Huang[2], C. T. Tsai[2], and C. W. Liang[2]

[1]*Department of Electronics Engineering, National Chiao Tung University, Taiwan* [2]*United Microelectronics Corporation (UMC), Taiwan*

Abstract- The variation of saturation drain current ($I_{d,sat}$), induced by the random dopant variation (RDF), has been extensively studied by a new multivariate analysis method. It was found that the variation of $I_{d,sat}$ is originated from $V_{th,sat}$ and saturation velocity(V_{sat}), while the variation of $V_{th,sat}$ comes from the drain induced barrier lowering(DIBL). However, the experimental results shows that V_{sat} dominates the variation of $I_{d,sat}$. From the transport theory, V_{sat} is further decomposed into V_{inj} and B_{sat}, showing that V_{inj} is the dominant factor of $I_{d,sat}$ variation. The faster the V_{inj} is, the less the $I_{d,sat}$ variation becomes. If one improves the injection velocity, then the variation of $I_{d,sat}$ can be suppressed. This has been one of the significant benefits of strained silicon technology in CMOS device scaling.

1. Introduction

As CMOS devices are scaled further, the RDF induced variation increases. In more recent years, most of the studies were focused on the threshold voltage and linear drain current variations [1]. However, almost none has been paid on the random dopant effect in the quasi-transport regime, until recently we have some preliminary results on the importance of this RDF induced variations in the saturation regime [2]. By considering the transport effect, the saturation drain current of a device is governed by: (Fig. 1, Table 1)

$$I_{d,sat}= Q_{eff}V_{sat}(V_{gs} - V_{th,sat})= Q_{eff}V_{inj}B_{sat}[V_{gs} - (V_{th,lin}-DIBL)]$$

where V_{sat}, V_{inj}, and B_{sat} are the saturation velocity, injection velocity, and ballistic efficiency respectively. Here, V_{inj} is the injection efficiency of carriers from the source to the channel, and B_{sat} represents how many carriers will be transported through the channel toward the drain side. The higher the B_{sat} is, the larger the $I_{d,sat}$ becomes. In this work, we will first develop a new method, called multivariate analysis, to identify the major parameters which will affect the RDF induced I_{dsat} variation and demonstrate it with advanced strained CMOS devices.

2. Device Preparation

The poly-Si gate CMOS devices made on an advanced foundry platform, with EOT(SiON)=12Å, were prepared (Fig. 2), including SiC S/D-Extension strained nMOS devices [3], raised SiGe S/D strained pMOS devices [4], and conventional bulk-Si devices. Several areas of devices have been selected to gather the standard deviation of V_{th} and I_d. Devices with different areas were used to calculate the V_{th} and I_{dsat} variations.

3. Results and Discussion

A. $I_{d,sat}$ Variations and Multivariate Analysis

The experimental result in Fig. 3 is the Pelgrom plot for $V_{th,sat}$ and $V_{th,lin}$, in which $V_{th,lin}$ is inversely proportional to the square root of the device area, but $V_{th,sat}$ no longer obeys the rule! The difference comes from the DIBL effect. Therefore, the $I_{d,sat}$ variation can be separated into $V_{th,sat}$ and V_{sat} terms, in which $V_{th,sat}$ is represented by $V_{th,lin}$ and DIBL, different from the $V_{th,lin}$ variation. On the other hand, we may utilize the saturation velocity model to extract the transport parameters. Fig. 4 shows the V_{sat} characteristics of the experimental devices in Fig. 2. But how these variables, DIBL, V_{sat}, and

$V_{th,sat}$, affect the variation of $I_{d,sat}$, i.e., the percentage of the influences from these variables are unknown? To quantify the contribution of each component to I_d variation, a statistics tool, Multivariate Analysis (MA), is introduced, Table 2. By performing MA, we can decouple I_d variation, into several multiple variables, in terms of its *variance*, defined as the *square of variation*. By judging the components of those variables, we can identify the major source of I_d variation. Table 3 lists the notations of variables to provide a complete and quantified result of the I_d variation.

B. $I_{d,sat}$ Variation from the Ballistic Transport Theory

First, we evaluate the contributions of $V_{th,lin}$ and DIBL to $V_{th,sat}$, utilizing: $V_{th,sat}= V_{th,lin}-$ DIBL for strained and control devices by MA method(Fig. 5). It shows that DIBL dominates $V_{th,sat}$ variation, since as gate length shrinks, $V_{th,sat}$ rolls off seriously, which further fluctuates $V_{th,sat}$, called *short channel variation*. Furthermore, we decouple $I_{d,sat}$ variation into two parts, V_{sat} and $V_{th,sat}$, for the comparisons of nMOS splits and pMOS splits, respectively (Figs. 6 and 7). Results show that: V_{sat} is dominant of the $I_{d,sat}$ variation, while $V_{th,sat}$ is a supporting role. This is because, in the saturation region, the channel barrier is rolled-off by the drain bias and results in a lower barrier height such that it will be favorable for the carrier transporting through the barrier. With this huge disturbance of the carriers through the channel barrier, it reflects a large V_{sat}. In comparison, $V_{th,sat}$ also becomes increasingly important as channel length reduces. In short, V_{sat} dominates the $I_{d,sat}$ variation while short channel effect induced variation is important as device is further scaled.

Moreover, it was noted that, in strained nMOS devices, as gate length reduces, V_{sat} variation increases and results in a more serious $I_{d,sat}$ variation than that of the control one. To rule out the factors which cause the variation, we further decouple V_{sat} into V_{inj} and B_{sat} for nMOS and pMOS, respectively. (Figs. 8 and 9) We found that the velocity(V_{inj}) is the dominant factor of V_{sat} variation, in which we can see that : (1) V_{sat} is dominated by the injection velocity, V_{inj}, (2) from the comparison of Fig. 6(left) and Fig. 7(left), the Vsat variation of nMOS is larger than that of pMOS since nMOS has a higher V_{inj}, (3) nMOS exhibits an increase of B_{sat} when channel length reduces which gives rise to a larger variation of V_{sat}. This can be explained by the carbon out-diffusion from the source/drain in strained nMOS devices[5] which led to more $I_{d,sat}$ variation by comparing to the control ones. Fig. 10 shows a summary of the comparison between conventional and strained nMOS and pMOS. In particular, I_{dsat} variation of nMOS is smaller than that of pMOS, different from the common understanding that $V_{th,lin}$ variation in nMOS is always larger than the pMOS ones. (Fig.10)

In conclusion, the origins of $I_{d,sat}$ variation can be identified by a simple statistic method, multivariate analysis. It can be reasonably explained by the interactions between the carriers and the channel barrier. If the channel barrier is lower, the interaction between the carriers and barrier is weak, and $I_{d,sat}$ variation will be weakened; if the velocity of carriers is slower, the interactions between the carriers and barrier will be stronger, and $I_{d,sat}$ variation is larger. As a result, two guidelines can be drawn to suppress the $I_{d,sat}$ variation in designing CMOS devices with good variability by: (1) improving the carriers velocity by the strained silicon technology, and (2)having a good short channel effect control. These provide us valuable information in designing high performance and good variability CMOS devices in the future.

978-1-4673-0996-7/12 $31.00 © 2012 IEEE

part by the National Science under contract NSC100-2221-E009-016-MY3.

References:

[1] W. Lee al., in *Symposium on VLSI Tech.*, p. 112, 2009.

[2] E. R. Hsieh et al., to appear in *Symposium on VLSI-TSA*, 2012.

[3] S. Chung et al., in *Symposium on VLSI Tech.*, p. 158, 2009.

[4] C. H. Tsai et al., in *Symposium on VLSI Tech.*, p. 188, 2006.

[5] E. R. Hsieh et al., in *Tech. Dig. IEDM*, p. 779, 2009.

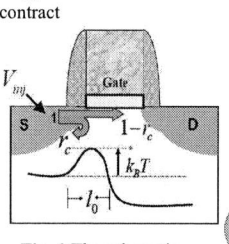

Fig. 1 The schematic showing the carrier transport behavior. r_c is the reflection coefficient; k_bT is the barrier; V_{inj} is the injection velocity.

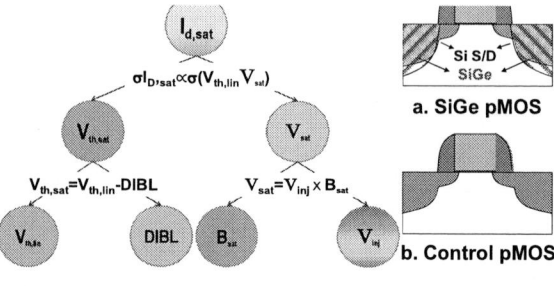

Table 1 $I_{d,sat}$ fluctuation can be decoupled into $V_{th,sat}$ and V_{sat}; $V_{th,sat}$ can be decoupled into $V_{th,lin}$ and DIBL; V_{sat} can be further decoupled into B_{sat} and V_{inj} where $B_{sat}= (1- r_c)/(1 + r_c)$.

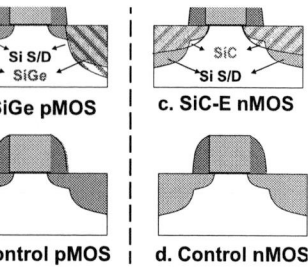

Fig. 2 The structures of the tested devices, including the strained pMOS and control pMOS (a,b) ; the strained SiC-E nMOS and control nMOS (c,d).

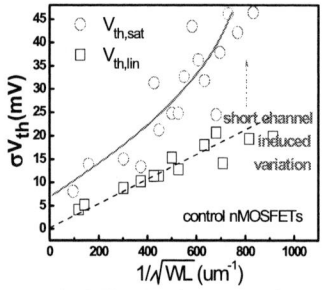

Fig. 3 The V_{th} Pelgrom plot of nMOSFETs. Note $\sigma V_{th,sat}$ is much larger than $\sigma V_{th,lin}$ due to the short channel induced variation.

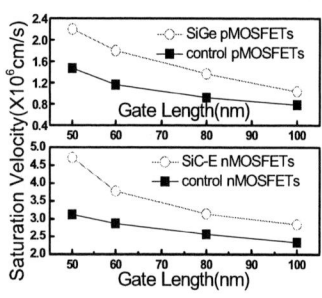

Fig. 4 The comparison of saturation velocities as a function of gate length in pMOSFET splits *(top)* and nMOSFETs splits *(bottom)*.

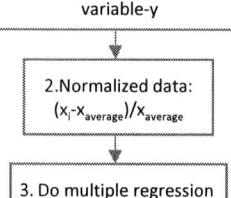

1. Define the independent variable-x&u and dependent variable-y

2. Normalized data: $(x_i-x_{average})/x_{average}$

3. Do multiple regression analysis : $y= ax+bu$

4. Analyze the variance of regression equation $\sigma^2 y= a^2\sigma^2 x+ b^2\sigma^2 u$

Table 2 Multivariate Analysis: The independent and dependent variables are chosen; the multiple regression analysis is performed. For example, the variance of y can be decoupled into two independent variables x, u.

Experiments	Variables of Multivariate analysis		
	$\sigma^2 y$	$a^2\sigma^2 x$	$b^2\sigma^2 u$
$V_{th,sat}=V_{th,lin}$-DIBL	$\sigma^2 V_{th,sat}$	$\sigma^2 V_{th,lin}$	σ^2 DIBL
$\sigma I_{d,sat}\propto\sigma(V_{th,sat}V_{sat})$	$\sigma^2 I_{d,sat}$	$\sigma^2 V_{th,sat}$	$\sigma^2 V_{sat}$
$V_{sat}=B_{sat}V_{inj}$	$\sigma^2 V_{sat}$	$\sigma^2 B_{sat}$	$\sigma^2 V_{inj}$

e.g., $\sigma^2 I_{d,lin}$: the normalized of variance for $I_{d,lin}$

Table 3 The symbols of variables used in Multivariate analysis corresponding to these experiments. For instance, $\sigma^2 tV_{th,sat}$ can be obtained as a function of $\sigma^2 V_{th,lin}$ and σ^2DIBL.

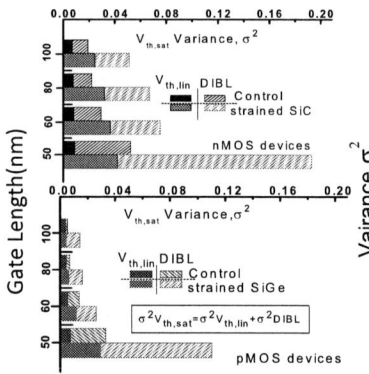

Fig. 5 The comparisons of the variance for $V_{th,sat}$, $V_{th,lin}$, and DIBL of n- and pMOSFETs. Note that $\sigma^2 V_{th,sat}$ is dominated by σ^2DIBL, the short channel variation.

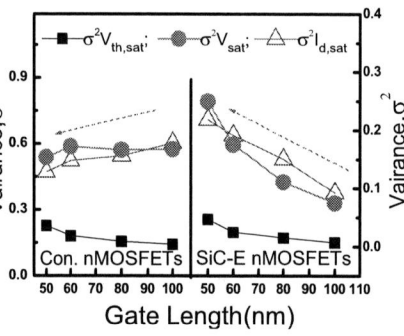

Fig. 6 The comparison of the variances (σ^2) for control *(left)* & strained nMSOFETs *(right)*. Note that $\sigma^2 I_{d,sat}$ and $\sigma^2 V_{sat}$ follow the same trend. $\sigma^2 I_{d,sat}$ decreases for control nMOSFETs *(left)* but increases for strained one *(right)* as gate length reduces.

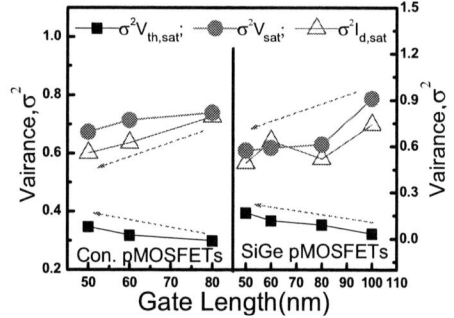

Fig. 7 The comparison of the variances (σ^2) for control *(left)* and strained pMSOFETs *(right)*. Note that $\sigma^2 I_{d,sat}$ and $\sigma^2 V_{sat}$(saturation velocity) follow the same trend, which decreases with the shrinking gate length.

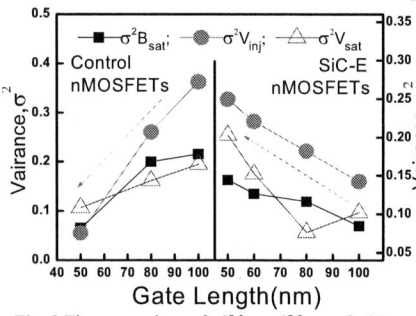

Fig. 8 The comparison of $\sigma^2 V_{inj}$, $\sigma^2 V_{sat}$, and $\sigma^2 B_{sat}$ for control *(left)* & strained one *(right)*. Note $\sigma^2 V_{inj}$ and $\sigma^2 B_{sat}$ increases with reducing gate length due to carbon out-diffusion.

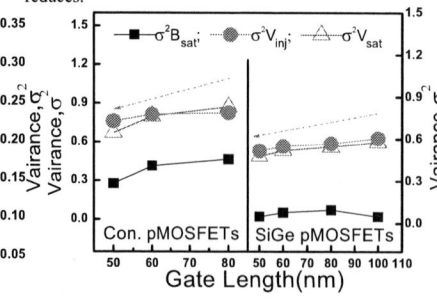

Fig. 9 The comparison of $\sigma^2 V_{inj}$, $\sigma^2 V_{sat}$, and $\sigma^2 B_{sat}$ for control *(left)* & strained one *(right)*. Note $\sigma^2 V_{inj}$ and $\sigma^2 V_{sat}$ follow the same trend and decrease with shrinking gate length.

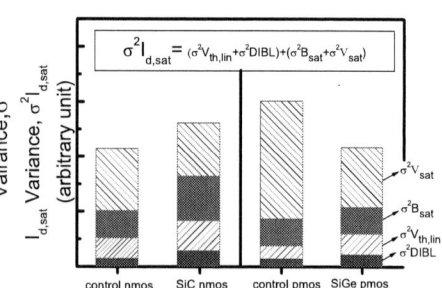

Fig. 10 The comparison of $I_{d,sat}$ for n- and pMOS devices. $I_{d,sat}$ fluctuation of nMOS devices is smaller than that of pMOS devices, but $I_{d,stat}$ fluctuation of strained nMOS devices is larger than that of pMOS

978-1-4673-0996-7/12 $31.00 © 2012 IEEE

On the Statistical Trap-Response (STR) Method for Characterizing Random Trap Occupancy and NBTI Fluctuation

Jibin Zou[1], Changze Liu[1], Runsheng Wang[1]*, Xiaoqing Xu[1], Jinhua Liu[2], Hanming Wu[2], Yangyuan Wang[1], Ru Huang[1]*

[1]Institute of Microelectronics, Peking University, Beijing 100871, China, *E-mail: ruhuang@pku.edu.cn; r.wang@pku.edu.cn
[2]Semiconductor Manufacturing International Corporation (SMIC) Beijing 100176, China

Abstract

In nanoscale devices with only a few oxide traps, characterization of trap response during NBTI stress is challenging due to the stochastic nature of trapping/detrapping behavior. This paper successfully extends the statistical trap-response (STR) method from DC to AC device operation, for getting a full understanding of the trap occupancy probability and the aging-induced dynamic variations under DC and AC NBTI. The AC trap response and the AC NBTI fluctuations are found largely deviating from the DC case, indicating different physical mechanisms.

Introduction

As devices scaling down, only a few traps exist in the gate dielectric of MOSFETs, leading to random trap occupancy behavior and the resulted stochastic degradation of NBTI, which has been paid increasing attention recently [1-10]. In these studies, the characterization method plays an important role. Since the direct measurement of trapping event during ultra-fast BTI stress is challenging, and the trap behavior should be investigated statistically due to the stochastic NBTI nature, thus a better way to study the random trap occupancy and NBTI degradation is statistically taking lots of step-like I_D or V_{th} relaxation traces originated from few trap detrapping events after short BTI stress. We named this kind of method as the statistical trap-response (STR) technique [10]. The STR method can also be applied to study time-dependent NBTI fluctuation. As we demonstrated that during different device operation cycles (or periods), there is time-dependent cycle-to-cycle variation (CCV) of NBTI degradations [10], in addition to the traditional time-dependent device-to-device variation. Therefore, the STR method provides a useful characterization platform that links both random trap occupancy and NBTI fluctuation.

On the other hand, trap response and NBTI fluctuations may vary from DC to AC device operation, due to the "recovery" in each AC clock-cycle. However, whether the STR method can be extended to AC study needs to be further experimentally validated. In this paper, various configurations of both DC and AC STR method are comprehensively studied, with a full picture of random trap occupancy and dynamic NBTI variations obtained.

Experiments, Results and Discussion

The devices used in this work are pFETs with 0.9nm SiON gate dielectrics. The gate length L=60nm and width W=120nm. Both DC and AC STR methods are adopted for statistical CCV analysis. For DC case as shown in Fig. 1, the device is stressed shortly, and then fully recovered at a small $V_{g,recovery}$ to monitor ΔV_{th} relaxation trace (100 stress-recovery measurement cycles for statistics). AC case shown in Fig. 2 is similar to DC case, except that the stress signal is changed to AC. AC BTI stress is carried out at different frequencies f with $duty$=50%, t_{stress_AC}=2t_{stress_DC} for the same effective stress time as DC case. In addition, regarding the AC clock-period itself as a CCV source, on-line characterization of AC NBTI fluctuation is also adopted as shown in Fig. 3. The CCV statistics is extracted from the on-line I_D fluctuation of 100 cycles nearby a given time. Although this on-line method can clearly show NBTI fluctuation with stress time, it cannot comprehensively characterize NBTI degradation with one single measurement, as will be discussed later.

As shown in Fig. 4 and 5, discrete step-like V_{th} recovery resulted from single trap detrapping behavior can be observed in small devices for both DC and AC STR methods. It is noticed that not every trace is step-like; which indicates that a particular trap may not response to stress signal every measurement cycle. Therefore, a large number of repetitive recovery traces obtained under a fixed stress condition can provide a statistical method to characterize the trap-response property, in terms of trap occupancy probability (p). Fig. 6 further shows DC and AC results of single trap occupancy probability (for a fixed trap) extracted from these repetitive recovery traces, as a function of effective stress time. It is found that for both DC and AC cases, the trap occupancy probability obeys Weibull distribution. But p is obviously smaller in AC case, probably due to the relaxation effect in every AC clock-cycle when $V_{g,stress-low}$=0. If increasing f of AC signal, p is further reduced, confirming that trap occupancy behavior under AC case largely deviates from DC case.

Since that single-trap response behavior is linked to the stochastic nature of NBTI [2-5], it indicates that the NBTI fluctuation in time domain should vary from DC case to AC case. The CCV statistics of NBTI degradation are further extracted in terms of mean value $\mu(\Delta V_{th})$ and standard derivation $\sigma(\Delta V_{th})$, as shown in Fig. 7 (DC) and Fig. 8 (AC). The trends of μ under AC condition are similar with DC ones. However, μ of AC case at higher frequency is smaller than DC, which is also attributed to the recovery in each clock period of AC signal. The variation curves in both cases present a non-monotonic "kink" behavior, which can be understood by two competitive parallel mechanisms [10]: σ increases monotonically with the average number of activated traps per device (N), while it has a quadratic relationship with the effective trap occupancy probability (p_{eff}). Variation can be expressed by $\sigma^2 = \lambda p_{eff}(N^2 + 2N - p_{eff}N^2)$, where λ represents average ΔV_{th} induced by a single trap. Based on which, the statistical differences of NBTI fluctuation between DC and AC case can be explained. Besides, the frequency-dependent trend of AC results as shown in Fig. 8 is believed to be related to the frequency-dependent trap occupancy behavior as discussed above.

Moreover, the on-line characterization of AC NBTI fluctuation is also adopted for comparison. Fig. 9 shows the typical NBTI fluctuations within one clock period, as expected. And Fig. 10 intuitional shows the NBTI fluctuation varying between each AC-clock-cycle with stress time. Both μ and σ are further extracted in Fig. 11. The results indicate that μ varies largely when repeating the on-line characterizations, revealing that there is an additional NBTI fluctuation source between repetitive on-line measurements. For comprehensive evaluation, all the above 3 sources of random NBTI fluctuation must be accounted if applying on-line characterization method. Therefore, the AC STR method is more convenient than the on-line method for characterizing the total dynamic NBTI fluctuation. Besides, it can also be observed that the σ curves of on-line method present the same "kink" shape as AC STR method.

Summary

In this paper, by extending the STR method to AC configurations, the trap occupancy behavior and NBTI fluctuation under AC NBTI stress are comprehensively studied and found deviating from the DC case. The results indicate that the STR method provides a robust characterization platform for evaluating trap occupancy probability in nanoscale devices and predicting the NBTI-induced dynamic variation during circuit aging under both DC and AC conditions, from which the physical mechanisms of trap responses under stress and the NBTI fluctuations can be clearly obtained.

978-1-4673-0996-7/12 $31.00 © 2012 IEEE

Acknowledgements: This work was partly supported by the 973 Projects (2011CBA00601), the NSFC (61106085, 60625403), the National Science & Technology Major Project 02 (2009ZX02035-001).

References: [1] V. Huard, et al., *IRPS*, p. 289, 2008. [2] T. Grasser, et al. *IEDM*, p. 729, 2009. [3] T. Grasser, et al., *IRPS*, p. 16, 2010. [4] B. Kaczer, et al., *IEEE EDL*, p. 773, 2010. [5] B. Kaczer, et al., *IRPS*, p. 26, 2010. [6] S.E. Rauch III, *IEEE T-DMR*, p. 524, 2007. [7] M. Toledano-Luque, et al., *VLSI*, p. 152, 2011. [8] V. Huard, et al. *IRPS*, p. 33, 2010. [9] M. Toledano-Luque, et al., *IRPS*, p. 364, 2011. [10] C. Liu, et al., *IEDM*, p. 571, 2011.

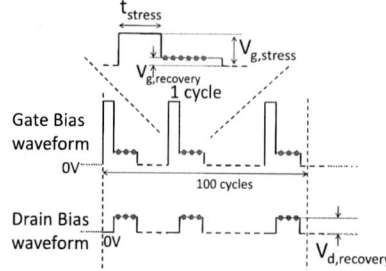

Fig. 1 The DC STR characterization method. The sampling rate is $10^6/s \sim 10^8/s$ for all the STR measurements.

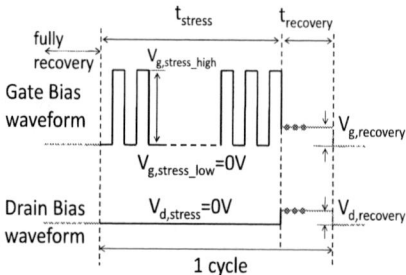

Fig. 2 The AC STR characterization method. It should be noticed that, the lower voltage of the AC signal during stress stage is pulled down to 0V.

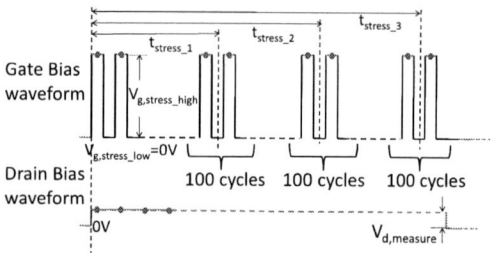

Fig. 3 The on-line AC characterization method. The device is stressed by AC signal for a relatively long time. The CCV statistics is extracted from the on-line I_D fluctuation of 100 cycles nearby a given time.

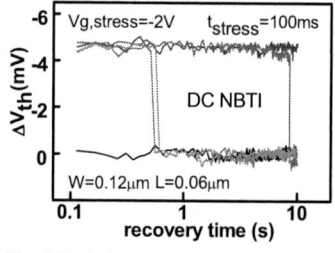

Fig. 4 Typical measured V_{th} recovery traces out of 100 traces by DC STR characterization. This is an example of single-trap relaxation. Trap occupancy probability p can be extracted.

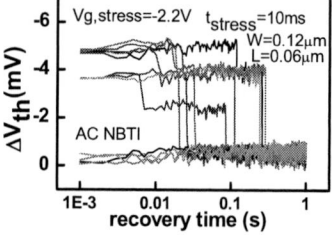

Fig. 5 Typical measured V_{th} recovery traces out of 100 traces for AC STR characterization. This is an example of multiple-trap relaxation.

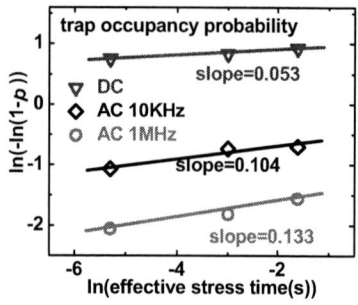

Fig. 6 Experimental results of the single trap occupancy probability (for a fixed trap) as a function of effective stress time. The results of DC and AC NBTI are compared.

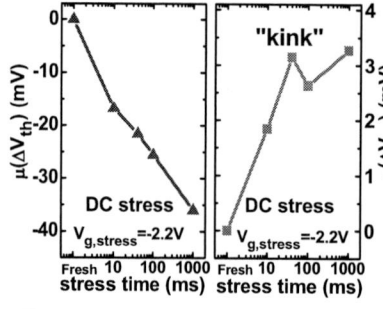

Fig. 7 Extracted μ and σ of CCV in DC NBTI. There is a non-monotonic "kink" behavior of σ: variation increases monotonically with N, while it has a quadratic relationship with p_{eff} [10].

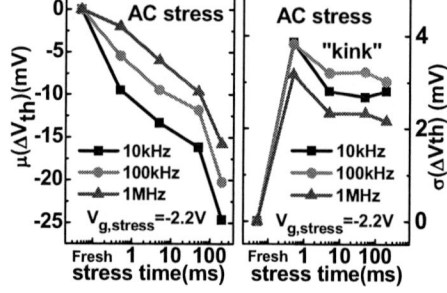

Fig. 8 Extracted μ and σ of CCV in AC NBTI. The results deviate from DC case, and the frequency-dependent NBTI fluctuation is related to the frequency-dependent trap response behavior as shown in Fig. 6.

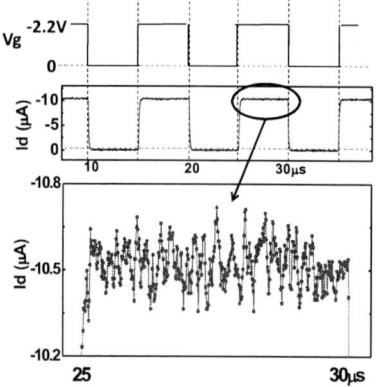

Fig. 9 Typical AC NBTI fluctuation in one AC cycle under the on-line measurement condition.

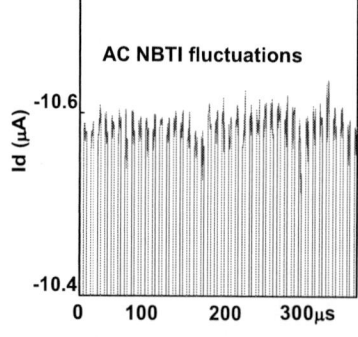

Fig. 10 The typical measurement of AC NBTI fluctuations with stress time by the on-line characterization method. AC NBTI fluctuation between AC clock-cycles is clearly shown.

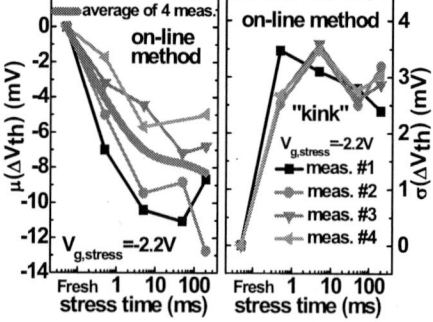

Fig. 11 The on-line AC NBTI fluctuation measurements at 100 kHz are repeated 4 times. μ varies severely when repeating measurements, revealing an additional NBTI fluctuation source. The trends of σ are similar to that of AC STR method.

978-1-4673-0996-7/12 $31.00 © 2012 IEEE

Statistical distribution of RTS amplitudes in 20nm SOI FinFETs

Xingsheng Wang[1*], Andrew R. Brown[2], Binjie Cheng[1], and Asen Asenov[1,2]

[1] Device Modelling Group, School of Engineering, University of Glasgow, Glasgow G12 8LT, U.K.
[2] Gold Standard Simulations Ltd., Rankine Building, Oakfield Avenue, Glasgow G12 8LT, U.K.
Tel: +44 (0)141 330 2964, *E-mail: Xingsheng.Wang@glasgow.ac.uk

Abstract

This abstract presents a comprehensive 3D simulation study on the impact of a single interface trapped charge in emerging 20nm gate-length FinFETs on an SOI substrate. The impact of the location of trapped charges on the Random Telegraph Signal (RTS) amplitudes is studied in detail. The RTS amplitude associated with particular trap position depends on the complex current density distribution in the Fin and is modified by 'native' statistical variability sources such as metal gate granularity (MGG), line edge roughness (LER), and random discrete dopants (RDD).

Introduction

Apart from the statistical variability inherent to fresh devices, random trapped charges at/near the oxide interface result in RTS and in a cumulative manner are responsible for the N/PBTI enhancement of the statistical variability [1,2]. Gigantic RTS amplitudes have been measured and reported in bulk transistors [3,4] and are associated with interactions between trapped charges and the underlying RDD [6]. Due to superior electrostatic integrity 3D FinFETs, pending introduction in the 22 nm CMOS technology, can tolerate very low channel doping concentration, resulting in reduced statistical variability [5]. This can also beneficially affect the RTS amplitudes, however detailed study of the RTS statistics in FinFETs is limited [5]. This abstract presents a 3D simulation study of the impact of the location of single trapped charges on FinFET characteristics, and the corresponding RTS distribution. FinFET specific sources of statistical variability are also taken into account.

Device and Simulation Methodology

The 20nm gate-length SOI FinFET design follows the ITRS 2010 featuring an EOT of 0.83nm and fin height/width of 25/10nm, as shown in Fig.1, achieving the required performance [5]. An individual interface trapped charge is resolved using the GSS 'atomistic' drift-diffusion simulator GARAND, including density-gradient quantum corrections, as demonstrated in Fig. 2. Random interface trapped charge is simulated in conjunction with the identified major statistical variability sources including MGG, LER of the gate (GER) and the fin (FER), and RDD. The statistical 3D simulations include ensemble of 1000 microscopically different FinFETs.

Results and Discussions

A. Location of individual trapped charge, and sensitivity

First we examine the threshold-voltage shift (ΔV_T) due to a single charge placed at the fin side. Fig.3 and Fig.6 show the ΔV_T, the bell shaped distribution corresponding to a differently location of the trapped charge. Trapped charge in the middle region of the fin results in a larger V_T-shift. Along the channel the most sensitive position in the middle corresponds to the peak in the source-to-drain barrier illustrated in Fig.4. With increased drain-bias the peak moves towards the source end, shifting accordingly the maximum of the sensitivity bell. However, in the vertical direction, due to quantum confinement and volume inversion in the subthreshold region, the maximum of the carrier distribution is in the middle, determining the vertical position of the bell maximum. Therefore a trapping charge located around half fin-height can cause the largest current reduction. The potential in the lateral direction and electron density in vertical direction result in strong correlations with ΔV_T in the respective directions, however this correlation does not hold over the whole side interface. The current density shows a high correlation with ΔV_T over the whole interface (Table 1). However, this influence of the trapped charge is limited to regions very close to the interface and has little effect on the current in the middle of the fin (Fig.5). Meanwhile, interface trapped charge located close to the drain end results in increased DIBL while trapping close to the source reduces DIBL, both contributing to DIBL variability (Fig.7).

B. Impact of trapping location on entire IV characteristics

Fig.8 and 9 shows the fractional gate voltage required in the presence of trapped charge to achieve the same current as the virgin FinFET. The subthreshold behaviour is explained above. In the overdrive region, more screening from strong inversion charges leads to less impact of the interface trapped charge [6]. In strong inversion conditions the highest density of mobile charge is in the upper corners of the fin (Fig. 10), screening the impact of the trapped charge there and requiring lower compensation voltage. Figs. 11 and 12 show the RTS/current-reduction magnitude. Similar to the previous two figures, trapping at middle of the FinFET side interface has the strongest impact on device characteristics in subthreshold.

C. Statistical Variability Enhances the Impact of Trapping

The distributions of fractional V_T change due to single random interface trapped charge are compared between the 'uniform' FinFET and 'atomistic' FinFET with MGG (5nm average grain-size), LER (3Δ of 2nm and correlation length 30 nm) and RDD. Fig.13 shows the complementary cumulative distribution function (CCDF), following $\propto \exp(-\Delta V_T / \eta)$ within the first 80%, with a little larger η of 2.61mV for atomistic devices compared with 2.56mV for the uniform device. In the tail, it is clear that charge trapping in the atomistic devices causes larger V_T shifts. Statistical variability sources result in current percolations in the sensitive middle region, and therefore enhance the RTS amplitude when the trapping occurs in a corresponding percolation path. As illustrated in Fig.14 the case with largest V_T shift happens to have a charge trapped in an enhanced current percolation path under a low workfuntion metal-grain and where the fin has narrowed due to FER.

Acknowledgement

This work is partially supported by EU ENIAC project MODERN and FP7 project TRAMS.

References

[1] B. Kaczer, et al., IRPS, 2010; [2] A. Brown, et al., T-ED, 57(9); [3] A. Ghetti, et al., T-ED, 56(8), 2009; [4] J. Franco, et al., IRPS, 2012; [5] X. Wang, et al., IEDM, 2011, p.103; [6] A. Asenov, et al., T-ED, 50(3), 2003.

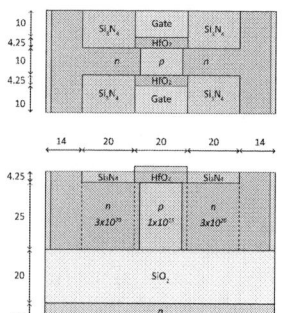

Fig. 1: The schematic view of 20nm gate-length SOI FinFET. The scale numbers are in the unit of nm.

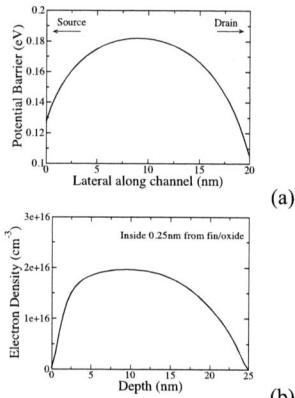

Fig. 2: The 3D device showing electron density in the fin. An electron is trapped at a side interface. The front slice is the potential, and the middle slice shows the current density. Vg=0.0V, Vd=0.05V.

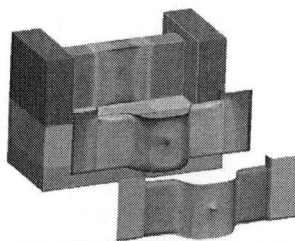
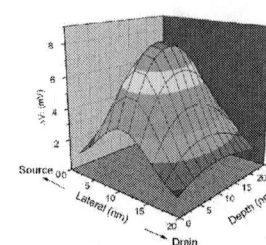

Fig. 3: Threshold-voltage shift due to a single trapping charge at different locations of fin side interface. It is seen that the sensitivity due to a single trapping is a bell-like shape; the most sensitive region is located in the middle of the side face at low drain-bias (a). With increasing drain-bias to saturation (b), the peak is reduced and also shifts towards the source end.

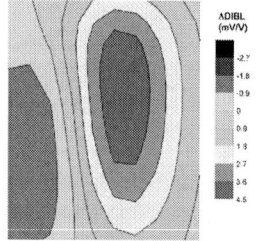

(a)

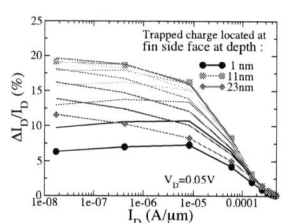

(b)

Fig. 4: (a) The potential barrier along channel-direction at half fin-depth and (b) the electron density along fin-depth in the middle-channel, near the fin side interface. Vg=0.2V, Vd=0.05V.

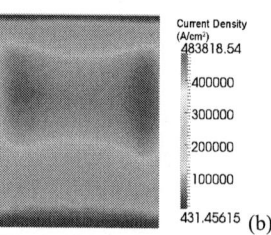

(a)

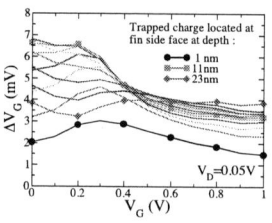

(b)

Fig. 5: The current density inside the fin with (a) top view at half fin-height, (b) vertical cross-section at half fin-thickness and (c) near the fin side interface parallel to fin side-face. Source/drain is on the left/right side. Vd=0.05V.

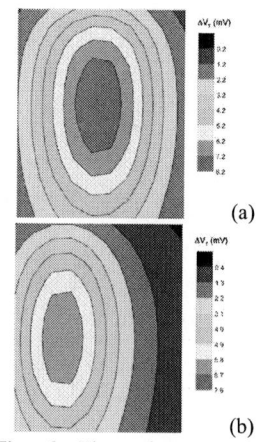

(c)

Table 1: correlations

Correlation Coeff.	ΔV_T
Potential (lateral)	0.92
eDensity (vertical)	0.94
Current Density (lateral/vertical)	0.96/ 0.94
Potential	0.57
eDensity	-0.41
Current Density	0.73

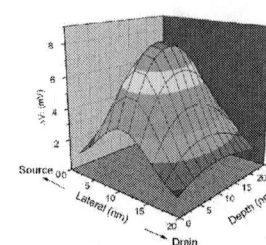

(a)

(b)

Fig. 6: The colour mapping sensitivities of threshold voltage to a single trapping charge at the fin side interface with (a) low and (b) high drain bias.

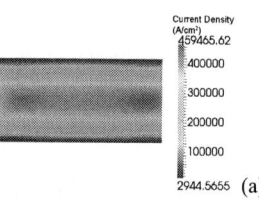

Fig. 7: The DIBL variation due to a single trapping charge at fin side interface.

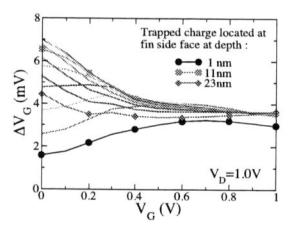

Fig. 8: The gate-voltage shift due to single trapped charge needed to give the same current as the original at low drain-bias.

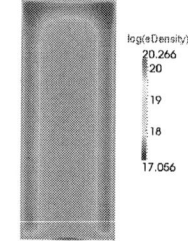

Fig. 9: The gate-voltage shift at high drain-bias is converged at high gate-voltage.

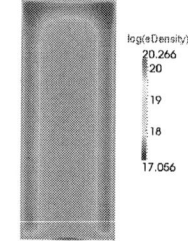

Fig. 10: The electron density inside the fin with Vg=1.0V, Vd=0.05V.

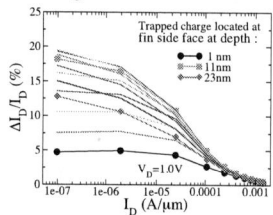

Fig. 11: The RTS magnitude due to differently located single trap on fin side interface. The relative magnitude is reduced with increasing gate-voltage and channel current.

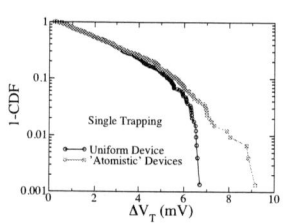

Fig. 12: The RTS magnitude due to differently located single trap on the fin side interface at high drain bias.

Fig. 13: The CCDF as single trapping charge induced threshold voltage shift for uniform device, and "atomistic" devices. Vd=1.0V.

Fig. 14: The largest Vt shift due to single charge trapping. Trapped charge is captured at the interface with fin-narrowing and under a metal grain of low workfunction.

978-1-4673-0996-7/12 $31.00 © 2012 IEEE

Self-Improvement of Cell Stability in SRAM by Post Fabrication Technique

Anil Kumar[1], Takuya Saraya[1], Shinji Miyano[2], and Toshiro Hiramoto[1]

[1]Institute of Industrial Science, The University of Tokyo, [2]STARC

4-6-1 Komaba, Meguro-ku, Tokyo 153-8505, Japan, Phone：+81-3-5452-6264, E-mail: anilkr@nano.iis.u-tokyo.ac.jp

Abstract

The post fabrication technique for self-improvement of SRAM cell stability is validated by experiment using 1k DMA SRAM TEG array. It is shown that the stability of unbalance cells is automatically improved by merely applying stress voltage to V_{DD} terminal. The mechanism of the phenomena is also analyzed by measuring V_{TH} of all transistors before and after stress and it is newly found that $|V_{TH}|$ of weaker PFET in the cell is selectively lowered by the self-improve mechanism.

Introduction

Increase in the V_{TH} variability of scaled transistors is a serious concern for the stability of SRAM cells [1-12]. Some unbalanced cells are likely to fail which has a severe impact on SRAM yield. Though there are some techniques, yet these techniques generally suffer from area penalty or speed loss. Recently, we have proposed a new post fabrication technique to improve SRAM cell stability by applying stress voltage [8-10]. In this technique, only stronger NFET and PFET in the cell are selectively stressed and their $|V_{TH}|$ is raised, resulting in self-improvement of cell stability.

In this paper, the post fabrication self-improvement technique is demonstrated using 1k SRAM DMA (device matrix array) TEG. All transistors in 1k cells after and before stress are measured, and the self-improvement mechanisms are also investigated. It is newly found that not only $|V_{TH}|$ of the stronger PFET is raised, but also $|V_{TH}|$ of the weaker PFET in the cell is selectively lowered, resulting in very effective self-improvement phenomena.

Self-Improvement Technique

In this technique, the stress voltage is applied to V_{DD} terminal of SRAM cell array. Figs. 1 & 2 show a schematic of 6T SRAM cell and method of the technique, respectively. V_{DD} is raised from 0V to the stress voltage (3.2V in this study), keeping WL at 0V. The scan time (stress time) is only several seconds. When V_{DD} is low enough, "high" or "low" of VL and VR nodes is determined by cell unbalance caused by the variability of 4 transistors.

Let us assume that VR is high which means TpR is strong (low V_{TH}) and TnL is also strong (Fig. 2). Here, we call pFET connected to the high node "p-high" and pFET connected to the low node "p-low". When V_{DD} is raised to 3.2V, BTI (bias temperature instability) stress is automatically applied to only "p-high" and "n-low" which are stronger, and these transistors are weakened. On the other hand, "p-low" and "n-high", which is weaker, are not stressed. As a result, the cell stability is improved [8-10].

Measurements

6T SRAM DMA TEG [10-12] analyzed was fabricated with 40nm planar bulk technology. Terminals for V_{DD}, WL, two BLs, and two storage nodes (VL and VR) can be accessed (Fig. 1), so that SRAM and all 6 transistors can be measured. Fig. 3 shows V_{TH} distribution of 1k driver NFETs and load PFETs. V_{TH} shows the normal distribution.

Figs. 4 & 5 show distributions of retention noise margin (RetNM) [4] and static noise margin (SNM) of 1k cells. Both RetNM and SNM are deviated from normal distribution [10,12]. Fig. 6 shows correlation between RetNM at 25°C and 100°C, showing insufficient correlation. Since the cell unbalance is more severe at higher temperature, the self-improvement stress is applied at 100°C and cell stability improvement at 100°C is targeted in this study.

Fig. 7 shows $|V_{TH}|$ shift of the TpL by stress. The shift is random: some cells show positive shift while some show negative shift. However, "p-high" show only positive shift as shown in Fig. 8, indicating the self-improvement mechanism works. This shift is caused by NBTI stress because "p-high" transistor is ON and the gate bias is low.

Moreover, as shown in Fig. 9, "p-low" shows only negative shift which is favorable for the self-improvement. The shift is even larger than that of "p-high". The "p-low" transistor is OFF: gate is high, source is high, and drain is low. Therefore, positive gate bias is applied between gate and drain and negative $|V_{TH}|$ shift (positive V_{TH} shift) is observed.

On the other hand, V_{TH} shift of both "n-high" and "n-low" by stress is very small (not shown).

Fig. 10 shows RetNM distribution before and after stress. Clear improvement of RetNM is observed, particularly in worst cells. Fig. 11 shows correlation between RetNM improvement and initial RetNM before stress. Apparently, negative correlation is observed, indicating that more unbalance cells are saved. Fig. 12 shows SNM distribution before and after stress. Clear improvement of SNM is also observed in worst cells, and the minimum operating voltage (Vmin) is apparently improved.

Conclusions

Post fabrication technique for self-improvement of SRAM cell stability is demonstrated by experiment. It is found that the V_{TH} shift of PFETs mainly contributes to the self-improvement mechanism. The time taken for the stress in this technique is very short; therefore this scheme is very useful to improve cell stability. Reliability and recovery after applying stress is being studied to analyze any penalty in performance and the net amount of improvement in variability.

Acknowledgement

This work was carried out as a part of the Extremely Low Power (ELP) project supported by METI and NEDO.

References

[1] A. J. Bhavnagarwala et al., IEEE JSSC, p.658, 2001.
[2] F. Tachibana, JJAP, 44, p.2147, 2005.
[3] L. Chang et al., VLSI Tech. Symp., p.128, 2005.
[4] A. Bhavnagarwala et al., IEDM., p.675, 2005.
[5] A. Asenov et al., VLSI Tech. Symp., p.87, 2007.
[6] K. J. Kuhn et al., IEDM., p.471, 2007.
[7] K. Takeuchi et al., IEDM., p.467, 2007.
[8] M. Suzuki et al. VLSI Tech. Symp., p.148, 2009.
[9] M. Suzuki et al. ISDRS, TP7-03, 2009.
[10] M. Suzuki et al. VLSI Tech. Symp., p.191, 2010.
[11] X. Song et al., IEDM, p.62, 2010.
[12] T. Hiramoto et al., IEEE TED, 58, p.2249, 2011.

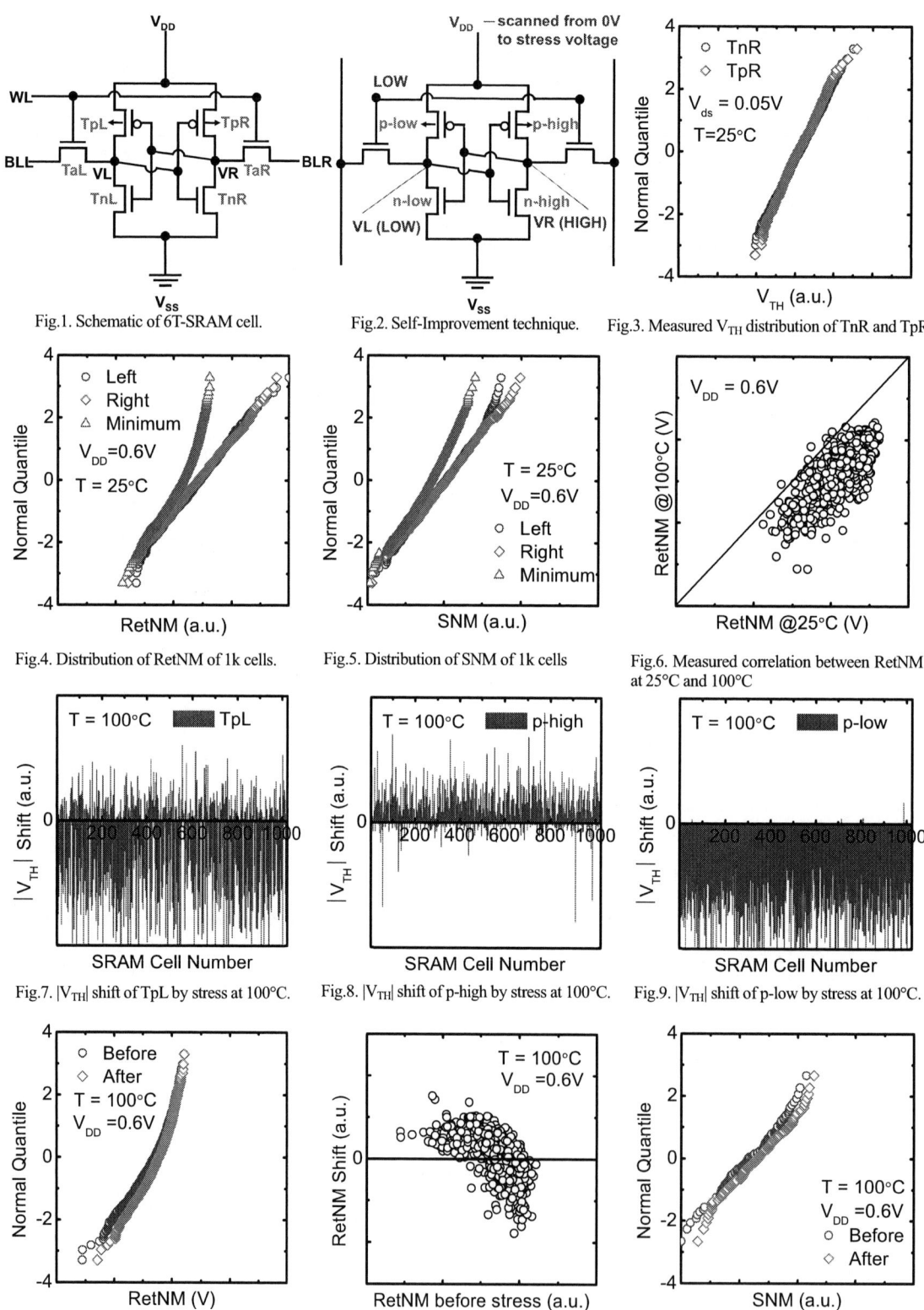

Fig.1. Schematic of 6T-SRAM cell.

Fig.2. Self-Improvement technique.

Fig.3. Measured V_{TH} distribution of TnR and TpR.

Fig.4. Distribution of RetNM of 1k cells.

Fig.5. Distribution of SNM of 1k cells

Fig.6. Measured correlation between RetNM at 25°C and 100°C

Fig.7. |V_{TH}| shift of TpL by stress at 100°C.

Fig.8. |V_{TH}| shift of p-high by stress at 100°C.

Fig.9. |V_{TH}| shift of p-low by stress at 100°C.

Fig.10. RetNM distribution of 1k cells before and after stress at 100°C.

Fig.11. RetNM change by stress at 100°C as a function of initial RetNM before stress.

Fig.12. SNM distribution before and after stress at 100°C.

978-1-4673-0996-7/12 $31.00 © 2012 IEEE

Low Standby Power Charge Trap Flash Memory with Tunneling Field Effect Transistor

Min Su Han, Jong Ho Lee[1], Dongsun Seo, Chong-Dae Park, Youngcheol Oh, and Il Hwan Cho

Department of Electronic Engineering, Myongji University
116 Myongji-ro, Cheoin-gu, Yongin, Gyeonggi-do, 449-728, Korea
Phone : +82-31-330-6380, Fax : +82-31-330-6977, E-mail : *ihcho77@mju.ac.kr*
[1]School of EECS and ISRC, Seoul National University, Seoul 151-742, Korea

I. Introduction

Recently, the mobile device market continually demands both excellent graphic and various functions on mobile device. To meet the specification of small mobile devices, low power consumption is necessary. The memory is one of the focused devices to reduce standby power and size. Standby power reduction can be realized by adoption of advanced technology and two of the device technologies were considered in this paper. The first technology is SONOS memories which is one of the candidate for low-power and small size nonvolatile memory [1-2]. The second one is the TFET which has p-i-n structure [3]. Though adoptions of advanced MOSFET technology to nonvolatile memory were common method [4-5], there was no research about memory with TFET.

In this paper, the memory characteristics of SONOS memory is investigated in terms of Fowler-Nordheim (FN) program efficiency, off state leakage current and disturbance characteristics .

II. Simulation Works

In this paper, two-dimensional simulation has been performed by ATLAS simulator [6]. A nonlocal band-to-band tunneling model has been utilized in device simulation which is normally used in previous works [3]. **Fig. 1** shows the simulated structure of the SONOS memory with the TFET. Abrupt source/drain doping profile has been assumed in this work. The p-type source, n-type drain and intrinsic region are doped at 10^{20} cm^{-3}, 10^{18} cm^{-3} and 10^{14} cm^{-3}, respectively The substrate is made of strained SiGe to increase on current and narrow the tunneling width. But It suffers from excessive off leakage current. To alleviate off leakage current, drain side doping concentration is lower than source [7]. Gate material is n+ polysilicon and a length of channel is 50 nm. Strained substrate t_{SSOI} is 50 nm. The thicknesses of tunneling oxide, nitride, and barrier oxide are 2 nm, 5 nm and 4 nm, respectively.

III. Results and Discussions

Fig. 2(a) shows simulated I-V characteristics of the proposed memory with FN program. The FN program voltage is 10 V on gate. Read operation is performed with 1 V drain bias. In this result, the proposed memory has similar memory characteristics with conventional SONOS memory. Off leakage current of the proposed memory is 1 fA/μm. Threshold voltage is measured with constant current method at $I_d = 10^{-6}$ A/μm. Threshold voltage is shifted by FN program voltage and time as shown in **Fig. 2(b)**. The proposed memory has huge threshold voltage change in accordance with FN program. At the point of program (10 V/100 μs), measured threshold voltage change is 2.65 V.

Fig. 3(a) is shows dimensional distribution of trapped charge in nitride with FN program time. Trapped charge in the nitride can be changed by FN program voltage and time. Since the proposed memory has p-i-n diode structure. Trapped charge concentration has asymmetrical profile in trap layer. However, the amount of trapped charge at the source and drain side is similar with etch side as shown in **Fig. 3(b)**.

Fig. 4(a) shows surface potential on the channel at 4 V of gate bias. It shows differences of potential distribution change with FN program. The surface potential is determined by the amount of trapped charge in the nitride layer. When the gate bias is 4V, surface potential at FN program (10 V/100 μs) is lower than others. Thus, energy barrier width for tunneling at source/channel junction is wide as shown in **Fig. 4(b)**. This result cause threshold voltage change under FN program.

To understand the origin of the threshold voltage change induced by trapped charge concentration, E-field and tunneling barrier width, tunneling current is approximately calculated by WKB method. The current is proportional to the electron tunneling probability T(E) calculated through the WKB method, given by the equation in (1) [8].

$$ I_{BTB} \propto T(E) \approx \exp\left(- \frac{4\sqrt{2m^*}E_g^{3/2}}{3|e|h(E)} \right) \quad (1) $$

Where m* is the carrier effective mass, e is the electron charge, E_g is the band gap, E is E-field at the source and channel junction, respectively.

Tunneling current is affected by tunneling barrier width at the source and channel junction. **Fig. 5** shows tunneling current and E-field relation. As shown in equation (1), |ln(I)| and electric field has linear relation. Different tunneling barrier width is induced by different programming time as shown **Fig. 6(a)**. **Fig. 6(b)** shows tunneling current and tunneling barrier width relation. E-field can be express E = $(E_g+\Delta\Phi)/\lambda$, where λ is tunneling barrier width at the source/channel junction, $\Delta\Phi$ is the energy range over which tunneling can happen [9].

Thus, tunneling current relation can be express $\ln(I) \propto \lambda$. If tunneling barrier width is narrow, tunneling current will be increased as shown in **Fig. 6(b)**.

Trapped charge can change the surface potential. It is accordance with equation $V_{Surface} = Q_N/C_{eff}$. Thus, the surface potential change is proportional to the amount of trapped charge concentration. It can be expressed as $\Delta V_{surface} \propto Q_N$. **Fig. 7** shows trapped charge concentration and surface potential change relation.

The proposed memory has p-i-n type is not suitable for general NAND flash memory structure which uses FN program. But NOR type array using FN programming is available for the proposed memory [10]. FN-NOR type memory array is shown in **Fig. 8**. Two cell array units share the same drain line. FN-NOR type memory array has advantages on low power operation and fast program speed [10]. The program disturbance on the same word line is considered as shown in **Fig. 9**. When the selected cell is programmed by 10 V of gate bias, unselected cell is biased by 10 V of gate bias and -7 V of source bias. The negative source bias affect source side charge injection. Thus, amount of trapped charge is smaller than drain side trapped charge.

Fig. 10 shows potential distribution on selected cell and unselected cell. The negative potential effect at source side bring out selected cell and unselected cell distinction at program state.

IV. Summary

SONOS memory with TFET is proposed to achieve off leakage current characteristics. SONOS memory with TFET exhibits extremely small off state leakage current, good FN program efficiency. Program characteristics and disturbance characteristics were investigated with device simulation. It is expected that SONOS memory with TFET can be a promising candidate for mobile devices with require low-power consumption.

Acknowledgement : This work was supported by the National Research Foundation of Korea (NRF) funded by the Ministry of Education, Science and Technology (MEST) under Grant 2010-0027704 (Mid-Career Researcher Program) and supported by the Basic Science Research Program through the NRF funded by the Ministry of Education, Science and Technology (2011-0005703).

Reference

[1] M. K. Kim, et al., *IEEE Trans.Nanotechnology*, vol.3, p417, 2004.

[2] P. Xuan, et al., *IEDM Tech. Dig.*, p26.4.1, 2003.

[3] W. Y. Choi, *Jpn. J. Appl. Phys.*.vol. 49. P04DJ12, 2010

[4] J. G. Yun, et al., *IEEE trans. Elec. Dev.Lett.*, vol. 56, p1721, 2009

[5] J. R. Hwang, et al., *IEDM Tech. Dig.*, p154, 2005

[6] ATLAS : Silvaco International 2010

[7] E. H. Toh et al., *J. Appl. Phys.*, vol.103, p104504, 2008.

[8] S. M. sze, "Physic of Semiconductor Devices", *John Wiley & Sons. Inc.*, 1981

[9] J. Knoch, et al., *IEEE Device Research Conf.*, p153, 2005

[10] M. Ohkawa, et al., *IEEE J. Sol. Sta. Cir.*, vol.31, p1584, 1996

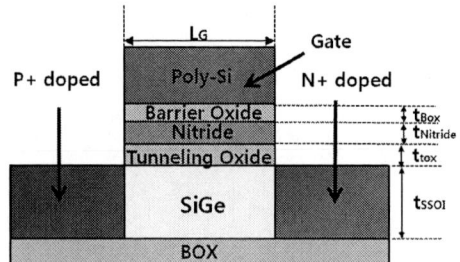

Fig. 1. The structure of SONOS memory with TFET on SSOI wafer.

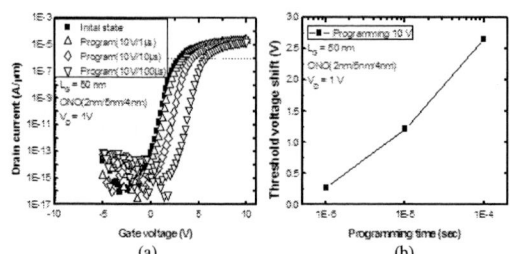

Fig. 2. (a) Current and voltage characteristics with programming time. (b) The shifted threshold voltage in accordance with programming time.

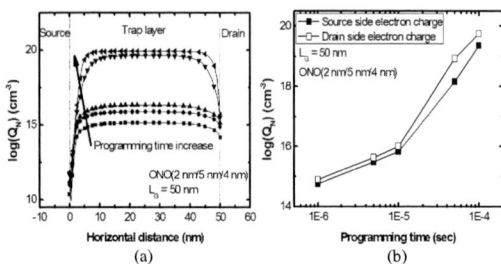

Fig. 3. (a) Trapped charge concentration at the interface between nitride and tunneling oxide. (b) Source and drain side trapped charge concentration difference in the nitride trap layer.

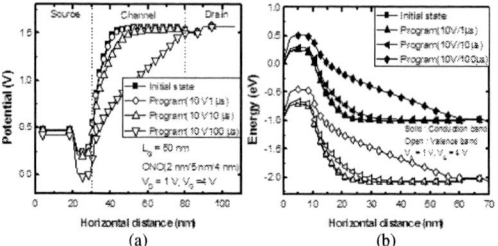

Fig. 4. Surface potential and energy band diagram at the 4 V of V_G. (a) Surface potential, (b) Energy band diagram.

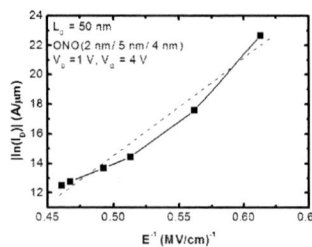

Fig. 5. Tunneling current and E-field relation at fixed programming voltage (10 V) and different programming time (1 μs, 5 μs, 10 μs, 50 μs, 100 μs).

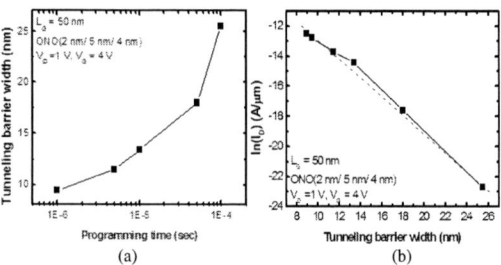

Fig. 6. (a) Tunneling barrier width at source/channel junction in accordance with different programming time (1 μs, 5 μs, 10 μs, 50 μs, 100 μs) when $V_G = 4$ V, $V_D = 1$ V are applied. (b) Tunneling current and tunneling barrier width relation at $V_D = 1$ V, $V_G = 4$ V.

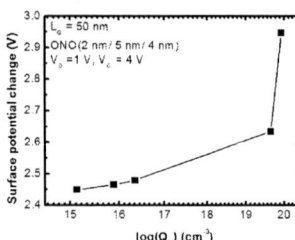

Fig. 7. Trapped charge concentration nitride trap layer and surface potential charge relation at $V_D = 1$ V, $V_G = 4$ V.

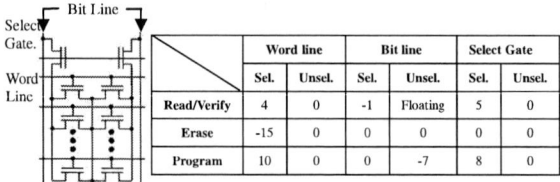

	Word line		Bit line		Select Gate	
	Sel.	Unsel.	Sel.	Unsel.	Sel.	Unsel.
Read/Verify	4	0	-1	Floating	5	0
Erase	-15	0	0	0	0	0
Program	10	0	0	-7	8	0

Fig. 8. FN-NOR type array circuit.

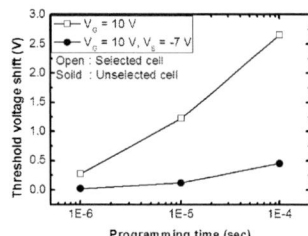

Fig. 9. The program disturbance on same word line in FN-NOR type array operation in accordance with programming voltage and different time.

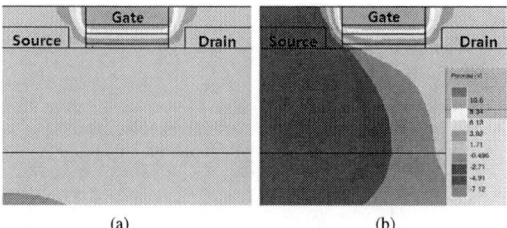

Fig. 10. The potential of selected cell and unselected cell programming at programming time 100 μs (a) Selected cell is programmed $V_G = 10$ V. (b) Unselected cell is biased $V_G = 10$ V and $V_S = -7$ V

978-1-4673-0996-7/12 $31.00 © 2012 IEEE

Charge-trap flash memory devices fabricated with nano-scale patterns on the Si_3N_4 trapping layer

Ho-Myoung An[1], Kyong Heon Kim[1], Hee-Dong Kim[1], Won-Ju Cho[2] and Tae Geun Kim[1*]

[1]School of Electrical Engineering, Korea University, Anam-dong 5-ga, Seongbuk-gu, Seoul 136-701, Korea.
[2]Department of Electronic Materials Engineering, Kwangwoon University, Wolgye-dong, Nowon-gu, Seoul 139-701, Korea
Phone: +82-2-924-5119 E-mail address: tgkim1@korea.ac.kr

1. Introduction

Recently, market demands for NAND flash memories with conventional floating gate (FG) structures have been increased due to their large-capacity storage applications [1]. Henceforward, the feature size reduction will be inevitable to accomplish ultra-high density and low-cost flash memories. However, the scale down of FG memory device is placed in an extreme situation due to the physical limits of gate length and tunneling oxide thickness [2]. In order to replace the traditional FG devices around the 30 nm or 2x nm technology node, the charge trap flash (CTF) memories composed of polysilicon/oxide/nitride/oxide/silicon (SONOS) structure are being developed for commercial applications. This is due to the merit of charge storage in discrete traps within silicon-nitride (Si_3N_4) layer, which prevents charge loss or leakage current through a tunnel oxide [3]. Although the fundamental feasibility of CTF NAND has been proven [4], further scaling of planar CTF NAND below 30 nm still faces extensive challenges. The most fundamental issue is the small number of stored electrons, because the dimension (both vertical and lateral size) of oxide-nitride-oxide (ONO) layer decreases. The few-electron storage threatens programming statistics as well as retention reliability.

In this work, we report improved memory characteristics of CTF memories by the application of nano-scale patterns on the Si_3N_4 trapping layer using nanosphere lithography (NSL).

2. Experiments

Figure 1 shows schematic diagrams of metal-aluminum-oxide-nitride-oxide-silicon (MANOS) capacitors fabricated with a flat surface and a patterned surface by NSL. Three different etching time (3, 6, 9 sec) was applied to the interface between the Si_3N_4 and the blocking layer for the patterned surface. All the processes were identical except the surface roughening process at the interface between the trapping layer and the blocking oxide layer. Firstly, a 5 nm-thick tunnel oxide layer was deposited on top of silicon at 750 °C by using the low-pressure chemical vapor deposition (LPCVD). And then a 10 nm-thick nitride layer of was deposited using LPCVD at 750 °C, by the reaction of dichlorosilane ($SiCl_2H_2$) and ammonia (NH_3) gases. Next, a 10 nm-thick blocking layer of Al_2O_3 was deposited for the oxide-nitride-aluminum oxide (ONA) stack using radio frequency (RF) sputter. After depositing the ONA layer, the gate electrode of aluminum (Al) with a 300 μm diameter was deposited on top of the blocking oxide layer using an electron beam evaporator.

3. Results and discussion

In order to investigate the increase of memory traps for different etching time (3, 6, 9 sec), the variation of memory windows (flatband voltage shift, ΔV_{FB}) as a function of etching time on the Si_3N_4 trap layer of MANOS capacitors, was measured from the C-V curves, as shown in Fig. 2. The memory windows increased from 2.88 V and 3.9 V when the etching time of the Si_3N_4 trap layer increased from 0 sec (for flat sample) to 3 sec (for patterned sample). The pattern depth of the Si_3N_4 trap layer after etching for 3 sec was measured to be ~ 4 nm by the AFM image, as shown in the inset of Fig. 2. For patterned samples with etching time from 6 sec to 9 sec, memory windows gradually decreased from 2.86 V to 2.5 V with increasing the etching time compared to the flat sample, despite the increased surface density area of the Si_3N_4 layer. Therefore, we confirm that the optimized etching time is to be 3 sec.

Fig. 3 shows program characteristics of the MANOS capacitors with a flat surface and the patterned surface of different depth, in order to demonstrate the effect of traps density on the device performance. It is found that the maximum V_{FB} shift of the capacitor with patterned surface is larger than that of the capacitor with a flat surface at a program voltage of 13 V. This result implies that the patterned nitride layer provides a large memory trap density at the interface between the Si_3N_4 trap layer and Al_2O_3 blocking layer and this, in turn, increases the V_{FB} shifts. Besides, the V_{FB} shift of 2.5 V, which is large enough for program operation, was obtained from the patterned structure at gate bias conditions of 12 V/50 ms. On the other hand, the same voltage shift was observed from the flat structure at gate bias conditions of 12 V/500 ms. However, as compared to the program result, nearly same V_{FB} shifts are observed for the flat and patterned samples under the same erase bias condition (not shown here).

In order to identify the difference in the program and erase performance in the MANOS capacitors with patterned surface, we estimated the electron and hole current density in the patterned local region during the applied positive and negative bias in the gate electrode, using a computer-aided design (TCAD) simulation, as shown in Fig. 4. The electron current density of 13.6 μA is much larger than that of hole current density of 4.3 μA at gate bias voltages of 12 V and -12 V, respectively. The current difference is larger in the positive bias than in the negative bias, probably due to by the electron mobility greater than hole mobility.

Fig. 5(a) shows room-temperature retention characteristics for the flat and patterned samples. As a result, the charge decay rates of the flat sample and the patterned samples were 40 mV/dec and 48 mV/dec at the programmed state, respectively, whereas those were 32 mV/dec and 73 mV/dec, respectively, at the erased state. Although the charge decay rates of the patterned sample is slightly larger than that of flat sample due to the rough interfaces between the Si_3N_4 trap layer and Al_2O_3 blocking layer, the patterned samples have large memory windows over a retention time of 10^9 s (or about 10 years). Finally, we evaluated the endurance properties with an operation window of 2.5 V by measuring the V_{FB} shifts against the program/erase (P/E) cycles as shown in Fig. 4(b). The memory windows of the two samples gradually decrease with increasing P/E cycling and are about 2 V after 10^4 P/E cycles. These results indicate that the incorporation of nanoscale patterns on the Si_3N_4 trap layer is useful to improve the endurance as well as retention properties of conventional CTF devices.

4. Conclusion

We proposed a novel CTF memory structure with surface patterned Si_3N_4 trap layers, in order to enhance the memory window and the performance for ultra-high density CTF devices. Due to the enlargement of surface memory-trap densities, the CTF devices with nano-scale surface patterns on the Si_3N_4 trap layer by NSL showed increased memory windows and improved program properties. In addition, the reasonable reliability, including data retention of 10 years and endurance of 10^4 P/E cycles, was obtained.

References

[1] S. Lai, in IEDM Tech. Dig., p. 11, 2008.
[2] S. Gerardin, IEEE Trans. Elect. Dev. vol. 57, p. 3016, 2010.
[3] M. H. White, IEEE Circuits Dev. Mag., vol. 16, p. 22, 2000.
[4] W. Kim, et. al., in Symp. on VLSI Tech. Dig., p. 188, 2009.

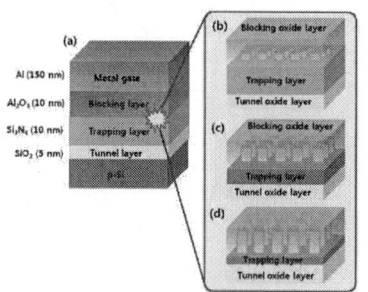

Fig. 1. Schematic daigrams of MANOS devices (a) with a flat surface and the patterned surface by NSL.

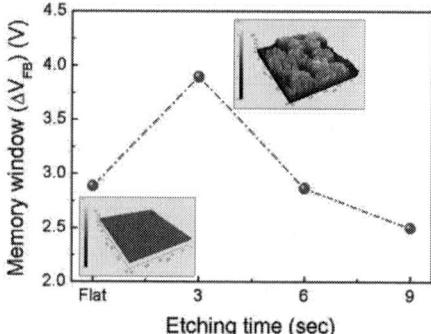

Fig. 2. The variation of memory windows as a function of etching time on the Si_3N_4 trap layer of MANOS capacitors, measured from the C-V curves.

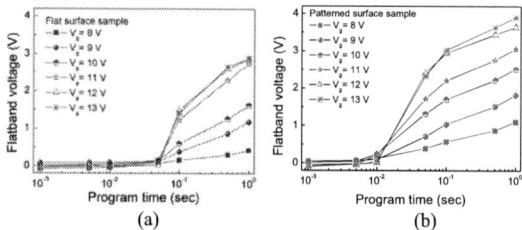

Fig. 3. Program characteristics of MANOS capacitors with flat surface and the patterned surface by NSL.

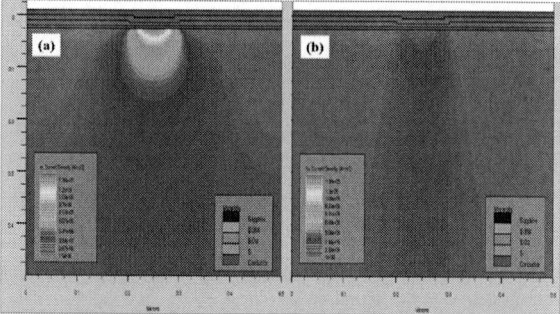

Fig. 4. Simulated results MANOS capacitors with a flat surface and the patterned surface by NSL. (a) Electron current density, (b) Hole current density.

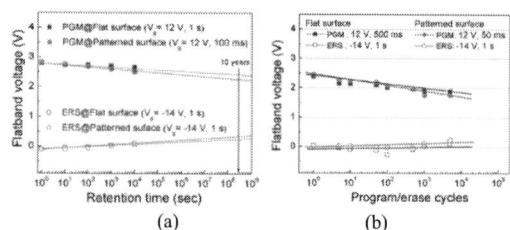

Fig. 5. Reliability characteristics of MANOS capacitors measured at room temperature with a flat surface and patterned surface. (a) Retention; (b) Endurance characteristics.

Simulation of Charge Trapping Memory with Silicon Nanocrystals Embedded in Silicon Nitride Layer

Yahua Peng, Xiaoyan Liu, Gang Du*, Yan Yang, Jinfeng Kang

Institute of Microelectronics, Peking University, Beijing, 100871, China.

* gangdu@pku.edu.cn

I. Introduction

Charge trap memory (CTM) has attained much attention recently due to its scalability, CMOS process compatibility, lower program/erase voltage and immunity from severe reliability [1-6]. To get faster program/erase speed, larger memory window and better retention characteristic, a charge trap layer with incorporating semiconductor or metal nanocrystals into Si_3N_4 or SiO_2 has been developed [7], [8]. Based on our previous numerical simulator for CTM [9], [10], we successfully develop a self-consistent method which has been calibrated and verified with experiment to evaluate the performance of device with Si nanocrystals (Si-NCs) embedded in Si_3N_4 layer. The critical models in CTM and Si-NCs are included to describe the mechanism of carriers transport across the multiple dielectric layers and charge trapping/detrapping in Si-NCs during the program/erase/retention operation. This method is a powerful approach for understanding the mechanism in CTM with Si-NCs embedded in the trap layer, predicting its failure mechanism, reliability and further device optimization. The results under different bias voltages, charge trap distributions, Si-NCs sizes, temperatures and gate dielectric layers' thicknesses are studied comprehensively.

II. Physical Model and Simulation Method

The physical models can be described in two aspects. The process of carriers tunneling across the multiple gate layers includes F-N tunneling, direct tunneling, trap assisted tunneling (TAT) and relaxation transport. The trapping/detrapping models of the trapped charge which mainly happened in the Si-NCs include charge capture, recombination and thermally excitation. Additionally, we add the effect of quantum confinement and coulomb blockade (Q&C) [11], [12] into our simulator which are indispensable especially when the nanocrystal size is smaller than 5nm.

The structure used in this work is the same as that in experiment work [13] which is shown in Fig. 1. The tunnel oxide and block oxide are set to SiO_2 which can be replaced by other materials. The tunnel oxide/trap layer/block oxide thickness is 2.5/13/7.5nm and Si-NCs' average diameter (D_{NC}) is 3.5nm except in special instructions. Our simulator has been calibrated and verified with the experiment data [13] as shown in Fig. 2 and the default parameters are listed in TABLE I. The I-V characteristic for the high threshold voltage (Vth) state and low Vth state is shown in Fig. 3 with scanning gate bias (Vg) is from -5V to 13V and drain bias (Vd) is 2V.

III. Results and Analysis

With the simulator, we evaluate the influence of physical models, bias voltage, charge trap distribution, Si-NCs size, temperature and gate dielectric layer's thickness which attract much interest. Q&C's effect couldn't be ignored especially for small nanocrystals because quantum confinement could shift the nanocrystals conduction band edge upward and coulomb blockade will raise its electrostatic potential. The effect of Q&C during program performance is illustrated in Fig. 4. Fig. 5 shows the comparison of results with and without TAT model. The program/erase speed is increased with TAT because it could accelerate the tunneling current. Fig. 6 shows the erase speed with different charge trap distributions including Uniform which is the default style in this work, Gaussian, Inverse Gaussian, Exponential Decrease and Exponential Increase. From the results we can see that more charge trap close to the substrate side in the trap layer corresponds to higher operation speed and this phenomenon could be more obvious in larger Si-NCs size. With 0V gate voltage the programmed state's Vth shift is smaller than 25% at 85℃ while time up to 10^7s. The retention characteristic under different temperatures is shown in Fig. 7. Fig. 8 shows the influence of tunnel oxide thickness on the program/erase speed. Thinner tunnel oxide corresponds to higher speed due to the increased tunneling probability. Fig. 9 illustrates the pulse time as a function of Si-NCs size while the Vth shifts up to 4V. Larger Si-NCs size with the same charge trap density could speed up the erase performance and make the program speed slow down which come within the reasonable range. Fig. 10 shows the retention characteristic under different Si-NCs sizes and tunnel oxide thicknesses at 85℃ with the same initial charge trap density, from which we can seen that the retention characteristic could be degraded with thinner tunnel oxide due to larger tunneling current.

IV. Conclusion

A simulation method for evaluating the performance of CTM with incorporating nanocrystals into the charge trap layer is presented and the effects of bias voltage, charge trap distribution, nanocrystal size, temperature and gate dielectric layer's thickness on program/erase/retention characteristic are studied. It can be a useful tool for designing nanocrystals based CTM.

Acknowledgement

This work is supported by the National Fundamental Basic Research Program of China (Grant No 2010CB934203 and 2011CBA00604)

Reference

[1] S.Lombardo, et al., *IEDM*, pp. 921-924, 2007. [2] G.Molas, et al., *IEDM*, pp. 22.5.1-22.5.4, 2010. [3] S.M.Amoroso, et al., *IEDM*, pp. 22.6.1-22.6.4, 2010. [4] Lee.Chang-Hyun, et al., *Appl. Phys. Lett.*, pp. 073510-073510-3, 2005. [5] Soon-Moon Jung et al., *IEDM*, pp. 1-4, 2006. [6] Hang-Ting Lue, et al., *IEDM*, pp. 547-550, 2005. [7] Jong Jin Lee, et al., *TED*, pp.507-511, 2005. [8] G.Molas, et al., *IEDM*, pp. 453-456, 2007. [9] Y.C.Song, et al., *SISPAD.*, pp. 41-44, 2008. [10] Y.C.Song, et al., *SNW.*, pp. 1-2, 2008. [11] C.Bostedt, et al., *Appl. Phys. Lett.*, vol. 84, pp. 4056-4058, 2004. [12] R. F. Steimle, et al., *Proc. 4thIEEE Conf. Nanotechnol.*, pp. 290-292, 2004. [13] Sangmoo Choi, et al., *IEDM*, pp. 166-169, 2005.

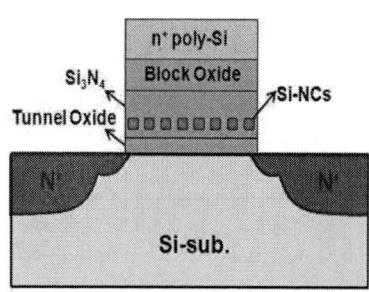

Fig. 1 Cross section of Charge Trap Memory device with Si-NCs in Si_3N_4 layer.

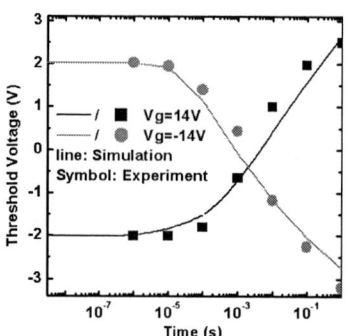

Fig. 2 Calibration of the simulation method and physical models. The symbols are from experiment data and lines are the simulation results.

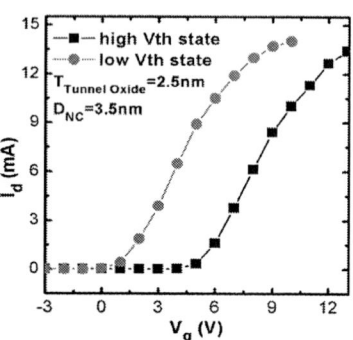

Fig. 3 Id-Vg characteristic for the high Vth state and low Vth state respectively which are corresponding to that in Fig. 2.

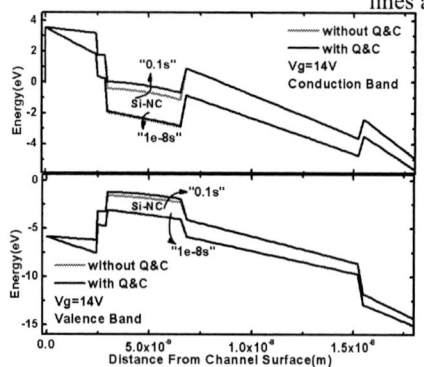

Fig. 4 Effects of quantum confinement and coulomb blockade on the Si-NCs' electrostatic potential during the program with Si-NCs size of 3.5nm.

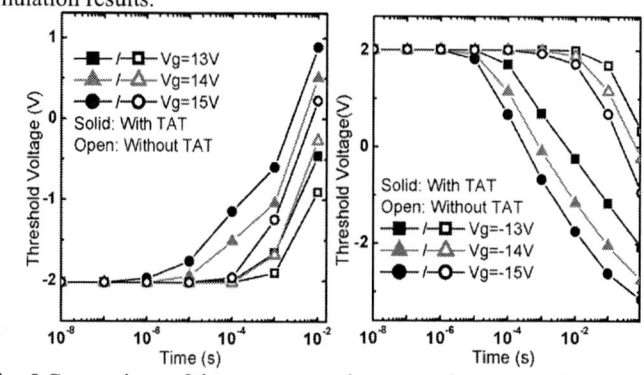

Fig. 5 Comparison of the program and erase performance with/without TAT model under various gate voltages for device illuminated in Fig. 1.

Parameter	Value
$\Phi_{n+\,poly-Si}$	4.37 eV
σ_e/σ_h	$2\times10^{-13}/2\times10^{-13}$ cm^2
$\Phi_{TAT,e}$	1.89 eV
$\Phi_{TAT,h}$	1.72 eV
$m^*_{e,Si3N4}$	0.41
$m^*_{h,Si3N4}$	0.8
$V_{d,read}$	2 V

TABLE I The default parameters used in this work

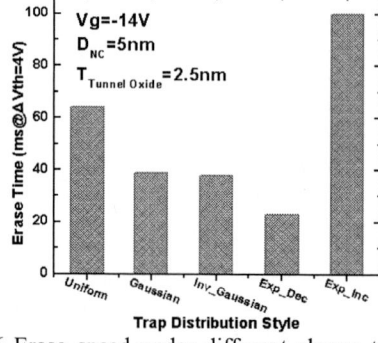

Fig. 6 Erase speed under different charge trap distributions while the other factors are the same.

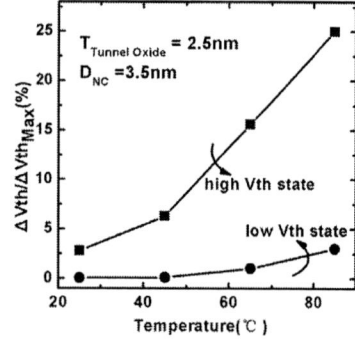

Fig. 7 Retention characteristic under different temperatures with time up to 10^7s.

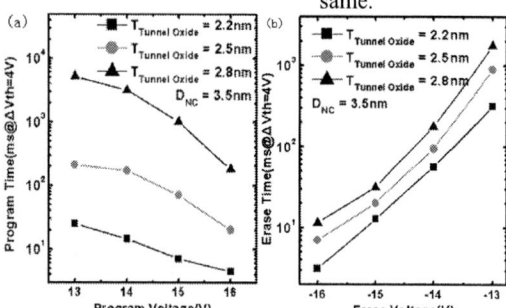

Fig. 8 Effect of tunnel oxide thickness on the device performance with 4V threshold voltage window under various gate voltages and the Si-NCs size is 3.5nm.

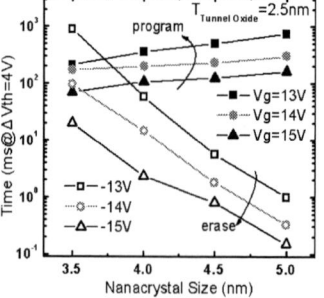

Fig. 9 Device performance as a function of Si-NCs size with the same tunnel oxide thickness and different gate voltages.

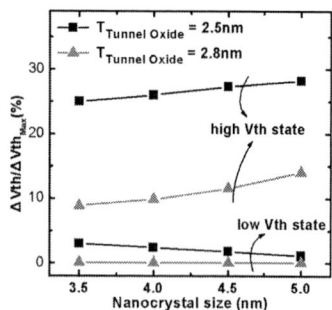

Fig. 10 Retention characteristic under different Si-NCs sizes and tunnel oxide thicknesses at 85℃ with time up to 10^7s.

978-1-4673-0996-7/12 $31.00 © 2012 IEEE

Nanodot-type Floating Gate Memory with High-density Nanodot Array Formed Utilizing *Listeria* Dps

Hiroki Kamitake[1,2], Kosuke Ohara[1], Mutsunori Uenuma[1,2], Bin Zheng[1,2], Yasuaki Ishikawa[1,2], Ichiro Yamashita[1,2,3], and Yukiharu Uraoka[1,2]

1 Nara Institute of Science and Technology, 8916-5 Takayamacho, Ikoma, Nara 630-0192, Japan

2 CREST, JST, 7 Goban, Chiyoda, Tokyo, 102-0075, Japan

3 ATRL, Panasonic Corporation, 3-4 Hikaridai, Seika, Kyoto, 619-0237, Japan

Tel: + 81-(743)-72-6064, Fax: + 81-(743)-72-6069 (e-mail:k-hiroki@ms.naist.jp)

Abstract — We formed a high-density two-dimensional nanodot array by utilizing Ti-binding Dps (TD) which is a *Listeria* Dps with Ti-binding peptides. A high-density nanodot array over 10^{12} cm^{-2} was formed on a SiO_2 at low temperature by specific adsorption of TD. The hysteresis of the MOS capacitor with nanodot array formed utilizing TD was larger than that of the MOS capacitor fabricated utilizing ferritin. This research contributes to realizing future memory devices.

I. INTRODUCTION

Nanodot-type floating gate memory (NFGM), which has a two dimensional array of nanodots as a floating gate, has attracted much attention owing to good reliability and high performance. Our research group has researched to develop new functional semiconductor devices by Bio-nano process (BNP) which is one of bottom-up processes [1], and fabricated NFGM utilizing ferritin [2]. Various types of homogeneous nanometer-sized particles (bio-nanodots: BND) can be formed in its cavity by biomineralization. We proposed the NFGM with a high-density nanodot-array utilizing *Listeria* Dps. In this study, we attempted to fabricate metal-oxide-semiconductor (MOS) capacitor with the high-density Co-BND array formed utilizing *Listeria* Dps modified Ti-binding peptides [3] on the surface Ti-binding Dps (TD). In addition, we compared the *C-V* characteristics of the fabricated MOS capacitor with that of the MOS capacitor with Co-BND array formed utilizing Ti-binding ferritin (TBF).

II. EXPERIMENTAL

Figure 1 shows Ferritin and *Listeria* Dps. We cleaned a Si substrate with a 3 nm-SiO_2 layer formed by rapid thermal annealing (RTA). Solution contained TD was dropped on the SiO_2 to adsorb TD on the SiO_2. Outer proteins of TD was eliminated by UV irradiation in an ozone atmosphere for 60 min (UV/ozone treatment), to form Co-BND array on the SiO_2. Then, a 15 nm-SiO_2 control oxide was deposited by plasma-enhanced CVD. Following the formation of Pt/TiN top electrode and Al bottom electrode by electron beam evaporation, the fabricated MOS capacitor was annealed in a reductive gas ($N_2:H_2$=9:1) at 450 °C to reduce the Co-BNDs from metal oxide nanodots to metal nanodots. Figure 2 shows the device structure of MOS capacitor embedded Co-BNDs.

III. RESULTS AND DISCUSSION

Figure 3 shows the SEM image of Co-BND array formed utilizing TD. We checked that Co-BNDs did not aggregate after UV/ozone treatment. The adsorption density of Co-BND array formed utilizing TD was 1.3×10^{12} cm^{-2}, which is higher than that of Co-BND array formed TBF as shown in Fig. 4. Figure 5-6 show the X-ray photoelectron spectroscopy (XPS) spectrums of carbon (C) and nitrogen (N). Figure 5-6 showed that the protein has been completely removed by UV/ozone treatment. We measured XPS spectrums of the Co-BNDs embedded in a 3 nm-SiO_2 to evaluate the reduction of Co-BNDs. Figure 7 shows the XPS spectrums of Co-BNDs before and after annealing in the reductive gas. In the case of the samples annealed in the reductive gas, a peak derived metal Co appeared. Figure 8, 9 show the *C-V* characteristics and memory windows of the fabricated MOS capacitors with Co-BND array. The band diagram of the devices was deduced as shown Fig. 10. The hysteresis of the MOS capacitor formed utilizing TD was larger than that of the MOS capacitor formed utilizing TBF. The enlargement of hysteresis was caused by increasing the amount of charge injected into Co-BND array because the adsorption density of Co-BND increased by utilizing Co-BND increased by utilizing TD.

IV. CONCLUSIONS

We increased the adsorption density of Co-BND array over 10^{12} cm^{-2} by utilizing TD. In addition, the *C-V* characteristics of MOS capacitor and the results of the XPS spectrums of C 1s and N 1s showed that the protein of TD was completely removed by UV/ozone treatment without influencing the device operation. The hysteresis of the MOS capacitor fabricated utilizing TD was larger than that of the MOS capacitor fabricated utilizing TBF. We confirmed that the high-performance memory can be fabricated utilizing *Listeria* Dps, compared with utilizing ferritin. These results contribute to realizing a next-generation memory.

ACKNOWLEDGMENT

This work was partially supported by AOARD.

REFERENCES

[1] I. Yamashita et al., *IEDM Tech. Dig.*, 447, (2006).

[2] K. Ohara et al., *Appl. Phys. Express*, 2, 9 (2009).

[3] T. Hayashi et al., *Langmuir*, 25, 18 (2009).

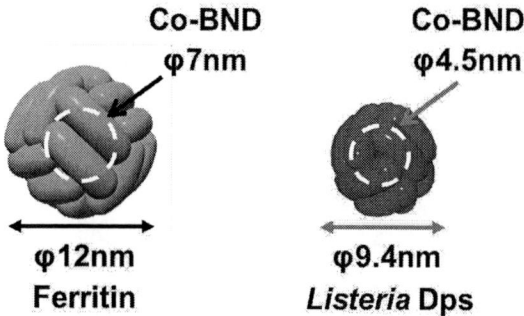

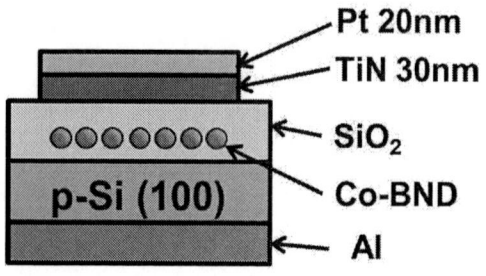

Fig.1 Ferritin and *Listeria* Dps.

Fig.2 Schematic illustration of MOS Capacitor structure.

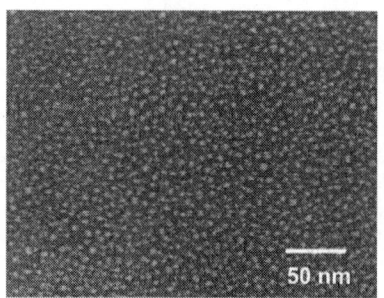

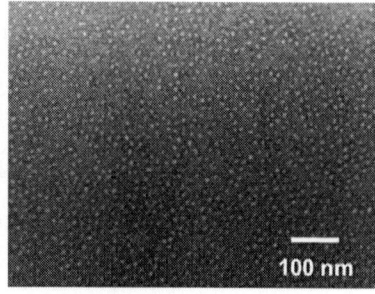

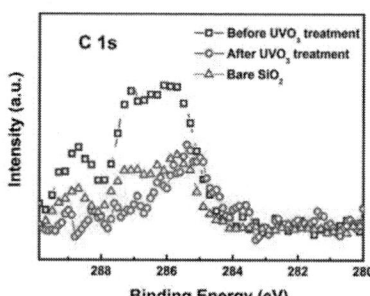

Fig.3 SEM images of Co-BND arrays formed utilizing TD.

Fig.4 SEM images of Co-BND arrays formed utilizing TBF.

Fig.5 XPS spectrums of C after eliminating protein by UV/ozone treatment.

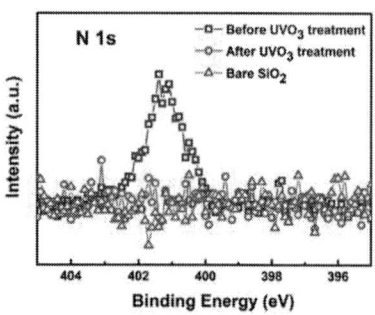

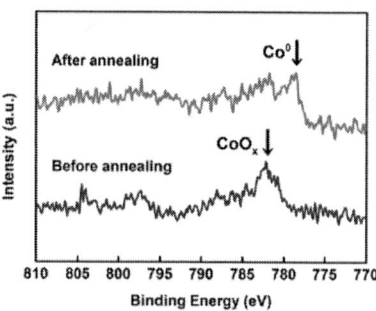

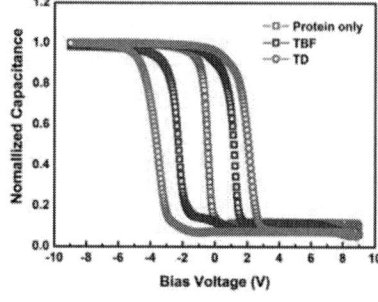

Fig.6 XPS spectrums of N after eliminating protein by UV/ozone treatment.

Fig.7 XPS spectrums of Co-BNDs.

Fig.8 *C-V* characteristics of the fabricated MOS capacitors.

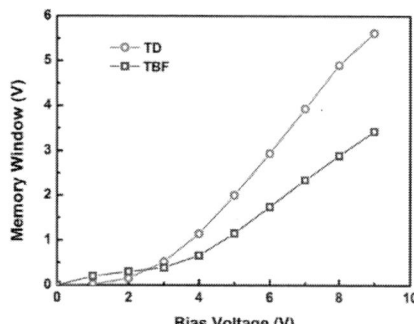

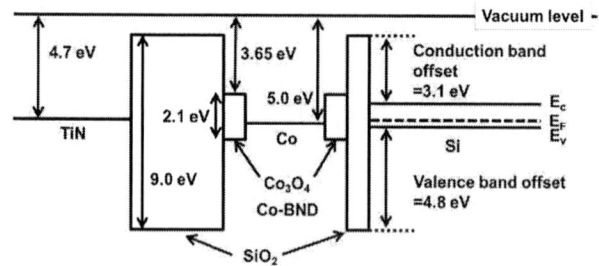

Fig.9 Memory windows of the fabricated MOS capacitors.

Fig.10 Band diagram of the MOS Capacitor fabricated utilizing TD.

978-1-4673-0996-7/12 $31.00 © 2012 IEEE

Impacts of Silicon Nanocrystal Incorporation on the Transfer Characteristics of Poly-Silicon Nanowire SONOS Devices

Ko-Hui Lee[1], Horng-Chih Lin[1,2*] and Tiao-Yuan Huang[1]

[1]Department of Electronics Engineering and Institute of Electronics, National Chiao-Tung University,
1001 Ta-Hsueh Road, Hsinchu, Taiwan 300, R.O.C.
[2]National Nano Device Labs.
*Phone:+886-35712121 ext.54193, Fax:+886-3-5724361, E-mail address: hclin@faculty.nctu.edu.tw.

Abstract — Gate-all-around poly-silicon nanowire (GAA poly-Si NW) SONOS devices embedded with silicon nanocrystals (Si-NCs) were fabricated and characterized. As Si-NCs are incorporated, the transfer characteristics show a large clockwise I_d-V_g hysteresis and a small kink under reverse sweep. Si dangling bonds located at SiNC/nitride interfaces are suspected to be responsible for the observations.

I. INTRODUCTION

Poly-Si-based memory devices are very attractive for future high-density memory applications due to its high feasibility for three-dimensional (3-D) electronics integration [1]. Multi-gated structure can enhance the device performance due to high gate controllability. Incorporation of nano-crystals (NCs) in the trapping layer is claimed to further improve the memory performance [2].

Several Si-NC formation methods have been reported, such as Si-precipitation [3, 4], aerosol deposition [5], and direct growth [6]. Among them, the direct growth technique has the advantage of easy integration with standard processes [7]. In this study, a technique similar to the direct growth was adopted for forming Si-NCs in silicon nitride layer [8]. In this paper, the impacts of the embedded Si-NCs on the fundamental characteristics of the fabricated devices are explored.

II. DEVICE FABRICATION

Si wafers capped with thermal oxide were used as the starting substrates. Figure 1(a) shows the 3-D structure of a GAA NW SONOS device. The fabrication processes are the same as those described in one of our previous papers [8]. Thicknesses of tunnel oxide and block oxide are 3 and 12 nm, respectively. $SiCl_2H_2$ and NH_3 gas sources were used for the deposition of silicon nitride. Figures 1(b) and (c) show the flow rate and schematic structure of the GAA SONOS device embedded with Si-NC layers. During the formation of Si-NCs, NH_3 gas source was switched off at 780 °C. Total nitride thickness is around 7 nm, and the NCs are located at 1.7 nm (1/4 of the thickness) away from the tunnel oxide. Control samples with the same total nitride thickness were also fabricated.

III. RESULTS AND DISCUSSION

Figures 2 (a) and (b) show the transfer characteristics of device (channel length = 0.4 μm) without and with embedded Si-NCs. The solid and dash lines represent the forward (increasing V_g) and reverse sweep (decreasing V_g). In control devices, negligible hysteresis is observed. However, for devices containing Si-NCs, a clockwise I_d-V_g hysteresis with window as large as 1.5 V is seen (Fig. 2(b)). For three consequent measurements, the I_d-V_g hysteresis curves repeat as shown in Fig. 3. These observations clearly indicate that fast electron trapping/de-trapping events associated with the incorporated Si-NCs occur during the measurements.

The power dependent micro-PL spectra of Si-NC embedded in a blanket silicon nitride excited by 325 nm He-Cd laser is shown in Fig. 4(a). It should be noted that, for the test samples without Si-NCs, negligible PL intensity is recorded (data not shown). The power dependence of the luminescence peak can be described by the equation $I_{PL} = \eta I_o^{\alpha}$, where I_{PL} is the PL intensity, I_o is the excitation power, α and η are the fitting parameters. The exponent α depends on the radiative recombination mechanisms, and is 0.61 for peaks centred at 550 nm (2.26 eV) as shown in Fig. 4(b). The extracted α of less than unity indicates that the peak is associated with free-to-bound transitions [9] which are presumably related to the traps introduced by the embedded Si-NCs.

Based on the above observation and the results reported in a previous study [10], the PL peak at around 2.26 eV is postulated to be related to the Si dangling bonds which are likely located at the interfaces between the Si-NCs and surrounding nitride. When the device is switched on, the channel electrons would tunnel through the tunnel oxide and get trapped by the dangling bonds. As a result, the threshold voltage increases. Moreover, the trapped electrons tend to return to the channel as the device is turned off. This explains the occurrence of the hysteresis seen in Figs. 2(b) and 3. Note that the I_d-V_g characteristics show that there exists a small kink at around 10^{-10} A under the reverse sweep. Such a kink is postulated to be related to the onset of the de-trapping process.

IV. CONCLUSIONS

In this paper, we have investigated the electrical characteristics of GAA poly-Si NW SONOS devices with or without incorporating Si-NCs. An obvious clockwise hysteresis is seen in the transfer characteristics as Si-NCs are contained. Si dangling bonds presenting at Si-NC/nitride interfaces are postulated to be the sites for trapping/de-trapping of electrons during the measurements.

978-1-4673-0996-7/12 $31.00 © 2012 IEEE

Acknowledgment We would like to thank the National Nano Device Laboratories (NDL) and Nano Facility Center (NFC) for assistance in device fabrication. This work was supported in part by the Ministry of Education in Taiwan under ATU Program, and the National Science Council under contract No. NSC 99-2221-E-009 -172 and No. NSC 99-2221-E-009 -167 -MY3.

REFERENCES

[1] Yung-Chun Wu. Po-Wen Su, Chin-Wei Chang, and Min-Feng Hung, IEEE Electron Device Lett., vol. 29, p. 1226 (2008).

[2] J. Fu1et al., in IEDM Tech. Dig., p. 79 (2007).

[3] A. Meldrum et al., Adv. Mater., vol. 13, p.1431 (2001).

[4] L. A. Nesbit, Appl. Phys. Lett., vol. 46, p. 38 (1985).

[5] M. Ostraat, J. De Blauwe, M. Green, D. Bell, H. Atwater, and R. Flagan, J. Electrochem. Soc., vol. 148, p. 265 (2001).

[6] Y. Shi et al., in Proc. 5th ICSSICT, p. 834 (1998).

[7] Jan De Blauwe, IEEE Trans. on nanotechnology, vol. 1, p. 72 (2002).

[8] Cheng-Wei Luo et al., IEEE Trans. Electron Devices., vol. 58, p. 1879 (2011).

[9] T. Schmidt, K. Lischka and W. Zulehner, Phys. Rev. B, vol. 45, p. 8989 (1992).

[10] John Robertson, et al., Appl. Phys. Lett., vol. 44, p. 415 (1984).

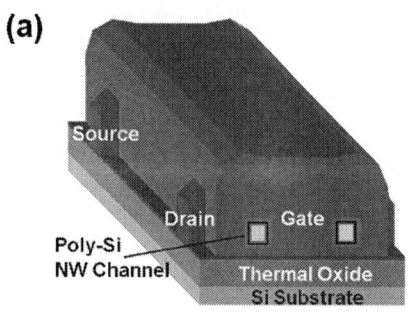

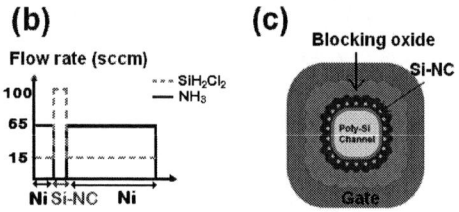

Fig. 1 (a) 3-D structure of the GAA NW SONOS device. (b) Flow rates of $SiCl_2H_2$ and NH_3 for incorporating the Si-NCs in the nitride. (c) Schematic structure of the ONO with SiNCs surrounding the poly-Si NW channel.

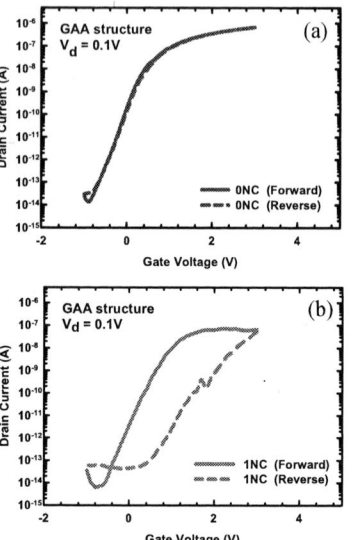

Fig. 2 Transfer curves of GAA NW SONOS TFTs **(a)** without **(b)** with embedded Si-NCs.

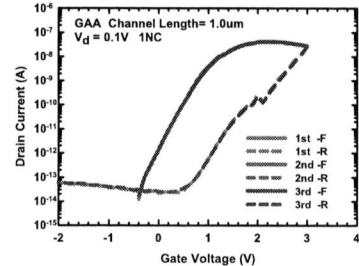

Fig. 3 Transfer curves of three consecutive sweepings for a GAA NW SONOS device (with NC) showing reproducible hysteresis characteristics.

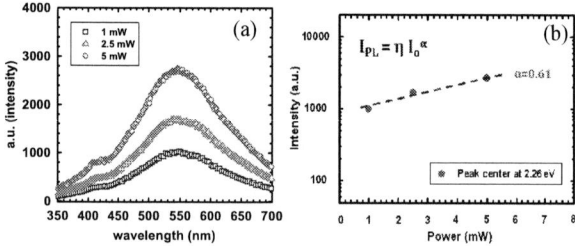

Fig. 4 (a) Micro-photoluminescence spectra of a nitride layer conating Si-NCs under various excitation power. (b) Power dependence of the luminescence peak at 550 nm.

3-D Stacked NAND Flash Memory Having Lateral Bit-line Layers and Vertical Gate

Ju-Wan Lee, Min-Kyu Jeong, Byung-Gook Park, Hyungcheol Shin and Jong-Ho Lee

School of EECS and ISRC, Seoul National University, Seoul 151-742, Korea
Phone: +82-2-880-1727; Fax: +82-2-882-4658; E-mail: jhl@snu.ac.kr

1. Introduction

Recently, demand for non-volatile memory have increased since digital applications like MP3, digital camera, solid state disk (SSD) and so on using non-volatile memory have increased in market. 3-D memories have been gathering increasing attention as future ultra-high density memory technologies because they increase bit density and reduce bit cost. As a candidate, tunneling oxide, poly-Si body, and tunneling dielectric are formed sequentially in through-holes [1]-[3]. However these structures have a limitation in the number of vertical bit cell and large V_{th} distribution in a bit line because through-hole size becomes narrow as the number of bit cell increases [1], [2]. Metal gate structure makes bad retention characteristics in high temperature operation due to stress of high thermal expansion coefficient [3]. Thus we require 3-D stacked NAND flash string to solve the problems mentioned above.

In this work, we study a new 3-D stacked NAND flash memory structure and explain the fabrication sequence. The three layers of stacked poly-Si bodies are used as bit-lines. The trench etch is used to define active layers. The side of stacked bodies is used as a channel and the method to improve the device characteristics is proposed. Finally, key features of fabricated device are investigated.

2. Fabrication

Fig. 1 shows the schematic view of fabrication sequence for proposed 3-D stacked NAND flash memory device. On the Si substrate, SiO_2 and poly-Si are alternately deposited and the SiO_2 layers are formed by using MTO (Medium Temperature Oxide) (a). The thicknesses of SiO_2 and poly-Si are 40 nm and 100 nm, respectively. The body is doped with boron ion implantation. To dope the vertically stacked bodies, ion implantations with three different energies were performed. If the body width becomes nano-scale, the body can be doped by a tilt ion implantation after etching of the body stack. The trench is formed by dry etching using a hard mask for active layers (b). Here, the vertically stacked three layers of poly-Si are bit-lines (BLs) and the BLs are separated by SiO_2. The side of poly-Si is used as a channel. Tunneling oxide-nitride-blocking oxide (ONO) stack was formed by a consecutive deposition process. After the formation of an ONO layer, doped poly silicon for the word-line (WL) was deposited and filled the trench out (c). After that, the doped poly-Si layer is etched to pattern WLs (d). Then, side n^+ regions were formed by ion implantation in the bodies except the regions overlapped by the WLs. For the n^+ regions, arsenic ions were implanted being tilted by 30° with two rotations (0° and 180°) to form the n^+ regions on the sides of the etched poly-Si layers. To anneal the implanted ions, the samples were annealed at 750 °C for 30 min and 950 °C for 5 s. The blocking oxide and nitride layers between WLs are selectively removed by wet etch process in order to separate charge trapping layers. To form reasonable contact holes at both ends of the bottom and center BLs, the top BL layer near the contact holes is etched firstly, and the center BL near the contact holes for the bottom BL is

etched (e). Then n^+ doping is performed on both ends of a BL body to provide contact regions for BL wiring. A layer of oxide is deposited using HDPCVD and contact hole is etched. The contact holes are formed on the n^+ region at one end of the BL, and on both n^+ body and p-body regions at the other end of the BL to achieve grounded body. Finally, metal layers (Ti/TiN/Al/TiN) are deposited using endure sputter and are etched for metal pad (f).

3. Experimental Results and Discussion

The inset of Fig. 2 shows the SEM image of fabricated 3-D stacked NAND flash memory device. Three body layers are vertically stacked and they are separated by SiO_2 layer. After the active trench etch process, the side of bodies is revealed and the slope is about 80 degree. The ONO gate stack is deposited on the side of stacked bodies as a thickness of 3/6/9 nm, respectively.

Fig. 2 shows program and erase characteristics (bit-line current (I_{BL}) versus control gate bias (V_{CG})) of fabricated 3-D stacked NAND flash memory device. The measured device is located on the top BL layer and the gate length (L_g) is 0.5 μm. The p-body and side n^+ doping concentrations are 5×10^{16} cm^{-3} and 5×10^{17} cm^{-3}, respectively. For the program and erase, 19 V and -18 V are applied, respectively. The threshold voltage shift (ΔV_{th}) is 2.4 V when the bit-line current (I_{BL}) is 10 nA. Since the channel is formed on the side of etched poly-Si body, the device characteristics are not good.

Fig. 3 shows the retention characteristics of a device. No serious charge loss is detected up to 10^4 s. The V_{th} for erase state is nearly constant over time, but the V_{th} for program state decreases. The ΔV_{th} of 2.44 V at the initial time was decreased to 2.1 V after 10^4 s and will be decreased to 1.84 V after 10^6 s. Fig. 4 shows the endurance characteristic of a device. We can observe similar V_{th} behavior with P/E cycling for program and erase states. Although the V_{th}s for program and erase states are fluctuated, the ΔV_{th} is 1.58 V after 10^4 cycles.

To improve the quality of the side channel, we adopted chemical dry etch (CDE) after etching the BL stack. The inset of Fig. 5 shows the SEM image of fabricated 3-D stacked NAND flash memory device where the side of the etched poly-Si is slightly dented due to the CDE process. By using the CDE process, we can significantly reduce the etch damage which comes from etch of BL stack. Fig. 5 compares measured I-V characteristics with and without the CDE process after etching the BL stack. The open and closed symbols represent I-V characteristics of fabricated devices without and with CDE process, respectively. Thanks to the CDE process, on- and off-current are increased and decreased, respectively, significantly. The ΔV_{th} is 2.4 V at the I_{BL} of 10 nA.

4. Conclusion

In this paper, we have studied a new 3-D stacked NAND flash memory structure and explained the fabrication sequence and key features of fabricated devices. Reasonable operation of the devices was shown in terms of ΔV_{th}, retention and cycling characteristics. Moreover, the

978-1-4673-0996-7/12 $31.00 © 2012 IEEE

device characteristics were quite improved by removing the etch damage on the side surface (channel) of poly-Si BL layers when CDE process was adopted after etching the BL stack.

Acknowledgement

This work was supported by "Development of novel 3D stacked devices and core materials for the next generation Flash memory sponsored by Ministry of Knowledge Economy" in 2012.

References

[1] R. Katsumata et al., *VLSI Tech. Dig.*, p. 136, 2010.
[2] S. J. Whang et al., *IEDM Tech. Dig.*, p. 668, 2010.
[3] J. Jang et al., *VLSI Tech. Dig.*, p. 192, 2009.

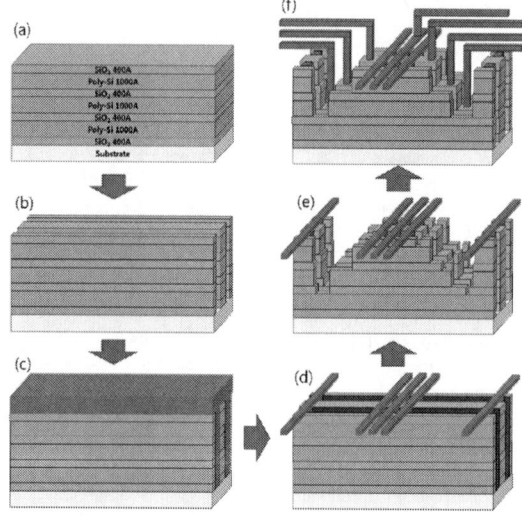

Fig. 1. The schematic view of fabrication sequence for proposed 3-D stacked NAND flash memory device.

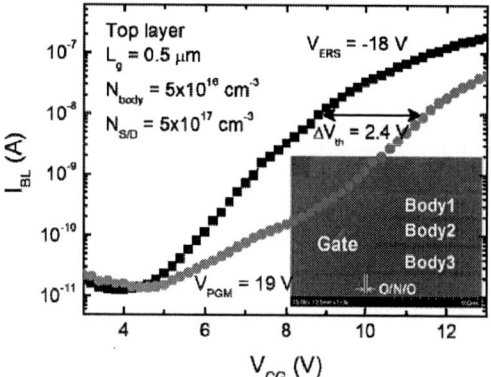

Fig. 2. *I-V* characteristics with program and erase of a fabricated 3-D stacked NAND flash memory device. The top layer of a single gate device having L_g of 0.5 μm is measured. For the program and erase, 19 V and -18 V are applied. Inset: SEM image of fabricated 3-D stacked NAND flash memory device.

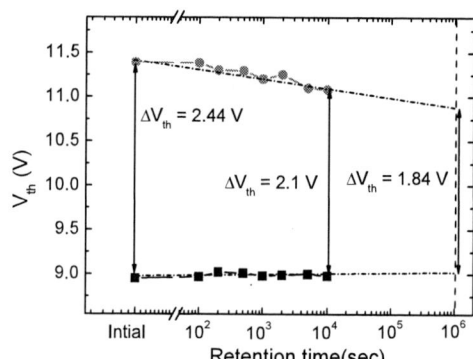

Fig. 3. Retention characteristics of a device. The ΔV_{th} after 10^6 sec is expected as 1.84 V.

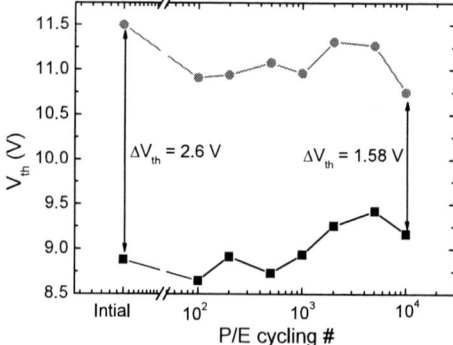

Fig. 4. Endurance characteristic of a device. Although the V_{th}s for program and erase states are fluctuated, the ΔV_{th} is 1.58 V after 10^4 cycles.

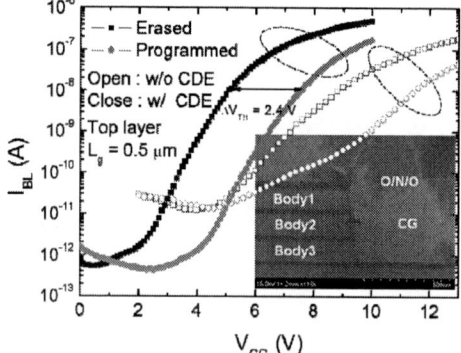

Fig. 5. Comparison of measured *I-V* characteristics with program and erase between without and with CDE process after etching the BL stack. Inset: SEM image of fabricated 3-D stacked NAND flash memory device which CDE process is added after etching the BL stack.

978-1-4673-0996-7/12 $31.00 © 2012 IEEE

Effect of Cu Insertion Layer between Top Electrode and Switching Layer on Resistive Switching Characteristics

Sunghun Jung, Jeong-Hoon Oh, Kyung-Chang Ryoo, Sungjun Kim, Jong-Ho Lee, Hyungcheol Shin, and Byung-Gook Park[*]

Inter-university Semiconductor Research Center (ISRC) and

School of Electrical Engineering and Computer Science, Seoul National University,

San 56-1, Sillim-dong, Gwanak-gu, Seoul 151-742, Republic of Korea.

Tel.: +82-2-880-7279, Fax: +82-2-882-4658, E-mail address: jsh1127@snu.ac.kr

Abstract

By inserting copper (Cu) metal layer between platinum (Pt) and titanium dioxide (TiO_2), we have observed both unipolar and bipolar resistive switching characteristics in $Pt/Cu/TiO_2/Pt$ stacked RRAM cell. In order to analyze the conduction mechanism, we have conducted I-V fitting. And based on measurement results of bias polarity dependency, we have found that copper plays a role as oxygen reservoir. It can explain redox mechanism in bipolar resistive switching cell.

1. Introduction

Resistive Random Access Memory (RRAM) has been regarded as an alternative of NAND flash memory due to its high scalability and low power consumption [1]. In order to achieve better memory performance, many research groups have studied various materials such as NiO_x, WO_x, and TiO_x [2]. Among them, TiO_x-based RRAM can have good compatibility with selection devices because it shows unipolar and/or bipolar resistive switching characteristics according to measurement conditions and combinations of stacked layers. However, its conduction and switching mechanisms are still controversial [3, 4].

In this paper, relationship between conduction and switching mechanism is investigated in $Pt/Cu/TiO_2/Pt$ stacked RRAM showing both unipolar and bipolar resistive switching characteristics. And the effect of Cu layer is also investigated through I-V fitting and measurement of bias polarity dependency.

2. Fabrication

Fig. 1 (a) shows SEM image of fabricated $Pt/Cu/TiO_2/$ Pt RRAM cell. Fabrication process is as follows. First, a 70 nm thick SiO_2 and 90 nm thick platinum (Pt) bottom electrode were deposited on Si wafer step by step. Then, TiO_2 with 40 nm thickness was deposited by plasma-enhanced atomic layer deposition (PEALD). In order to form the top electrode, a 10 nm thick copper and 50 nm platinum layer were deposited by sputter and patterned by using a shadow mask. A diameter of top electrode is 80 μm. Fig. 1 (b) shows measurement setup. We applied positive or negative bias to the top electrode in the I-V sweep mode while the bottom electrode was grounded as common.

3. Results and Discussion

Fig. 2 shows the I-V curves which indicate the resistive switching behavior of $Pt/Cu/TiO_2/Pt$ stacked cells. Both bipolar (a) and unipolar (b) resistive switching characteristics were observed [5].

First of all, in order to analyze conduction and switching mechanisms of bipolar resistive switching, we conducted I-V fitting in a double logarithmic plot. While the first reset operation (2nd curve in Fig. 2 (a) and Fig. 3 (a)) shows abrupt transition from low resistance state (LRS) to high resistance state (HRS), the second reset transition (6th curve in Fig. 2 (a)) occurs gradually as shown in Fig. 3 (b). Difference of two reset operations results from the LRS of a RRAM cell. In case of LRS current flow following Ohm's law (Fig. 3 (a)), abrupt change occurs. On the other hand, in case of LRS current flow following space charge limited conduction (SCLC), smooth transition takes place. After reset operation, as shown in Fig. 4, both cells follow SCLC regardless of the reset transition type [3, 6]. And then, after bipolar set operation, it is found that the slope of I-V curve in LRS cell shows a deviation from 1 as shown in Fig. 5. It is different from LRS cell operating in unipolar mode (Fig. 6).

In order to explain the switching mechanism in detail, we measured the dependence of bias polarity in pristine state before forming process. When the voltage sweep level increases more and more toward positive direction, there is no current change as indicated in Fig. 7 (a). However as shown in Fig. 7 (b), in case of negative direction sweep, noticeable current drop occurs. In control group, which is $Pt/TiO_2/Pt$ RRAM without Cu layer, there is no current change compared with the cell inserted Cu layer during negative direction sweep. When only negative voltage is applied to top electrode in $Pt/Cu/TiO_2/Pt$, current reduction is observed. Such bias polarity dependency can be considered as the effect of Cu layer insertion between the top electrode and the switching layer.

The effect of Cu layer can be explained as illustrated in Fig. 8. When the negative bias is applied to the top electrode, negatively charged oxygen ions existing near the Cu/TiO_2 interface move toward the anode. And then oxygen ions recombine the oxygen vacancies [7]. Consequently, the current decreases because oxygen vacancy is regarded as the current path. In reset operation of bipolar resistive switching, the same phenomenon contributes to the current reduction. Therefore, bipolar resistive switching behavior in $Pt/Cu/TiO_2/Pt$ stacked RRAM can be well achieved in negative bias reset scheme.

4. Conclusion

In this paper, both unipolar and bipolar resistive switching characteristics in $Pt/Cu/TiO_2/Pt$ system are observed. Reset operation type (sharp and gradual reset) is determined according to conduction mechanism of low resistance state. Through the measurement for bias polarity dependency, we have found that the Cu layer can help the current be reduced under only negative bias. From this result, positive set and negative reset operation can be explained.

Acknowledgement

This work was supported by the Smart IT Convergence System Research Center funded by the Ministry of Education, Science and Technology as Global Frontier Project. And the authors would like to thank Prof. Cheol Seong Hwang for experimental support.

References

[1] W. C. Chien, *et al.*, *IEDM Tech. Dig.*, (2010), pp. 440-443.
[2] Kyung-Chang Ryoo, *et al.*, *Japanese Journal of Applied Physics*, **50**, (2011), 04DD15.
[3] Jeong-Hoon Oh, *et al.*, *International Conference on Solid State Devices and Materials*, (2011), pp. 154-155.
[4] Lin Yang, *et al.*, *Appied Physics Letters*, **95**, (2009), 013109.
[5] B. Gao, *et al.*, *IEDM Tech. Dig.*, (2011), pp. 417-420.
[6] Kyung Min Kim, *et al.*, *Nanotechnology*, **22**, (2011), 254010.
[7] N Xu, *et al.*, *Semicond. Sci. Technol.*, **23**, (2008), 075019.

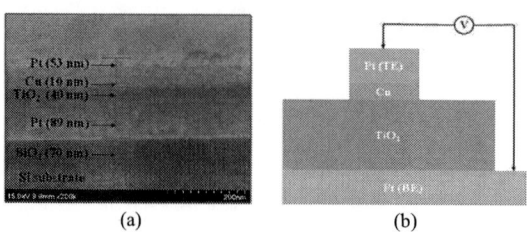

Fig. 1. (a) SEM image and (b) measurement system of Pt/Cu/TiO$_2$/Pt stacked RRAM.

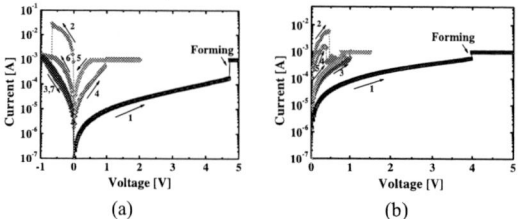

Fig. 2. (a) Bipolar and (b) unipolar resistive switching characteristics in Pt/Cu/TiO$_2$/Pt stacked RRAM.

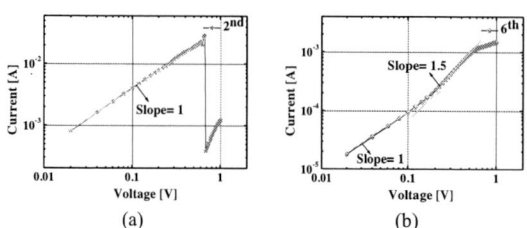

Fig. 3. Reset operation types of (a) sharp and (b) gradual transitions in case of bipolar resistive switching.

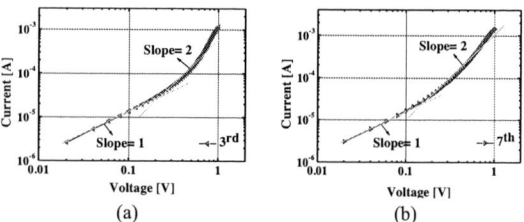

Fig. 4. High resistance state (HRS) fitting results after reset operation in case of bipolar resistive switching.

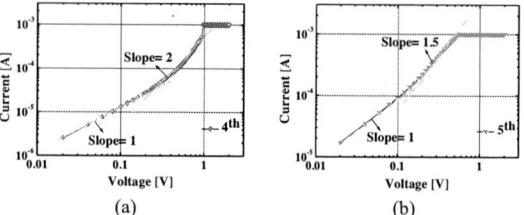

Fig. 5. (a) Set operation and (b) Low resistance state (LRS) after set operation in case of bipolar resistive switching.

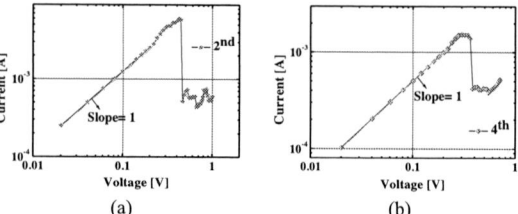

Fig. 6. Reset operations by Joule heating in case of unipolar resistive switching.

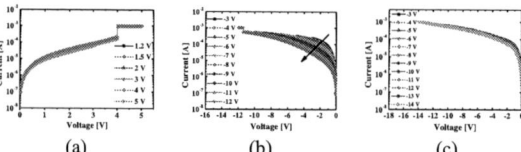

Fig. 7. (a) Positive and (b) negative voltage sweep curves of pristine cells in Pt/Cu/TiO$_2$/Pt RRAM and (c) negative voltage sweep curves of pristine cell in Pt/TiO$_2$/Pt RRAM.

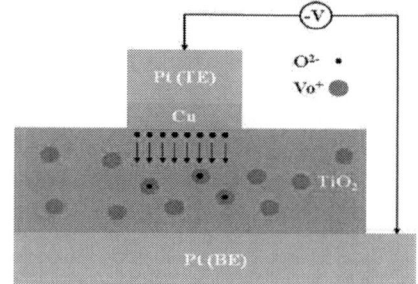

Fig. 8. Schematic diagram for explaining the current reduction when negative bias is applied to the top electrode.

978-1-4673-0996-7/12 $31.00 © 2012 IEEE

Self- compliance Unipolar Resistive Switching and Mechanism of Cu/SiO$_2$/TiN RRAM Devices

D. Yu, L.F. Liu[*], P. Huang, F.F. Zhang, B. Chen, B. Gao, Y. Hou, D.D Han[*], Y Wang, J.F. Kang, X. Zhang

Institute of Microelectronics, Peking University, Beijing 100871, China

[*]E-mail: lfliu@pku.edu.cn; handd@ime.pku.edu.cn

Abstract

CMOS compatible Cu/SiO$_2$/TiN-based resistive random access memory (RRAM) was fabricated and investigated. Unique self-compliance unipolar resistive switching (RS) was observed, as well as good retention and uniformity of resistance states. A physical model based on formation and rupture of Cu conductive filament (CF) is proposed, considering both thermal and electrical effect, and verified by experiments.

Index Terms: unipolar switching, RRAM, SiO$_2$

Introduction

RRAM has been widely studied for several strong points，such as fast switching speed, high integration, low power consumption and so on[1,2], especially when traditional memory faces severe scale-down problems. Recently, much attention has been paid to RRAM devices based on solid electrolyte such as SiO$_2$, due to ultra low power consumption, high scalability and good compatibility with CMOS process[3,4],and several researches have been done to improve performance like retention, uniformity and so on[5,6]. In our study, CMOS compatible RAAM devices of Cu/SiO$_2$/TiN structure were observed to show self-compliance unipolar resistive switching, as well as good retention and uniformity of resistance states. A physical model for unipolar RS based on Cu CF(s) is proposed and verified.

Experimental

Fig.1.(a) and (b) show the schematic structure and fabrication process of CMOS compatible Cu/SiO$_2$/TiN devices, respectively. Electrical measurements were performed by Keithley4200 analyzers with BE grounded.

Results and Discussion

The fresh cell shows initially high resistance state (HRS), and an electrical forming process with a 2.5~3.5V positive bias is necessary. Unique self-compliance unipolar resistive switching was observed in Cu/SiO$_2$/TiN devices with a high switching current (Ireset≈10mA) and big resistance ratio(>10^3,read at 0.5V), as shown in Fig2(a). A good linearity between bias and current in low resistance state (LRS), showed in inset figure, suggests a metallic CF between electrodes. Fig.2(b) and (c) dispaly bipolar RS and unipolar RS under bidirectional sweep without set current compliance, respectively, showing the co-existence of self-compliance bipolar and unipolar RS.

Fig.3(a) and (b) show distribution of reset voltage and LRS resistance during 90 RS cycles under both 25℃ and 85℃, while good distribution of LRS resistance is observed. The reduced reset voltage found at 85℃ is probably induced by diffusion of Cu atoms in CF to ambient. Fig.4 shows the distribution of HRS resistance. HRS becomes stable and decreases with improved uniformity as RS goes on and temperature increases,

related to diffusion of Cu probably. Both HRS and LRS are stable more than 5000s at 85 ℃ ,as shown in Fig.5,implying the good retention performance.

Division still exists regarding the switching mechanism for unipolar RRAM[7,8]. We believe that the metallic Cu CF(s), formed by electrochemical reaction[3], is responsible for RS in our device. Formation and rupture of Cu CF(s) result in LRS and HRS respectively. High LRS current is the key factor for unipolar RS, considering thermal power's square dependence on local current. In addition, the migration of Cu under electrical field does have a big effect on device resistance, responsible for reforming of Cu CF(s) and set process.Fig.6 shows the schematic of unipolar resistive switching –(a):LRS with Cu CF between electrodes；(b): thermal diffusion and migration of Cu under a small electrical field, while the former is dominate, causing the shrink of CF; (c):rupture of CF because of more and more Cu atom loss to ambient SiO$_2$; (d):migration of Cu becomes dominate under a comparably high bias ,with weak thermal diffusion caused by low current; (e):reforming of CF-1) migration keeps dominate if bias is high enough, thus LRS is stable ;2) migration is weaker than diffusion again, thus LRS is unstable, but finally stable, attributed to migration's exponent dependence on bias (see Fig.8).

Fig.7 shows the influence of reset current compliance on reset process, proving the key role of high switching current in unipolar RS. Occasional temporary resistance sticking in LRS (soft error) occurs during sweep, which can be solved, shown in Fig.9, by changing sweep voltage to a smaller one. All above is consistent with our model, while self-compliance is resulted from contradictory influence between thermal and electrical effect on CF, preventing draft increase of current in LRS.

Conclusion

Self-compliance unipolar RS of Cu/SiO$_2$/Pt devices with large resistance ratio, good retention and uniformity of LRS and HRS have been observed and studied. A physical model based on Cu CF is proposed, taking both thermal and electrical effect into consideration. Different testing measures were conducted to verify this model.

Acknowledgement

This work was supported in part by 973 and NSFC (Grant Nos. 2011CBA00600, 2010CB934203, 60906040, and 60925015).

References

[1] Y.S. Chen et al, IEDM2009, p105; [2] C.H.Cheng et al, VLSI2010,p85; [3]C. Schindler, et al, TED. 54, p2762, (2007);[4] C. Schindler, et al, APL. 92, p122910, (2008); [5] K.-L. Lin et al., J. Appl. Phys., 109, p084104, (2011); [6] B.J.Choi et al, Adv. Mater. 23, p3847,(2011);[7] Y S Chen, et al. J. Phys. D: Appl. Phys. 45, p065303(2012);[8] Ugo Russo，et al., EDL, 56,p186,(2009).

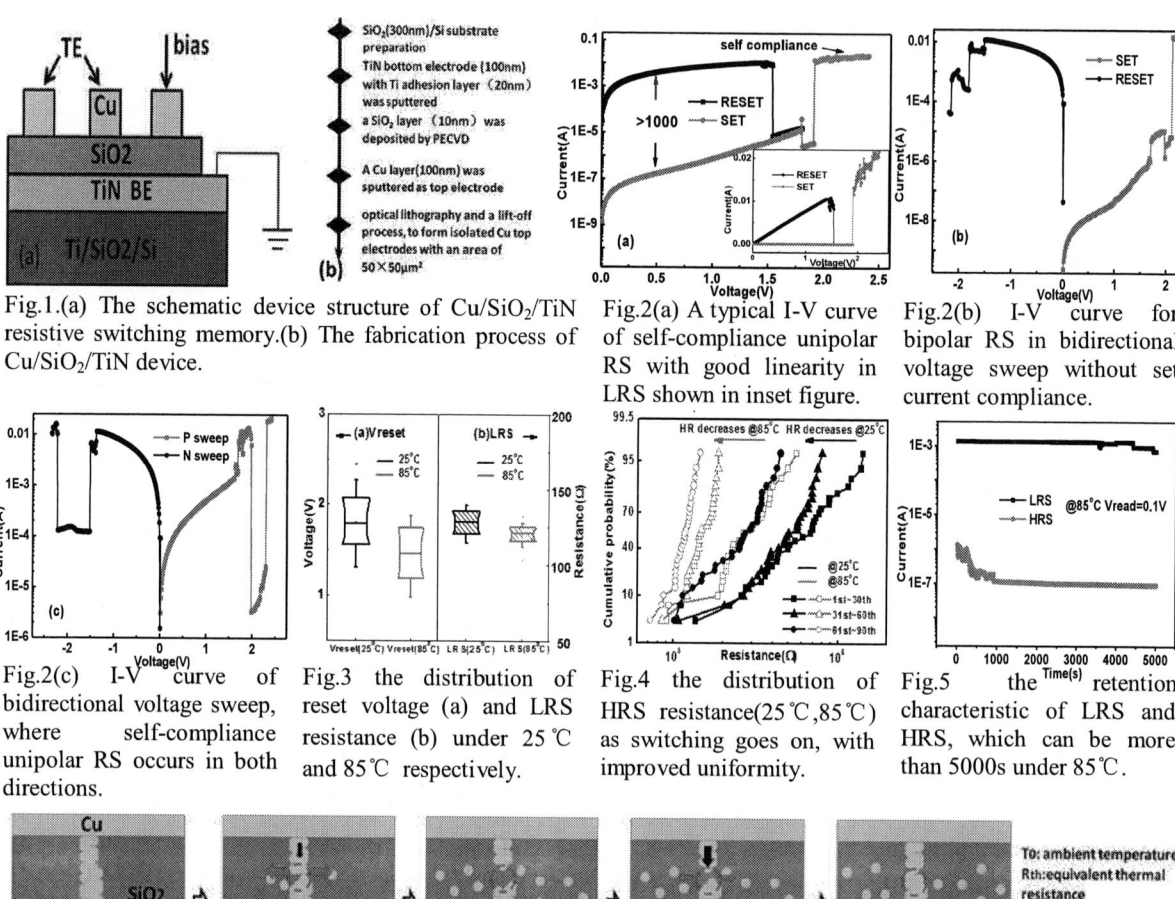

Fig.1.(a) The schematic device structure of Cu/SiO₂/TiN resistive switching memory.(b) The fabrication process of Cu/SiO₂/TiN device.

Fig.2(a) A typical I-V curve of self-compliance unipolar RS with good linearity in LRS shown in inset figure.

Fig.2(b) I-V curve for bipolar RS in bidirectional voltage sweep without set current compliance.

Fig.2(c) I-V curve of bidirectional voltage sweep, where self-compliance unipolar RS occurs in both directions.

Fig.3 the distribution of reset voltage (a) and LRS resistance (b) under 25℃ and 85℃ respectively.

Fig.4 the distribution of HRS resistance(25℃,85℃) as switching goes on, with improved uniformity.

Fig.5 the retention characteristic of LRS and HRS, which can be more than 5000s under 85℃.

Fig.6 schema of unipolar resistive switching: (a):LRS with Cu CF between electrodes; (b): shrink of CF because of thermal diffusion, stronger than migration of Cu under small electrical field; (c):rupture of CF because of more and more Cu atom loss to ambient SiO₂; (d):migration of Cu becomes dominate under a comparably high bias ,with weak thermal diffusion caused by low current; (e):reforming of CF.

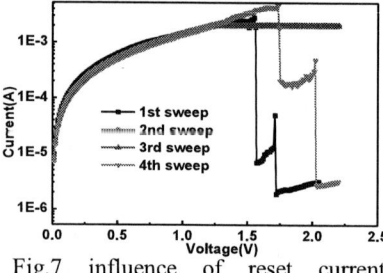

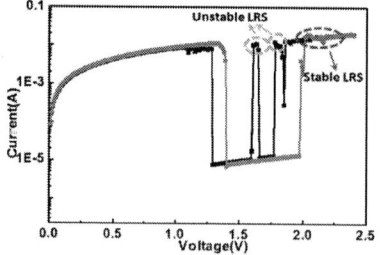

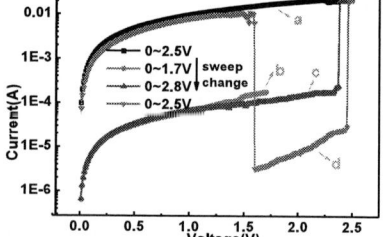

Fig.7 influence of reset current compliance on reset process, where 2mA current compliance is applied in 2nd and 3rd sweep, causing reset failure.

Fig 8 unstable LRS during set process, induced by competition between thermal diffusion and electrical migration of Cu.

Fig.9 temporary resistance sticking in LRS(curve a) and solution-curve b: device resistance switches to and sticks in a higher level resistance after several RS cycles using 0~1.7V sweep; curve c:0~2.8V sweep is applied for set; curve d:normal RS occurs again.

*** Formatting Issue - Best Available Paper/Graphic ***

Rectifying Characteristics and Implementation of n-Si/HfO$_2$ based Devices for 1D1R-based Cross-Bar Memory Array

F. F. Zhang, P. Huang, B. Chen, D. Yu, Y. H. Fu, L. Ma, B. Gao, L. F. Liu, X. Y. Liu, *J. F. Kang

Institute of Microelectronics, Peking University, Beijing 100871, China, *E-mail: kangjf@pku.edu.cn

Abstract

Excellent rectifying characteristics are demonstrated in the fab-friendly n-Si/HfO$_2$/Ni/TiN devices with rectification ratio of >10^7 and the driving current of 1mA as a 1D-like selector. The rectified unipolar switching behaviors are demonstrated in the 1D1R cell structured with a diode-like device of n-Si/HfO$_2$/Ni/TiN (1D) and a unipolar RRAM of n$^+$-Si/HfO$_x$/Ni/TiN (1R). Based on the measured I-V characteristics, these excellent selection behavior can be implemented in the cross-bar memory array of >64K bits RRAM with large read margin.

Index Terms: rectifying, resistive switching, 1D1R

Introduction

One diode-one resistor (1D1R) has been extensively investigated to solve the sneak path current problem in the crossbar memory array of unipolar-type resistive switching memory [1, 2]. However, conventional pn junction- or Schottky-diodes usually suffer from significantly reduced forward current with scaling area, making the driving current lower than the SET/RESET current. Therefore, it is required to explore the new diode-like devices with high driving current and high rectification ratio for the 1D1R application of crossbar memory array. In this paper, a novel n-Si/HfO$_2$ based device using fab-friendly materials is reported and demonstrated with the rectifying characteristics including large rectification ratio of >10^7, higher and area-independent forward current than 1mA@1V, and robust switching uniformity and reliability. Based on the excellent selection behaviors, a crossbar memory array of >64K bits with large read margin can be realized.

Experiment

The n-Si substrate (ρ: 0.05-0.2Ω·cm) was used for the study. Phosphorus ions implantation were performed on with energy of 40KeV and dose of 5×10^{15} cm^{-2} to form an n$^+$ area on some part of the substrate. After SC1&DHF cleaning, 23nm HfO$_2$ film was deposited by rf sputtering, followed by a furnace annealing in O$_2$ at 450℃ for 30min. Then a 6nm Ni layer and 100nm TiN top electrode were deposited and patterned by the areas of 20×20 μm^2, 50×50 μm^2, 100×100 μm^2. 200nm Al was deposited on the back of the substrate as the bottom electrode. Finally, a post annealing process in N$_2$ at 430℃ was performed for 30min. Fig.1 schematically shows the structure of the Si/HfO$_2$ based device. Electrical measurements were performed by Agilent4156C and Keithley400 analyzers.

Result and Discussion

The fresh device is in high resistance state as the black current-voltage (I-V) curve and transferred into low resistance state with a pronounced rectifying effect after a forming process of 12V as shown in Fig.2. Fig.3 shows the measured I-V curves of the devices with various areas. The almost consistent rectifying behaviors with high rectification ratio and forward driving current are observed in the devices with different areas, which is suitable for 1D1R crossbar array of unipolar RRAM [3].

Fig.4 shows the measured I-V curves of the 100× 100 μm^2 device after 1st, 50th, and 500th sweeping cycles from -2V to +2V. Excellent switching uniformity is observed. No observable degradation of the ON-OFF current is measured as shown in Fig.5, indicating good retention performance of the fabricated device.

Fig.6 shows measured area dependence of the forward and reverse current read at 1V, indicating the excellent scalability of the n-Si/HfO$_2$ based device as a diode-like selector. The diode-like I-V curves without area-dependence also imply that the filament-like [4, 5] Schottky emission dominated the current behavior of the device. Good agreement between the measured I-V data and Schottky emission model under biases from -2 to 0 and 0 to 0.5V as shown in Fig.7 supports the assumption. Fig. 8 illustrates the conduction mechanism of the rectifying effect of the devices. Conductive filaments (CFs) consisted of oxygen vacancies in HfO$_2$ are generated during the forming process. The Schottky-like barrier in Si/HfO$_2$ interface is attributed to the offset between Si Fermi level and energy level of oxygen vacancies in HfO$_2$[6].

A unipolar resistive device n$^+$-Si/HfO$_2$/Ni/TiN fabricated on the same wafer is connected with the device to identify the performance of the n-Si-HfO$_2$ based selector. The typical unipolar I-V behavior is measured in the n$^+$-Si/HfO$_2$ based devices as shown in Fig. 8. Fig. 9 shows the measured I-V curves of the structured 1D1R cell. The rectified resistive switching characteristics are demonstrated. Using the extracted data from Fig. 9, we calculated the read margin dependence on the number of word lines when the device is implemented to 1D1R crossbar structure using the method presented in ref.7 and 8. The calculation shows that a crossbar memory array of >2^{16} (64K bits) with large read margin can be realized.

Conclusion

An n-Si/HfO$_2$/Ni/TiN device is proposed and excellent rectifying characteristics with more than 10^7 rectification ratio, sufficient high forward driving current and excellent uniformity and scalability are demonstrated. The excellent selection behavior of the n-Si/HfO$_2$ based devices is preferable for the application of 1D1R-based RRAM crossbar array.

Acknowledgement

This work is partly supported by the 973 and NSFC Programs (2011CBA00600, and 60906040).

Reference

*** Formatting Issue - Best Available Paper/Graphic ***

[1] W.Y.Park et al, Nanotenology 21, p195201(2010); [2] K.W. Zhang et al, Sci China Tech Sci 54, p811 (2011); [3] X. A. Tran et al, EDL 32, p396 (2011); [4] B. Chen, et al, EDL, 32, p282(2011); [5] B. Gao et al, IEDM2011, p417; [6]K. Xiong at al, APL 87, p183505 (2005); [7] J.L. Liang et al, TED 57, p2531(2010); [8] J. J. Huang et al, IEDM2011, p733

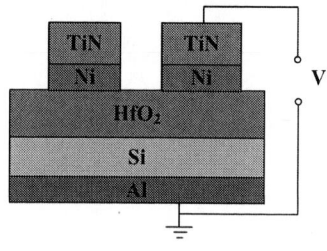

Fig.1 Schematic structure of the Si/HfO$_2$-based device with fab-friendly materials.

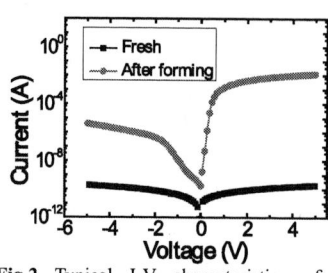

Fig.2 Typical I-V characteristics of the Si/HfO$_2$-based device. The rectifying behavior is observed after a forming process.

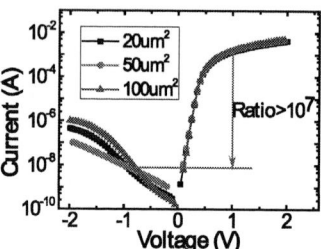

Fig.3 Typical I-V curves of the Si/HfO$_2$-based devices with different areas. The forward driving current larger than 1mA without area-dependence is measured.

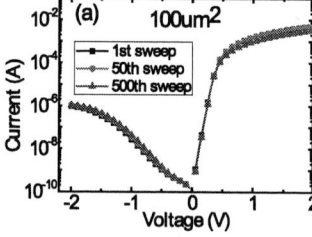

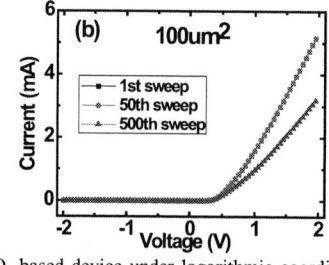

Fig.4 The switching behaviors of the Si/HfO$_2$-based device under logarithmic coordinates (a) and linear coordinates (b). Excellent uniformity of the switching behavior with a threshold voltage of 0.45V is measured. The characteristics show that the devices have high potential application to be used as a selector.

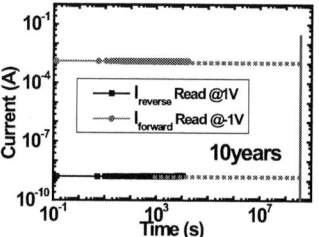

Fig.5 Retention characteristics of a rectifying device measured at RT. No obvious degradation of ON-Off current is observed.

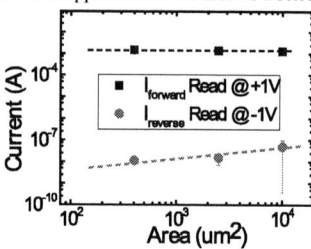

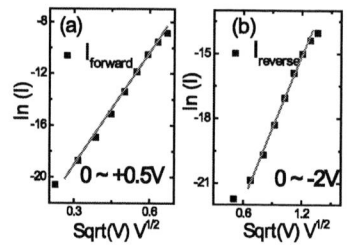

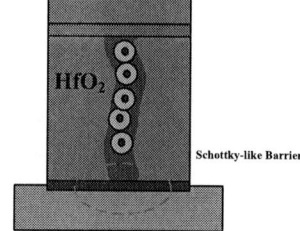

978-1-4673-0996-7/12 $31.00 © 2012 IEEE

Oxygen-induced High-k Degradation in TiN/HfSiO Gate Stacks

Takuji Hosoi, Yuki Odake, Keisuke Chikaraishi, Hiroaki Arimura, Naomu Kitano, Takayoshi Shimura, and Heiji Watanabe

[1] Graduate School of Engineering, Osaka University, 2-1 Yamadaoka, Suita, Osaka 565-0871, Japan

Email: hosoi@mls.eng.osaka-u.ac.jp

Abstract — We have investigated the diffusion kinetics of Hf in TiN/HfSiO gate stacks. The Hf upward diffusion is found to be independent of interfacial SiO_2 growth, but depends on the amount of oxygen in the gate stacks. It is also revealed that Hf diffusion into TiN electrode occurs at above 650°C and leads to high-k degradation.

I. Introduction

Effective work function control and equivalent oxide thickness (EOT) scaling are the major concerns for implementing metal/high-k gate stacks with gate-first process. In addition to interfacial SiO_2 growth, it has been reported that metal elements such as Hf and La atoms in high-k layers diffuse into gate electrode after high-temperature activation annealing [1-4]. Consequently, a loss of leakage reduction merit due to permittivity lowering of Hf-based high-k layer has been pointed out [4]. We have also found that metal inserted poly-Si stack (MIPS) structure can suppress both SiO_2 growth and Hf upward diffusion [4]. These observations suggest that the Hf diffusion correlates with SiO_2 growth, but the detailed mechanism is unclear. In this study, we systematically investigated the driving force for the Hf upward diffusion into TiN gate electrode in TiN/HfSiO gate stacks.

II. Experimental

TiN/HfSiO/SiO_2 gate stacks were fabricated on p-Si substrates by utilizing *in situ* physical vapor deposition (PVD) based process [5]. HfSiO dielectrics were formed by solid phase interface reaction (SPIR) between 0.5-nm-thick metal Hf and 1.8-nm-thick SiO_2. Then, 10- or 100-nm-thick TiN was also deposited without breaking vacuum. To examine the Hf diffusion by high-temperature annealing, we performed PDA at 600-1000°C for 30 s in N_2 ambient. All the samples were subjected to forming gas annealing (FGA) at 450°C for 30 min in 3%-H_2/N_2 ambient, followed by XPS analysis using monochromatic Al Kα source at 1486.6 eV. For electrical characterization, MOS capacitors were also fabricated. The EOT and flatband voltage (V_{fb}) were extracted from C-V measurements.

III. Results and Discussion

As shown in Figs. 1(a,b), Si $2p$ and Hf $4f$ XPS spectra taken from HfSiO surface of 10-nm-thick TiN/HfSiO stack with PDA at 900°C and subsequent TiN removal indicates the Hf diffusion into TiN electrode and interface SiO_2 growth. Surprisingly, even in 100-nm-thick TiN/HfSiO stack, both Si and Hf atoms are detected from the TiN surface (Figs. 1(c, d)).

Since SiO_2 growth occurs with Si emission from the SiO_2/Si interface [6], emitted Si atoms were considered to diffuse together with Hf atoms during PDA. In order to evaluate Hf and Si distribution in TiN, XPS analyses were performed for 100-nm-thick TiN sample after every TiN wet etching cycle (Fig. 2). The obtained profiles revealed that the Hf concentration gradually decreases toward the TiN surface, while Si concentration is almost constant within TiN layer and is highest at the TiN surface. This implies the higher diffusion rate of Si than that of Hf at 900°C, as illustrated in Fig. 3.

Next, we focus on the role of oxygen in Hf diffusion. In order to exclude residual oxygen in PDA ambient, we used MIPS structures and oxygen was intentionally incorporated into TiN by low-pressure oxidation of TiN surface at 650°C before poly-Si deposition (Fig. 4). O $1s$ spectra taken from TiN surfaces indicate that the amount of incorporated oxygen can be controlled by changing the oxygen partial pressure (Fig. 5). The EOT increase after PDA is found to depend on the amount of incorporated oxygen (Fig. 6). As shown in Figs. 7(a, b), Hf peak taken from HfSiO surface reduced with increasing oxygen partial pressure due to Hf upward diffusion, while no change in Si-O peak. This clearly indicates that Hf diffused into TiN during *in situ* low-pressure oxidation at 650°C without SiO_2 regrowth, and Hf diffusion is independent of Si diffusion. However, a slight increase in Si-O peak due to SiO_2 growth was observed after PDA at 700°C (Figs. 7(c, d)). This means that the temperature required for Hf diffusion and that for SiO_2 growth are close to each other. It is found that the amount of diffused Hf atoms crucially depends on the oxygen partial pressure even at 1000°C (Fig. 7(f)). Since the Si-O peak is almost identical regardless of the oxidation pressure (Fig. 7(e)), difference in EOT observed after high-temperature PDA in Fig. 6 is mainly due to permittivity lowering of high-k layer.

This work was partly supported by the Industrial Technology Research Grant Program in 2007 from the NEDO of Japan.

References

[1] T. Matsuki *et al.*, Jpn. J. Appl. Phys. **46**, 1921 (2007).
[2] S. Sakashita *et al.*, Jpn. J. Appl. Phys. **46**, 1859 (2007).
[3] H. Shinohara *et al.*, SSDM2009, 789.
[4] H. Arimura *et al.*, Appl. Phys. Lett. **99**, 142907 (2011).
[5] H. Watanabe *et al.*, Jpn. J. Appl. Phys. **46**, 1910 (2007).
[6] H. Kageshima *et al.*, Jpn. J. Appl. Phys. **38**, L971 (1999).

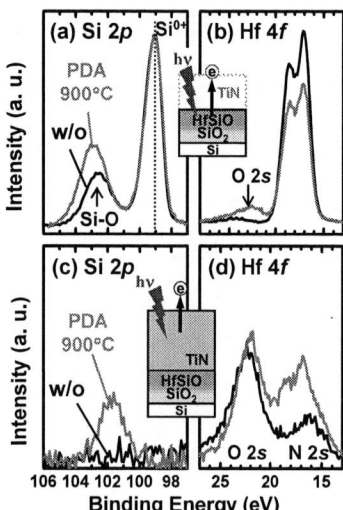

Fig. 1 Si 2p and Hf 4f core-level spectra taken from (a, b) HfSiO surfaces of 10-nm-thick TiN/HfSiO gate stacks and from (c, d) TiN surfaces of 100-nm-thick TiN/HfSiO gate stacks with and without PDA at 900°C for 30 sec.

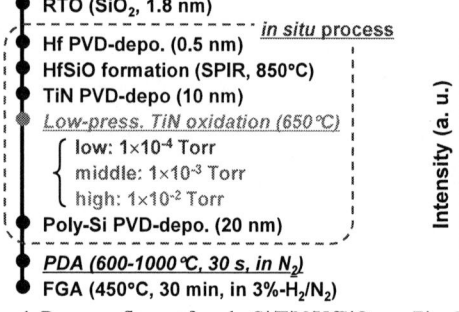

Fig. 4 Process flow of poly-Si/TiN/HfSiO gate stacks with low-pressure oxidation of TiN surfaces. HfSiO layer was formed by solid phase interface reaction (SPIR) between Hf metal and underlying SiO$_2$ under low O$_2$ pressure at 850°C.

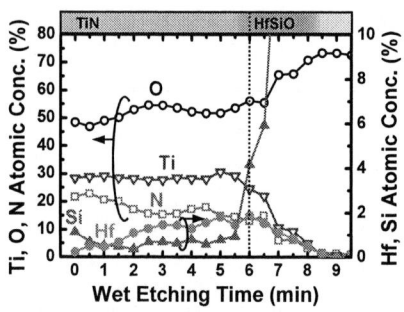

Fig. 2 Atomic concentration profiles in 100-nm-thick TiN/HfSiON stacks with PDA at 900°C obtained by repeating XPS analyses. XPS spectra were acquired after H$_2$O$_2$ wet etching of TiN layer for every 30 sec.

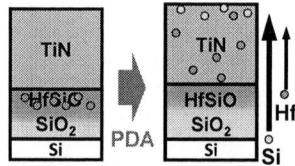

Fig. 3 Schematics of Hf and Si upward diffusion after high-temperature PDA. Hf diffusion and SiO$_2$ growth result in low permittivity of gate dielectric layer.

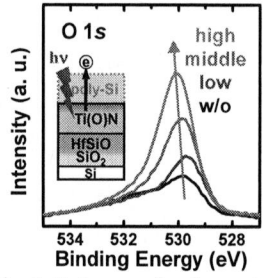

Fig. 5 O 1s core-level spectra taken from oxidized TiN surfaces of poly-Si/TiN/HfSiO stacks after a removal of poly-Si layer by wet etching.

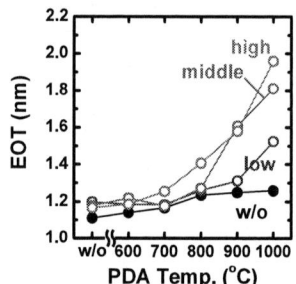

Fig. 6 Changes in EOT of poly-Si/TiN/HfSiO stacks with low-pressure oxidation of TiN surfaces depending on PDA temperature in N$_2$.

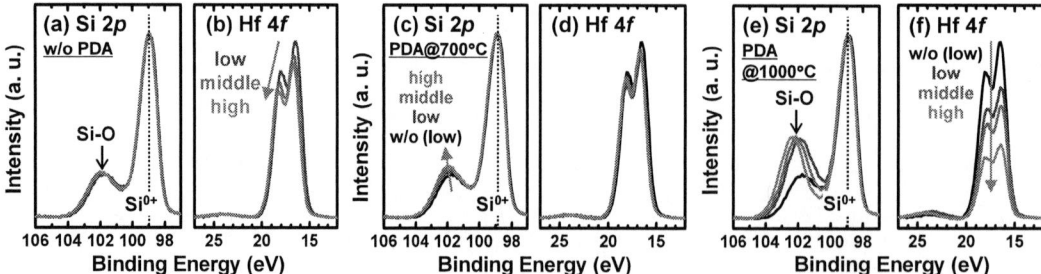

Fig. 7 Si 2p and Hf 4f core-level spectra taken from HfSiO surfaces of 10-nm-thick poly-Si/TiN/HfSiO gate stacks and (a, b) without and with PDA at (c, d) 700°C and (e, f) 1000°C for 30 sec.

978-1-4673-0996-7/12 $31.00 © 2012 IEEE

Metal/Ge Schottky Barrier Modulation With C-Containing Layer by Chemical Bath

Wei Wang[a], Jing Wang[b], Mei Zhao, Renrong Liang and Jun Xu

a) Institute of Microelectronics, Tsinghua University, Beijing 100084, China, Email:wwang05@mails.thu.edu.cn
b) Corresponding author: Institute of Microelectronics, Tsinghua University, Beijing 100084, China. Fax: 86-10-62771130 Tel: 86-13911020082

Abstract

We inserted a C-containing layer in a metal/Ge structure, using a chemical bath. This layer enabled the Schottky barrier height (SBH) to be modulated. The chemical bath with 1-octadecene and 1-dodecene were performed separately with Ge substrates. The ultrathin C-containing layer stops the penetration of free electron wave functions from the metal to the Ge. Metal-induced gap states are alleviated and the pinned Fermi level is released. The SBH is lowered to 0.17 eV. This new formation method is promising and much less complex than traditional ones.

Introduction

Ge is a promising substitute for Si as its high carrier mobility and compatible process. In Ge n-MOSFETs, traditional dopants such as P, As, and Sb have low solid solubility, increased diffusion, and are difficult to activate.[1,2] Metal S/D and Schottky contacts is a substitute. But there is strong Fermi-level pinning just above the valence band edge of Ge, near the charge neutrality level, caused by metal-induced gap states (MIGS) [3,4], which leads a high SBH.

In this study, we introduced a C-containing monolayer as the blocking layer by applying a chemical bath to the Ge substrate. The effective barrier height ($\varphi_{b,eff}$) can be modulated as a result of Fermi-level de-pinning.

Experiments and Results

The chemical bath was applied to n-type (Sb-doped) (100) Ge wafers with resistivities of 24.5–27.5 Ω cm. The non-thermionic emission part of the current could be greatly suppressed during the tests by using a low-doped substrate. Surface cleaning was carried out as shown in in Fig. 1. After Ge–H bond formation on the surface, the substrates were immersed in the chemical baths. 1-Octadecene and 1-dodecene were used individually in the baths. Chemical baths were applied at 160 °C for 2 h. After removal from the baths, the samples were rinsed with alcohol and de-ionized water for three cycles. Other experiments with different conditions also performed for comparison. All conditions and corresponding sample indexes are listed in Table I.

The formation process of monolayer is similar as hydrogermylation [5] and shown in Fig.2. Unsaturated alkene accept electron during the reaction. The Ge-C gives twice bounding energy of Ge-H. The unsaturated bounds receive electrons from Ge-H, and form Ge-C bounds on the surface of substrate [6].

Representative X-ray photoelectron spectroscopy (XPS) data obtained from these samples are shown in Figs. 3 and 4. Fig. 3 shows that the C 1s spectra of samples 1 and 2 have obvious peaks at 283.4 eV; these are below the main C–C peaks and can be ascribable to Ge–C surface bonds. Sample 2 has a third peak at 286.3 eV, which indicates that the surface may not be fully covered by the C-containing layer. The peaks at 30.2 eV in the Ge 3d spectra in Fig. 4 prove the presence of Ge–C. In Fig. 4(b), there is a Ge–O peak at 33 eV which is not obvious in Fig. 4(a). This is the result of oxidation of unreacted Ge–H bonds at a later stage in the process [7]. Higher temperatures ensure increased coverage of the surface by the blocking layer and lead to better results.

The thickness of the blocking layer for sample 1 can be determined by the cross-sectional transmission electron microscopy (TEM) image as Fig.5. A 2.2nm result was obtained which indicated the average thickness per CH$_2$ in the alkyl chain was 1.22Å.

Effective barrier height $\varphi_{eff,b}$ can be extracted from the current density versus voltage (J–V) characteristics which are shown in Fig.6. Samples without blocking layer gives a typical rectification characteristic, which indicates a high SBH. By inserting the layer, current density got significant improvement. The extracted results suggest a pure SBH of 0.62 eV for the untreated sample (Sample 5). It is due to the Fermi level pinning just above 0.04 V from the top of the Ge valence band. Fig.7 shows current density of different samples at V = 1V. $\varphi_{eff,b}$ for different samples is shown in Fig.8. All samples with blocking layer achieved a lower $\varphi_{eff,b}$. The thickness of the blocking layer is quite critical for the current. The best result is given by sample 1 at 0.46 eV. The sample with native oxide (Sample 6) gives a worse result due to the too thick layer and dominating tunneling barrier.

The transport mechanism of these samples is thermionic emission mixed with tunneling [8]. So $\varphi_{eff,b}$ is consisted with Schottky barrier and tunneling barrier. The tunneling barrier height is related to the thickness of blocking layer. The derived SBH of sample 1 is at about 0.17 eV [9].

Conclusions

In summary, chemical baths consisting of different liquid alkenes were applied to <100> Ge substrates under various conditions. Alkyl monolayers were formed on the surface; these layers blocked MIGS penetration to the substrates and released the Fermi-level pinning. The derived $\varphi_{b,eff}$ is about 0.46 eV and the SBH is 0.17 eV. Compared with other blocking-layer formation methods, this is a very easy and promising way of forming layers to lower the effective barrier height.

Acknowledgement

The work is supported by Chinese National Key Basic Research Program (No. 2011CBA00602) and National Key Scientific and Technological Projects (2009ZX02035-004-02 and 2011ZX02708-002).

References

[1] C. Chui et al., APL, 83, 3275, 2003 [2] C. Chui et al., APL, 87, 091909, 2005 [3] A. Dimoulas et al., APL, 89, 252110, 2009 [4] M. Kobayashi et al., VLSIT, Symposi, 54, 2008 [5] K. Choi et al., Langmuir, 16, 7737, 2000 [6] D. Lide, "CRC Handbook of Chemistry and Physics", CRC Press, p 9.64, 2010 [7] R. Chen et al., Chem. Mater. 18, 3733, 2006 [8] D. Schroder, "Semiconductor Material and Device Characterization, 3rd ed", Wiley, p 154, 2007 [9] Y. Liu et al., ChemPhysChem, 3, 799, 2002

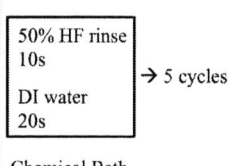

FIG. 1. Substrate-cleaning process. A clean surface with Ge–H bonds is critical for use in the chemical bath.

TABLE. 1. Conditions for different experiments. Sample 5 was cleaned, but a chemical bath was not applied. Sample 6 was as-received 3.7-nm oxide

Sample	Environment	Bath Time	Temperature
1	1-octadecene	2h	160°C
2	1-octadecene	2h	80°C
3	1-octadecene	1h	80°C
4	1- dodecene	2h	160°C
5	Clean, No layer	N/A	N/A
6	Native Oxide	N/A	N/A

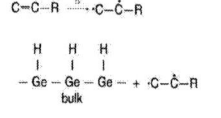

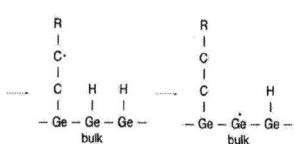

FIG. 2. The processing of reaction is similar as hydrogermylation. Once the reaction begins, it will continue easily.

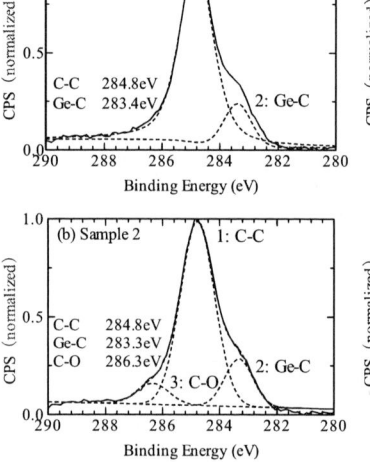

FIG. 3. Detailed XPS spectra of C 1s. (a) The peak at the binding energy lower than that of C–C represents electrons shifting to C, which suggests a Ge–C bond. (b) The third peak indicates a C–O bond, as a result of the lower temperature and short bath time. The surface was not completely covered.

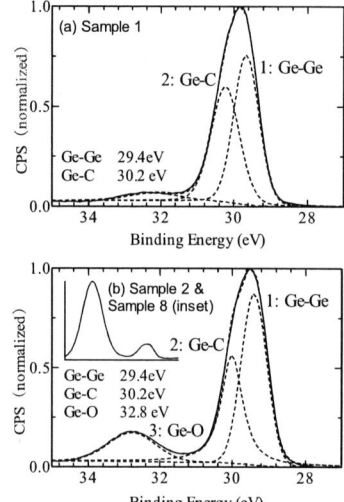

FIG. 4. Detailed XPS spectra of Ge 3d: (a) for sample 1, the higher binding energy peak at 30.2 eV is assigned to Ge–C and (b) there is an obvious Ge–O peak for sample 2.

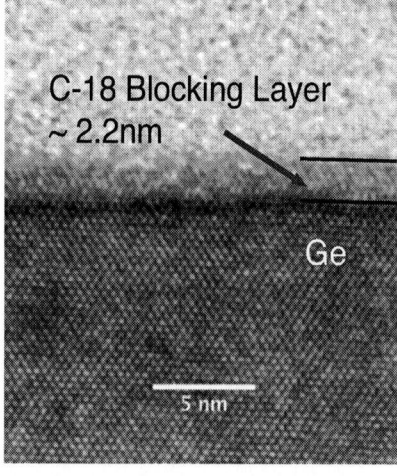

FIG. 5. Cross-sectional TEM result for sample 1. A blocking layer of thickness about 2.2 nm is formed. With high temperature and long bath time, the uniformity and quality of the layer is quite good.

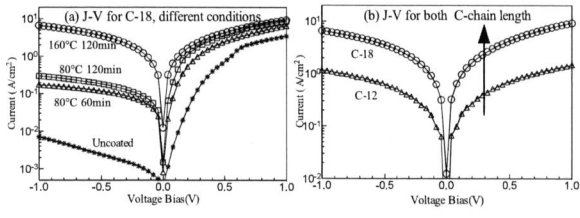

FIG. 6. (a) J–V results for samples prepared in 1-octadecene with different temperatures and durations. The results for the sample without a blocking layer show a rectification characteristic. All the other results show significant increases in both biases. (b) J–V results for the same conditions (160 °C and 120 min) but different bath environments (C-12 and C-18).

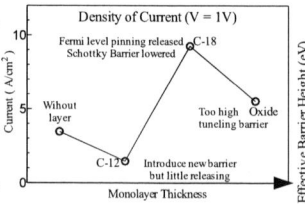

FIG. 7. Current density changes with layer thickness at high positive voltage biases. An optimal point can be obtained near C-18.

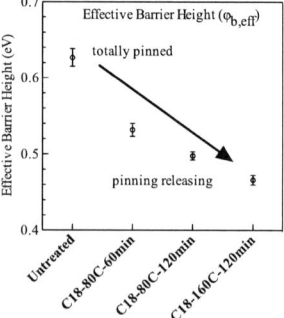

FIG. 8. Extracted $\varphi_{b,eff}$ values for some samples. These effective barrier heights consist of SBHs and tunneling barrier heights.

978-1-4673-0996-7/12 $31.00 © 2012 IEEE

Orientation and Size Effects on Ballistic Electron Transport Properties in Gate-All-Around Rectangular Germanium Nanowire FETs

Seigo Mori[*], Naoya Morioka, Jun Suda, and Tsunenobu Kimoto

Department of Electronic Science and Engineering, Kyoto University, Kyoto 615-8510, Japan
[*]*Electronic mail: mori@semicon.kuee.kyoto-u.ac.jp   Phone/Fax: +81-75-383-2302/2303*

Introduction

Germanium nanowires (GeNWs) are promising as a superior channel material for both n- and p-channel field-effect transistors (FETs), compared with silicon channels. Electronic states in GeNWs have strong dependence on the nanowire (NW) orientation and the cross-sectional shape/size owing to the anisotropic valleys in the conduction band of the bulk structure [1, 2]. Because device scaling should proceed from FinFETs to NW FETs in future, the fundamental understanding of rectangular NWs is particularly important. In this study, we calculate the conduction band structure of rectangular GeNWs with different orientation and size, and analyzed transport properties of gate-all-around (GAA) GeNW FETs. Two key parameters of drive current and injection velocity under ballistic electron transport are discussed.

Calculation Model

We calculated the conduction band structure of GeNWs using a nearest-neighbor $sp^3d^5s^*$ tight-binding (TB) method [3], neglecting spin-orbit coupling. The dangling bonds were passivated by giving excess energy to sp^3 hybridized bonds [4]. The lattice constant (a_0) was assumed to be the same as bulk (5.646 Å). The structures of calculated nanowires (NWs) have orientation and sidewalls of [001]/(100)/(010), [110]/(001)/($1\bar{1}0$), [111]/($\bar{1}\bar{1}2$)/($1\bar{1}0$), and [112]/($\bar{1}\bar{1}1$)/($1\bar{1}0$) as shown in Fig. 1. These NWs have vertically long rectangular cross section with the constant width of 2.0 nm. When the electronic states in GeNWs were calculated, we considered the structures surrounded by 0.6-nm-thick SiO_2 and metal gate and solved the TB and the 2D-Poisson equations self-consistently. The transport characteristics of GeNW FETs were simulated by a semiclassical top-of-the-barrier ballistic model [5]. The drain voltage was 0.6 V.

Results and Discussion

Figures 2-5 depict the conduction band structure of [001], [110], [111], and [112] GeNWs at high gate voltage ($V_G = 0.6$ V). The zero point of energy is the Fermi level. All dispersions have two main valleys at the Γ point (Γ valley) and the Brillouin zone edge (off-Γ valley). Only the Γ valley of [001] NW originates from bulk Δ valleys and the others from bulk L valleys. Regardless of the substrate face, the conduction band minimum (CBM) of [001] and [110] NWs is always located at Γ valley (Fig. 2, Fig. 3) and that of [111] NWs is off-Γ valley (Fig. 4). On the other hand, both Γ and off-Γ valleys can be CBM in [112] NWs, depending on the substrate face (Fig. 5). Comparison of the calculated effective mass at the CBM of rectangular GeNWs is summarized as follows: [112]/($1\bar{1}0$) < [110]/(001) < [111]/($\bar{1}\bar{1}2$) < [110]/($1\bar{1}0$) < [112]/($\bar{1}\bar{1}1$) < [111]/($1\bar{1}0$) < [001].

The injection velocity, which determines intrinsic FET delay [6], is shown in Fig. 6 for calculated GAA GeNW FETs with a 2.0 nm×6 nm cross section at the same off-current of 2.0×10^{-10} A ($V_G = 0$ V). At the low gate voltage ($V_G < 0.2$ V), the effective mass at CBM determines the injection velocity. When the Fermi level reach and exceed the CBM, we have to consider the Fermi velocity defined by the gradient of each subband at the Fermi level. Because the slope of subband becomes higher at the wave number far from the bottom of subband, the injection velocity increases as the gate voltage.

Figure 7 shows the height dependence of the electron injection velocity in GeNW FETs. The injection velocity of [001] NW FETs and [110] NW FETs on the (001) face is less dependent on the height owing to little change of the band structure below the Fermi level (Fig. 2, Fig. 3 (a), (b)). In contrast, the injection velocity of [110] and [112] NW FETs is strongly dependent on the choice of the substrate plane.

We obtained the NW-height dependence of the drive current in GeNW FETs as shown in Fig. 8. In the case of square cross section, the drive current is the highest for [110] and [112] > [111] > [001] follow, being the same trend as that of the injection velocity (Fig. 7). This trend agrees with the calculation of circular NWs [1]. However, in the case of rectangular cross section, the orientation dependence of drive current is not always the same as that of injection velocity because ballistic drive current is determined by not only the injection velocity but also the density of states (DOS). Because a large DOS value is obtained when the effective mass and the number of subbands under the Fermi level are large, the drive current of [112] NW FETs on ($1\bar{1}0$) is low due to the small DOS (Fig. 5 (c)) in spite of the highest injection velocity. In contrast, [001] NW FETs have large DOS but low current because of their lowest injection velocity. [111] NW FETs exhibit low drive current because of the both small DOS (Fig. 4) and low injection velocity no matter which substrate face is used. [110] NW FETs on (001) achieve the highest drive current in whole height range owing to the both high injection velocity and large DOS (Fig. 3 (b)).

Conclusion

We calculated the conduction band structure of GeNWs by a tight-binding model and obtained the fundamental understanding of electron transport characteristics in [001], [110], [111], and [112] GeNW FETs. The simulation of ballistic electron transport revealed that [110] GeNW FETs on the (001) face achieve high drive current as well as high injection velocity, being the best choice for n-channel FETs.

[1] J. Wang *et al.*, *IEDM*, p. 530 (2005)
[2] M. Bescond *et al.*, *J. Comput. Electron.* **6**, p. 341 (2007)
[3] T. B. Boykin *et al.*, *Phys. Rev. B* **76**, 115319 (2007)
[4] S. Lee *et al.*, *Phys. Rev. B* **69**, 045316 (2004)
[5] A. Rahman, *et al.*, *IEEE Trans. Electron Devices* **50**, p. 1853 (2003)
[6] D. A. Antoniadis *et al.*, *IBM J. Res. Develop.* **50**, p. 363 (2006)

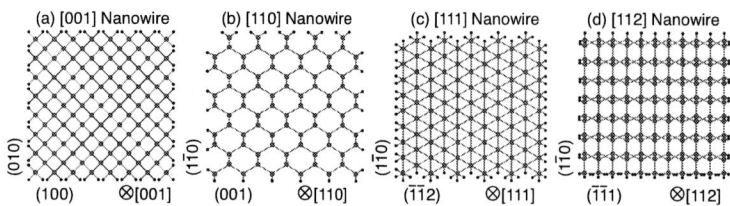

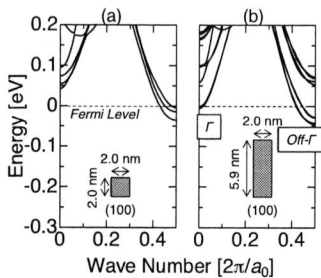

Fig. 1: The ball-stick structures of calculated GeNWs oriented to (a) [001], (b) [110], (c) [111], and (d) [112]. All the structures illustrate 2.0 nm×2.0 nm NWs. The surface stick with a small black ball indicates a passivated sp^3 bond.

Fig. 2: The calculated conduction band structure of [001] GeNWs with (a) 2.0 nm×2.0 nm and (b) 2.0 nm×5.9 nm cross section at high gate voltage (V_G = 0.6 V).

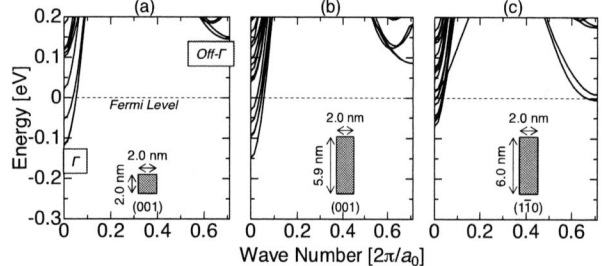

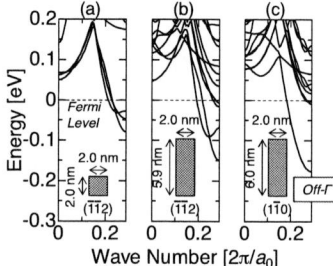

Fig. 3: The calculated conduction band structure of [110] GeNWs with (a) 2.0 nm×2.0 nm and (b) 2.0 nm×5.9 nm cross section on (001) and (c) 2.0 nm×6.0 nm cross section on ($1\bar{1}0$) at high gate voltage (V_G = 0.6 V).

Fig. 4: The calculated conduction band structure of [111] GeNWs with (a) 2.0 nm×2.0 nm and (b) 2.0 nm×5.9 nm cross section on ($\bar{1}\bar{1}2$) and with (c) 2.0 nm×6.0 nm cross section on ($1\bar{1}0$) at high gate voltage (V_G = 0.6 V).

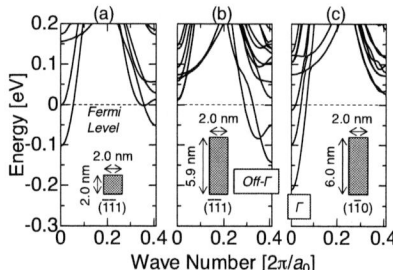

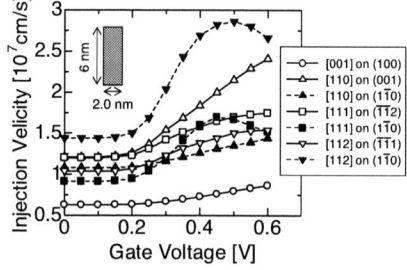

Fig. 5: The calculated conduction band structure of [112] GeNWs with (a) 2.0 nm×2.0 nm and (b) 2.0 nm×5.9 nm cross section on ($\bar{1}\bar{1}1$) and with (c) 2.0 nm×6.0 nm cross section on ($1\bar{1}0$) at high gate voltage (V_G = 0.6 V).

Fig. 6: The electron injection velocity versus gate voltage of 2.0 nm×6 nm cross-sectional GeNW FETs at the same off-current 2.0×10^{-10} A.

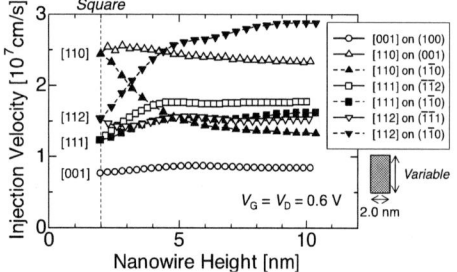

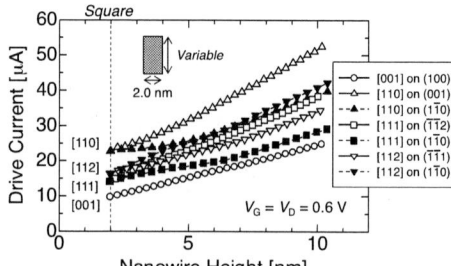

Fig. 7: The height dependence of the electron injection velocity in various GeNW FETs.

Fig. 8: The height dependence of the ballistic drive current in various GeNW FETs.

Quantum Transport Simulation of III-V MOSFETs based on Wigner Monte Carlo Approach

Yōsuke Maegawa[1*], Shunsuke Koba[1], Hideaki Tsuchiya[1,2], and Matsuto Ogawa[1]

[1]Department of Electrical and Electronic Engineering, Graduate School of Engineering, Kobe University
1-1, Rokko-dai, Nada-ku, Kobe 657-8501, Japan
Phone & Fax: +81-78-803-6082　*Email: 117t261t@stu.kobe-u.ac.jp

[2]JST CREST, Chiyoda, Tokyo 102-0075, Japan

III-V compound semiconductors are expected as a post-Si channel material, because they have higher electron mobility and lower effective mass than Si. Actually, the high performance of InGaAs MOSFETs with high-k gate dielectrics has been demonstrated [1,2]. On the other hand, due to a quasi-ballistic behavior of electron transport, III-V channel MOSFETs may be more vulnerable by quantum mechanical effects such as quantum reflection and tunneling, as compared to conventional Si-MOSFETs. In this paper, we investigate quantum transport effects in III-V channel MOSFETs by using a Wigner Monte Carlo (WMC) simulation [3,4], which can fully incorporate the quantum transport effects. As a result, we found that the quantum reflection reduces on-current, while the source-drain (SD) direct tunneling increases subthreshold current even as the channel length is larger than 10 nm.

Fig. 1 shows the device structure used in the simulation. We adopted a double-gate structure with the channel thickness of 5 nm and the gate oxide thickness of 1 nm. The channel material is InP and the channel length is set as 15 nm. The source/drain donor concentration is given as 2×10^{19} cm^{-3} [5,6], and the channel region is undoped. The electrical characteristics were computed by using the WMC device simulator developed in our laboratory [7,8] where we employed a mode-space expansion method as follows. The device is meshed into vertical slices in which the one-dimensional Schrödinger equation is solved to compute subband profiles and wave-functions for the Γ and L valleys of InP. Then, electron transport is simulated based on the Wigner or Boltzmann MC approach along the channel direction x for each valley and each subband. These two MC methods are self-consistently coupled with the two-dimensional Poisson's equation. We simulated electron transport in the Γ and L valleys as mentioned above, where the energy difference between them was taken as 0.832 eV. We considered phonon and impurity scatterings, while roughness scattering was ignored for simplicity.

Fig. 2 shows the drain current versus gate voltage (I_D - V_G) characteristics at $V_D = 0.5$ V, where the results for both the Wigner (quantum) and Boltzmann (classical) MC approaches are plotted. It is found that the subthreshold current increases and thus the subthreshold slope (SS) drastically increases in the quantum approach due to the SD direct tunneling. Incidentally, the current slopes at the minimum gate voltage are approximately 70 and 170 mV/dec for the Boltzmann and Wigner MC results, respectively. Next, Fig. 3 shows the drain current versus gate overvoltage (V_G - V_{th}) characteristics, where the threshold voltage V_{th} is defined as a gate voltage corresponding to $I_D = 0.03$ mA/μm. It is found that the drain current at high gate voltage is reduced in the quantum approach by approximately 10 ~ 20%, depending on the gate voltage. This is partly due to the quantum reflection inside the channel as discussed below. To exhibit the details of the above quantum phenomena, we plotted distribution functions in the phase-space computed by using the Wigner and Boltzmann MC methods in Figs. 4 and 5, respectively, where (a) and (b) correspond to V_G - $V_{th} = 0$ V (at threshold voltage) and V_G - $V_{th} = 0.5$ V (on-state), respectively. The contrast represents the number of electrons present in a cell of the phase-space. First, we can observe distinct oscillations in the Wigner distribution function at the on-state of Fig. 4(b), while they are not present in the corresponding Boltzmann distribution function of Fig. 5(b). In particular, oscillations are visible not only in the $k > 0$ region but also in the $k < 0$ region, which is the signature of quantum reflection [3,4,7-9]. They occur in the region where the potential abruptly drops between the barrier top and the drain-end of the channel, as shown on the upper side. As a result, the quantum reflection decreases the averaged electron velocity and contributes to the reduction of drain current at high gate voltage as compared to the classical one, as shown in Fig. 3.

Next, we examine the distribution functions at the threshold voltage. Comparing Figs. 4(a) and 5(a), we notice that tunneling electron trajectory appears in the Wigner distribution function as explained below. That is, a weak interference pattern is visible in the channel region of Fig. 4(a), but electron trajectory with acceleration is hardly observed. This means that the interference pattern in the channel region observed in Fig. 4(a) must be caused by the SD direct tunneling. In the meantime, the SS degradation due to the SD direct tunneling has been reported to become serious in channel lengths shorter than 10 nm for Si-MOSFETs [3,4,8,9]. In contrast, the SS degradation was demonstrated to be serious even as the channel length is larger than 10 nm for InP-MOSFET in this study. This is considered due to the lower effective mass of InP channel than that of Si channel. At the workshop, we will discuss the channel length dependence of quantum transport effects in III-V channel MOSFETs, and make a comparison between III-V and Si channel MOSFETs.

Acknowledgements This work was supported by a Grant-in-Aid for Scientific Research from the Japan Society for the Promotion of Science (JSPS), and the Japan Science and Technology Agency (JST)/CREST.

References [1] R. Terao et al., *APEX* **4** (2011) 054201. [2] S. Kim et al., *APEX* **5** (2012) 014201. [3] D. Querlioz et al., *IEEE-TED* **54** (2007) 2232. [4] D. Querlioz and P. Dollfus, *The Wigner Monte Carlo Method for Nanoelectronic Devices* (Wiley, New York, 2010) [5] H. Tsuchiya et al., *IEEE-EDL* **31** (2010) 365. [6] Y. Maegawa et al., *APEX* **4** (2011) 084301. [7] S. Koba et al., *JAP* **108** (2010) 064504. [8] S. Koba et al., *Proc. of SISPAD*, p. 79 (2011). [9] Y. Yamada et al., *IEEE-TED* **56** (2009) 1396.

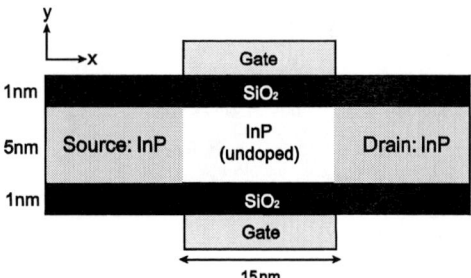

Fig. 1. Device structure used in the simulation. We adopted a double-gate structure with the channel thickness of 5 nm and the gate oxide thickness of 1 nm. The channel material is InP and the channel length is set as 15 nm. The source/drain donor concentration is given as 2×10^{19} cm^{-3}, and the channel region is undoped.

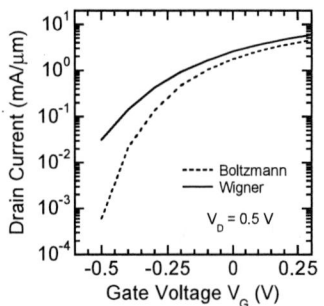

Fig. 2. Drain current versus gate voltage characteristics at $V_D = 0.5$ V, where results for both the Wigner and Boltzmann MC approaches are plotted.

Fig. 3. Drain current versus gate overvoltage ($V_G - V_{th}$) characteristics, where threshold voltage V_{th} is defined as a gate voltage corresponding to $I_D = 0.03$ mA/μm.

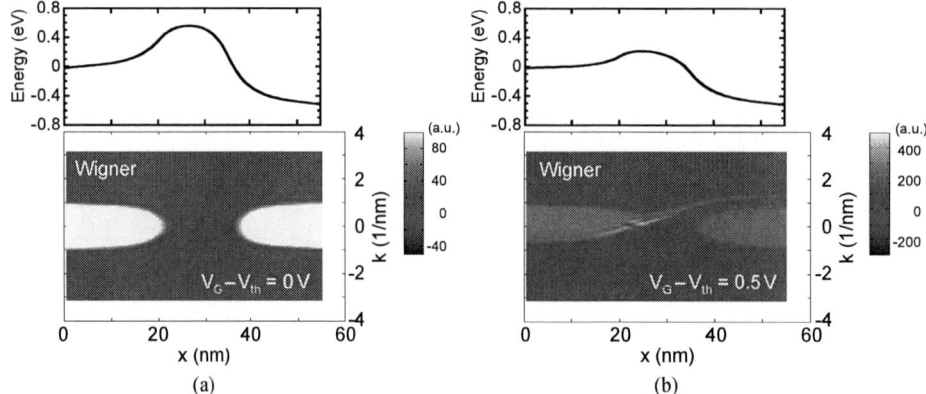

(a)   (b)

Fig. 4. Distribution functions in phase-space computed by using Wigner MC method. (a) and (b) correspond to $V_G - V_{th}$ = 0 V (at threshold voltage) and $V_G - V_{th}$ = 0.5 V (on-state), respectively. The contrast represents the number of electrons present in a cell of the phase-space. The upper figures indicate the corresponding potential distributions.

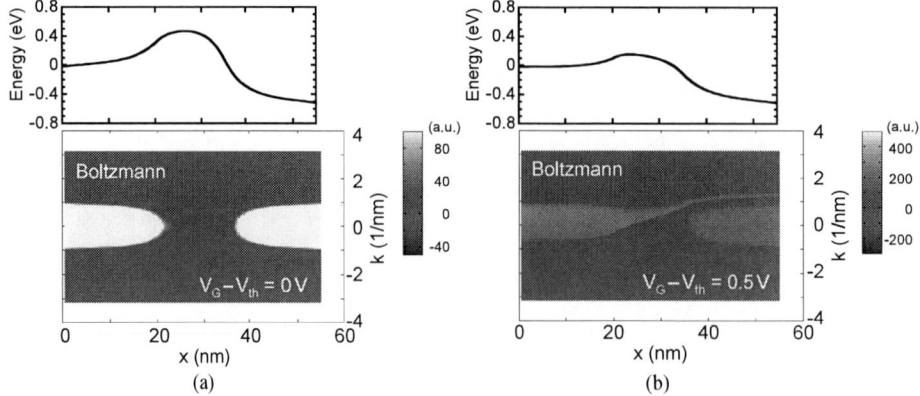

(a)   (b)

Fig. 5. Distribution functions in phase-space computed by using Boltzmann MC method. (a) and (b) correspond to $V_G - V_{th}$ = 0 V (at threshold voltage) and $V_G - V_{th}$ = 0.5 V (on-state), respectively.

978-1-4673-0996-7/12 $31.00 © 2012 IEEE

Mechanisms of Ambient Dependent Mobility Degradation in the Graphene MOSFETs on SiO₂ Substrate

[1]Y.G. Lee, [1]C.G. Kang, [2]C. Cho, [1]Y.H. Kim, [1]H.J. Hwang, [2]J.J.Kim, [1]U.J. Jung, [2]E.J.Park, [2]M.W.Kim, [1,2]B.H. Lee[*]

[1]School of Materials Science and Engineering, [2]Department of Nanobio Materials and Electronics
Gwangju Institute of Science and Technology, Oryong-dong 1, Buk-gu, Gwangju, Korea
Phone: +82-62-715-2308, Fax: +82-62-715-2304, E-mail: bhl@gist.ac.kr

Abstract: **Two different mechanisms affecting the device instability and mobility degradation at graphene MOSFET on SiO₂ substrate and their time constant, 40μsec and ~370μsec, have been identified. Oxygen/H₂O reaction at the surface of graphene was identified as a major source of device hysteresis causing mobility degradation and device instability.**

I. INTRODUCTION

Recently, graphene MOSFETs attracted a lot of attention, but soon, many pessimistic results associated with the device variability, performance degradation have been reported [1-3]. Large hysteresis, substrate dependence, and device scattering are representative challenges in addition to small band gap problem [4-7]. However, systematic research to investigate the origin of device instability has been scarce. As a result, there are many misunderstandings about the origin of device instability and performance degradation. For example, hysteresis of graphene MOSFETs was often attributed to charge trapping into surface adsorbates or a substrate [2]. In this work, a chemical reaction to generate/reduce hydroxyl bonds at the surface of graphene is identified as a major source of hysteresis using a quantitative study on the transient characteristics of graphene MOSFET in temporal domain.

II. EXPERIMENT

Fig.1 shows the schematic fabrication flow of graphene FET. Fig.2 shows the optical image of the graphene FET and Raman spectrum of monolayer graphene. After Al₂O₃ passivation and post deposition anneal, DC I-V and pulse I-V characteristics were measured in vacuum (~10^{-6} torr) and air ambient in sequence using a pulse measurement system.

III. RESULTS AND DISCUSSION

Fig. 3 shows typical DC I-V and pulse I-V curves measured at 300K-500K in vacuum. Hysteresis of DC I-V curves increased at high temperature, but that of pulse I-V didn't show an appreciable temperature dependence, indicating the presence of physical processes affecting the hysteresis with different time dependence (Fig.4(a)). Furthermore, temperature dependence of the hysteresis at DC I-V curves increased significantly in the air ambient (Fig.4(b)). In the air ambient, even pulse I-V curves showed a weak temperature dependence at 500K. For more quantitative analysis, the fast charging effect (ΔI) was measured at V_g-V_{Dirac}=25V for a fixed time, 1msec. Ambient dependence similar to that of hysteresis was confirmed at ΔI, but the amount was almost same for both electron and hole branch, indicating that the hysteresis with a fast time constant is independent of carrier type in the graphene (or bias polarity) (Fig.5). These results imply that the mechanisms causing the hysteresis have a strong dependence on the ambient, temperature and test speed.

For more detail study, the impacts of temperature and ambient on other device parameters were investigated. As expected, $g_{m,max}$ showed a strong temperature dependence in air while almost temperature independent in vacuum. As a result, the field effect mobility degraded drastically in the air ambient, especially in DC I-V measurement (Fig.6).

Time dependence of the reaction causing the hysteresis can be used to monitor the reaction speed under a constant bias (Fig.7). In vacuum, the time dependence can be explained with two different time constants. In the air ambient, very strong temperature dependence was observed, and the reaction became evident even within the pulse rise time (100μsec). However, mean values of time dependences could still be modeled with two different time constants (~40μsec and ~370μsec) similar to those of tests in vacuum (Fig.8). This means that the mechanisms governing the hysteresis in vacuum and air ambient are actually same, but the amounts of reaction are different due to the differences in the ambient.

Based on these observations, possible mechanisms for the hysteresis are suggested; chemical reaction and tunneling (Fig.9 (a)). Tunneling is a dominant mechanism with a fast trapping time constant < 50μsec. In addition, a chemical reaction involving the air ambient is introduced to explain the strong temperature dependence and slow time constants > 300μsec. As shown in Fig.9(b), the graphene can react with atmospheric oxygen and water molecules through the electrochemical reduction of $O_2 + 2H_2O + 4e = 4OH^-$ [8]. In terms of the contribution to the hysteresis, the chemical reaction has more dominant effects than the tunneling when the graphene is exposed to oxygen.

IV. CONCLUSION

Based on the ambient, temperature, test time dependence of device parameters, the origin of strong mobility degradation observed at the graphene MOSFET has been identified as an oxygen redox process. This finding provides a direction to improve the stability and performance of graphene MOSFET using oxygen deficient passivation surrounding the graphene.

Acknowledgement: This work was supported by SAMSUNG System LSI, WCU program through NRF grant funded by MEST (R31-10026,) and the IT R&D program of MKE/KEIT(10039174).

REFERENCES

[1] Z.Liu et. al., Nano Lett. 11, p.523, 2011.
[2] H.Wang et. al., ACS Nano 4, p.7221, 2010.
[3] I.Meric et. al., Nano Lett. 11, p.1093, 2011.
[4] C.G.Kang et. al., Nanotechnol. 22, p.295201, 2011.
[5] Y.G.Lee et. al., Appl. Phys. Lett. 98, p.183508, 2011.
[6] I.Meric et. al., IEDM Tech. Dig., p.556, 2010.
[7] J.Chen et.al., Nat. Nanotechnol. 58, p.206, 2008.
[8] C.G.Kang et. al., submitted for publication.

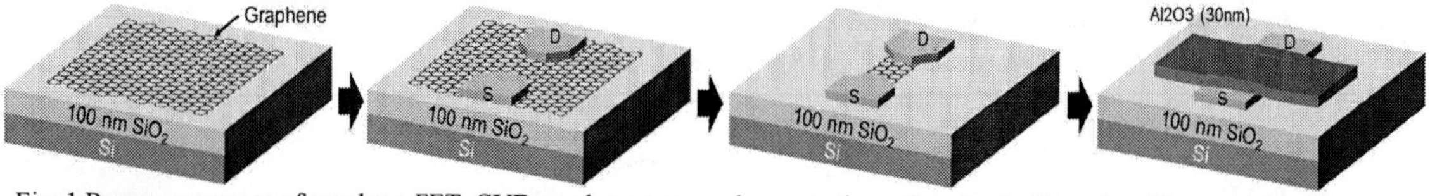

Fig. 1 Process sequence of graphene FET. CVD graphene was used as a starting substrate. Au/Pd=30nm/20nm, Al_2O_3=30nm.

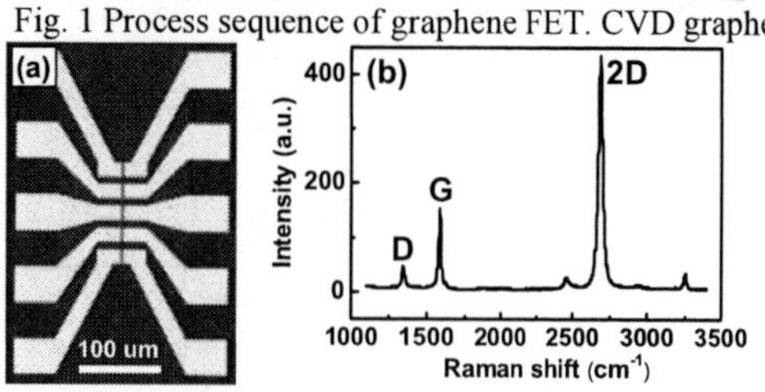

Fig. 2 (a) Optical image of graphene FET and (b) Raman spectrum of graphene showing monolayer thickness.

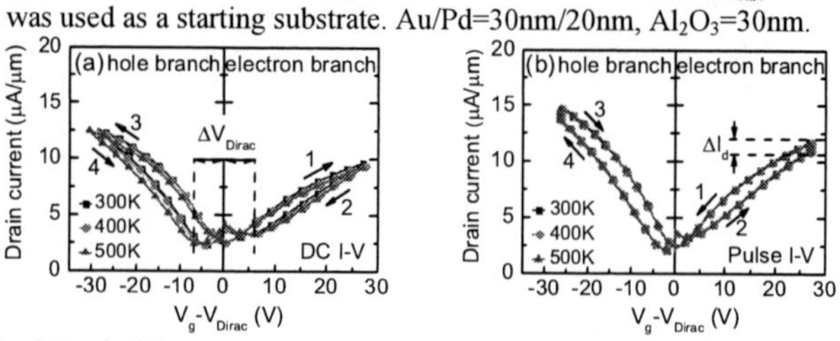

Fig. 3 Typical (a) DC I_d-V_g curves, (b) pulsed I_d-V_g curves of graphene FET at various temperature. V_d= 100mV. Pulse width=1msec.

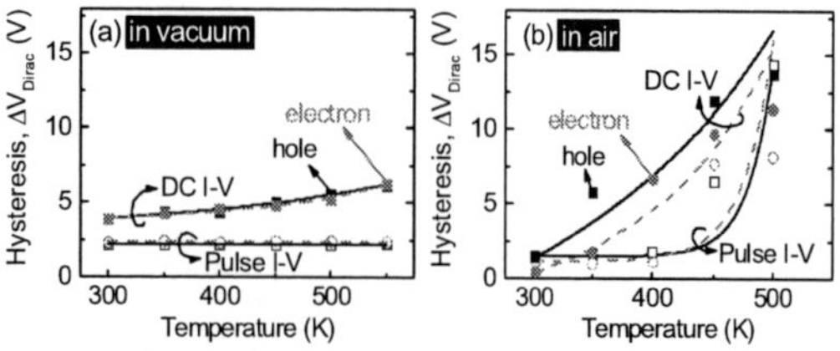

Fig. 4 Temperature dependence of hysteresis (ΔV_{Dirac}) measured in (a) vacuum, (b) air. Dashed line (electron), solid line (hole).

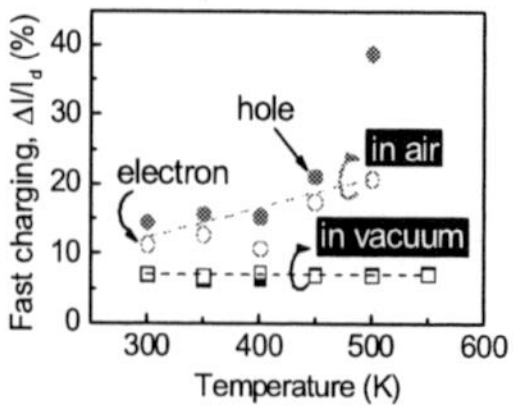

Fig. 5 Temp. dependence of fast charging within 1msec.

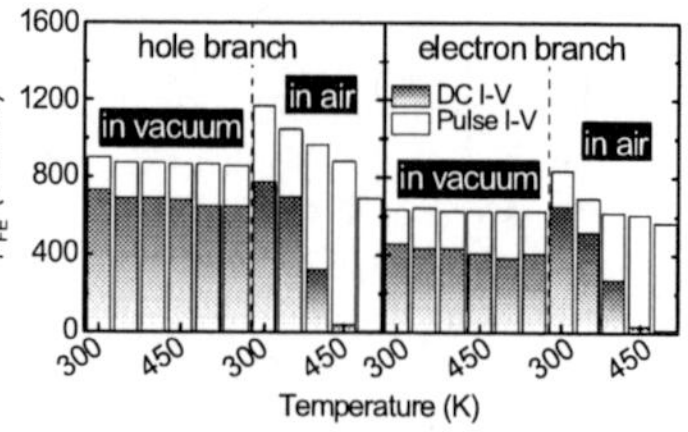

Fig. 6 Ambient dependence of field effect mobility from pulse I-V and DC I-V at various temperatures.

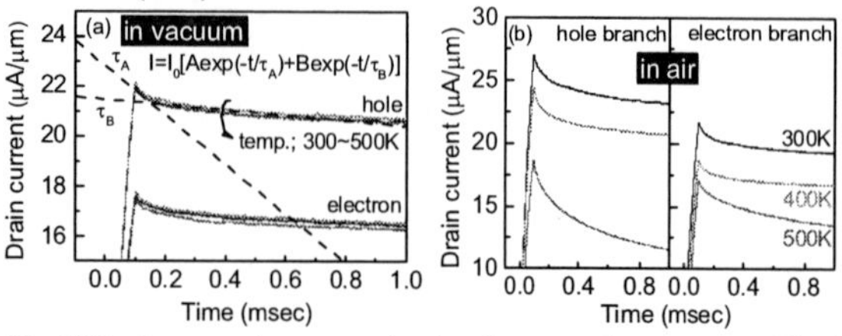

Fig. 7 Drain current measured using 1msec pulse at various temperatures (a) in vacuum, (b) in air. Two linear slopes shown by dashed line indicate the existence of two different time constants.

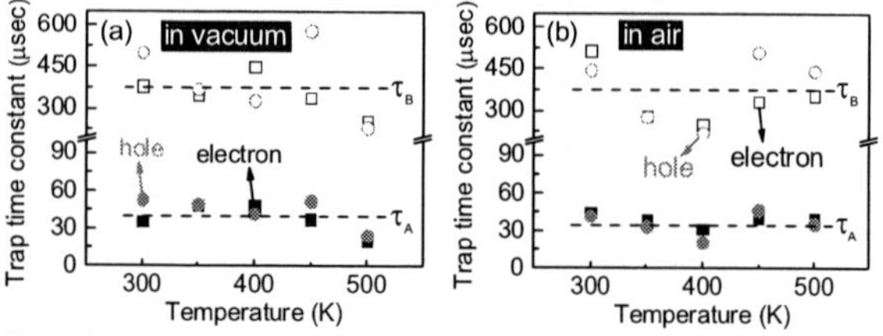

Fig. 8 Reaction time constants at various temperatures (a) in vacuum, (b) in air. τ_A and τ_B are the fast and slow time constants respectively. The equation of two trap model is $I=I_0[A\exp(-t/\tau_A)+B\exp(-t/\tau_B)]$.

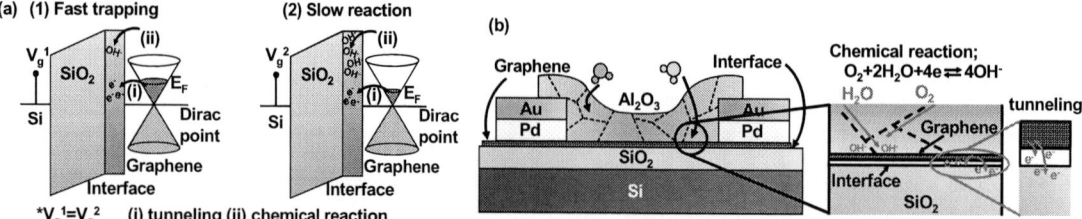

Fig. 9 (a) Mechanisms affecting the hysteresis of graphene FETs. Tunneling is the dominant mechanisms in the fast trapping while chemical reaction is the dominant mechanisms in the slow reaction. (b) The electrochemical reaction between the graphene and ambient can occur due to the molecules can penetrate the Al_2O_3 dielectric through the grain-boundary. The tunneling occurs from graphene to interface and SiO_2 layer.

Electronic Band Structures of Graphene Nanomeshes

Ryūtaro Sako[1*], Naomi Hasegawa[1], Hideaki Tsuchiya[1,2], and Matsuto Ogawa[1]

[1]Department of Electrical and Electronic Engineering, Graduate School of Engineering, Kobe University
1-1, Rokko-dai, Nada-ku, Kobe 657-8501, Japan
Phone & Fax: +81-78-803-6082   *Email: 119t226@stu.kobe-u.ac.jp

[2]JST CREST, Chiyoda, Tokyo 102-0075, Japan

Graphene nanomesh (GNM) is a highly interconnected network of graphene nanoribbons (GNRs) in which the size of nanoholes and the distance between them can be controlled down to the sub-10 nm scale [1]. GNM can open up a band gap in a large sheet of graphene to creat a semiconducting thin film. Actually, it was demonstrated that GNM-based transistors provide driving currents nearly 100 times greater than individual GNR devices, with a comparable on-off current ratio [1]. Furthermore, for practical use, GNM lattices should be much easier to produce and handle than GNRs. Therefore, the GNMs with variable periodicity and neck width are expected to offer a possibility of band gap engineering and graphene electronic applications [2]. In this study, we investigate the electronic band structures of GNMs with various geometric configurations based on a tight-binding approach [3], and examine the roles of the edge formation and neck width on the band gap opening.

Fig. 1 shows the atomic models for GNMs where the horizontal and vertical sides of the nanohole have an armchair-edge and a zigzag-edge, respectively. The number of carbon atoms in the neck of the zigzag-edge side, N_z, is fixed to be 4, while that in the neck of the armchair-edge side, N_a, is variable; N_a = (a) 2 and (b) 3. Since the unit cell used in the calculation is rectangle, which is denoted by the red dashed line in Figs. 1(a) and (b), the first Brillouin zone (FBZ) in the reciprocal space also becomes rectangle. We computed the electronic band structures for the two GNMs in the whole FBZ and their conduction bands are plotted in Figs. 2(a) and (b) for N_a = 2 and 3, respectively. Here, the horizontal and vertical axes indicate the k_x and k_y wavenumbers, respectively. First, it is found that the GNM with N_a = 2 has zero band gap at the K point along the k_y direction as denoted by the arrow, while the GNM with N_a = 3 has a band gap of approximately 0.8 eV at the FBZ edge of the k_y direction. The above results coincide with the well-known behavior of the band gap energies in the armchair-edged GNRs, that is, GNRs with $N_a = 3m + 2$ (m; a positive integer) group has zero band gap, while GNRs with $N_a = 3m$ group has a finite band gap [4,5]. Consequently, the electronic band structures for the present GNMs are considered to be strongly influenced by electronic states from the armchair-edge side of the nanoholes.

Next, to examine signature of the band structures, the dispersion curves along the k_x and k_y directions are plotted in Fig. 3. Note that for N_a = 2 shown in Fig. 3(a), the lowest dispersion curve in the conduction band touches the valence band at the K point along the k_y direction, which is to say, the band gap is zero. On the other hand, the dispersion curves with a finite band gap are confirmed in Fig. 3(b). However, it should be noted that even if the band gap is opened, the details of the dispersion curves including the effective mass of electrons significantly depend on the transport direction as found in Fig. 2(b) and 3(b). Therefore, a precise consideration of the geometric configurations, such as nanohole shape and periodicity, edge structure, neck width and current direction, will be important in a design of GNMs. At the workshop, we will further present the band structures of GNMs with larger nanohole sizes and shapes, and different edge structures, and then discuss anisotropic transport in GNMs.

Acknowledgements This work was supported by a Grant-in-Aid for Scientific Research from the Japan Society for the Promotion of Science (JSPS), and the Japan Science and Technology Agency (JST)/CREST.

References [1] J. Bai et al., *Nature Nanotech.* **5** (2010) 190. [2] V. H. Nguyen et al., *Nanotechnology* **23** (2012) 065201. [3] A. H. Castro Neto et al., *Rev. Mod. Phys.* **81** (2009) 109. [4] Y.-W. Son et al., *PRL* **97** (2006) 216803. [5] R. Sako et al., *IEEE-EDL* **32** (2011) 6.

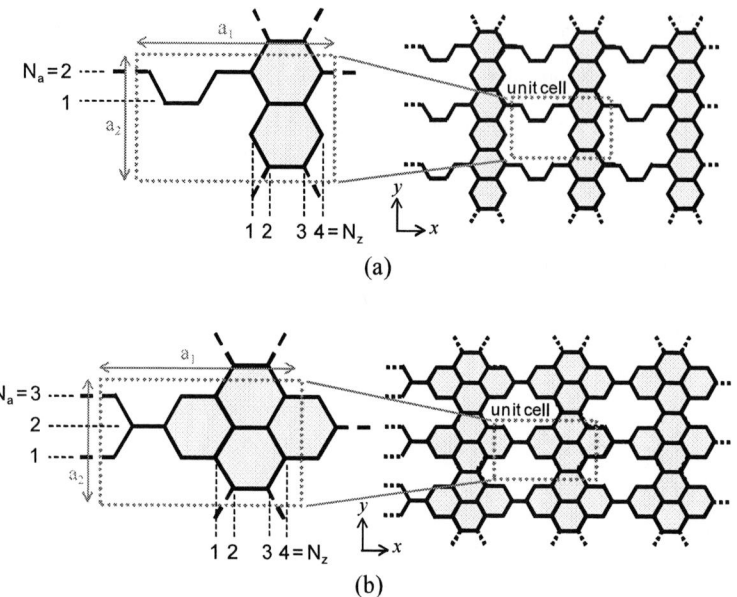

Fig. 1. Atomic models for GNMs where the horizontal and vertical sides of the nanohole have an armchair-edge and a zigzag-edge, respectively. The number of carbon atoms in the neck of the zigzag-edge side, N_z, is fixed to be 4, while that in the neck of the armchair-edge side, N_a, is variable; $N_a =$ (a) 2 and (b) 3. The red dashed lines represent the unit cells used in the calculation.

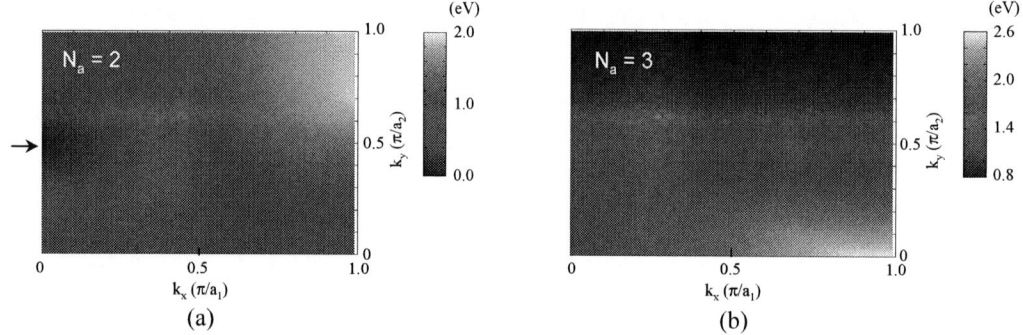

Fig. 2. Conduction band structures computed for the GNMs in the whole first Brillouin zone, where $N_a =$ (a) 2 and (b) 3. Here, the horizontal and vertical axes indicate the k_x and k_y wavenumbers, respectively. The arrow in (a) represents the position indicating zero band gap along the k_y direction.

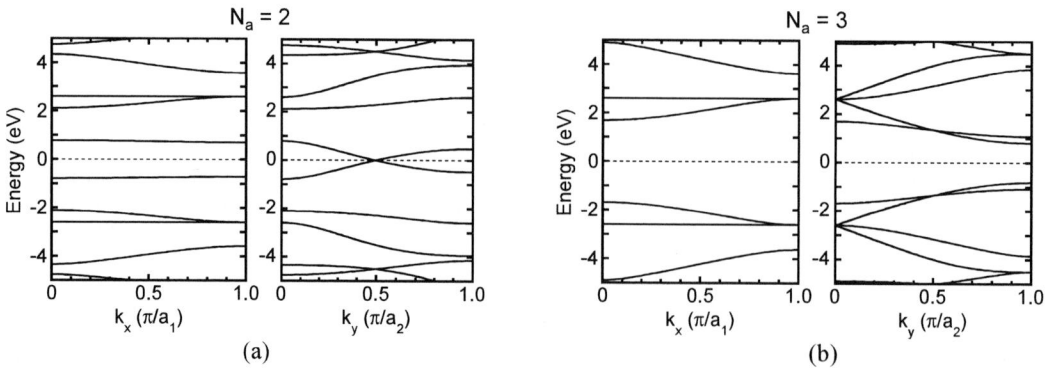

Fig. 3. Dispersion curves along the k_x and k_y directions for $N_a =$ (a) 2 and (b) 3. Note that each transverse wavenumber is taken to be zero, that is, $k_y = 0$ ($k_x = 0$) in the left (right) figure. In (a), the lowest dispersion curve in the conduction band touches the valence band at the K point along the k_y direction.

978-1-4673-0996-7/12 $31.00 © 2012 IEEE

Band Structure and Electron Transport in Multi-Junction Graphene Nanoribbons

Naomi Hasegawa[1], Ryūtaro Sako[1], Hideaki Tsuchiya[1,2*], and Matsuto Ogawa[1]

[1]Department of Electrical and Electronic Engineering, Graduate School of Engineering, Kobe University
1-1, Rokko-dai, Nada-ku, Kobe 657-8501, Japan
Phone & Fax: +81-78-803-6082　*Email: tsuchiya@eedept.kobe-u.ac.jp

[2]JST CREST, Chiyoda, Tokyo 102-0075, Japan

Graphene nanoribbons with armchair-edged configurations (A-GNRs) are expected to be a channel material for higher-speed operation of FETs, since they have an almost linear dispersion relation despite the opening of a finite band gap and hence high carrier velocity up to 5×10^7 cm/s is predicted [1,2]. However, in actually fabricated GNRs with the current technologies, the ribbon width and edge configuration may change along the length direction. On the other hand, the introduction of a heterojunction has been proposed to improve electrical characteristics of GNR tunneling FETs [3]. The A-GNRs are known to exhibit either semiconducting or metallic natures, depending on the number of carbon atoms in the width direction, which may enable us to form a potential barrier or a quantum well inside GNRs just by changing the transverse width along the length direction. In this study, we consider such an electrical heterojunction consisting of multi-connected semiconducting and metallic A-GNRs, and discuss its basic properties by performing the electronic band structure calculations.

Fig. 1 shows (a) the atomic model for multi-junction GNR (MJGNR) with armchair-edged structure. The wide and narrow GNR regions consist of $n = 3m+2$ (= 8; metallic) and $3m$ (= 6; semiconducting) configurations, respectively. Therefore, the wide region acts as a quantum well and the narrow one as a potential barrier, and then a periodic potential distribution is considered to be formed as shown in Fig. 1(b). The numbers of six-membered rings in the length direction of the wide and narrow GNRs are denoted as N_{3m+2} and N_{3m}, respectively. In this study, N_{3m+2} is fixed to be 5, which means that the length of the quantum well region is constant. On the other hand, N_{3m} is set to be variable, which aims to investigate the influences of electron coupling between neighboring quantum wells by changing the potential barrier thickness. We computed the electronic band structures for the atomic model shown in Fig. 1(a) on the basis of a tight-binding (TB) approach with a p_z orbital basis set [2], [4]-[6].

Fig. 2 shows the band structures of MJGNRs computed for various barrier lengths: N_{3m} = (a) 5, (b) 3 and (c) 1. It is found that the band gap energy only slightly decreases with N_{3m}, because the quantum well width is fixed. In contrast, the effective mass of the conduction band minimum significantly decreases with N_{3m}. To understand this behavior, we plotted the probability density distributions of the lowest-subband electrons in MJGNRs for N_{3m} = (a) 5, (b) 3 and (c) 1 in Fig. 3, where the wave number k was taken to be 0.5 π/a. Electrons are confined in the wide GNR region for (a) N_{3m} = 5 as expected, while they are spread into the narrow GNR region for (c) N_{3m} = 1. This is due to quantum tunneling through such an extremely thin potential barrier. Next, Fig. 4 shows the probability density distributions of the lowest-subband and the first-order higher subband for $N_{3m} = N_{3m+2}$ = 5 at k = 0.5 π/a. It is worth noting that the higher-subband exhibits the maximum distribution inside the narrow GNR region, contrary to the lowest-subband. This antisymmetric profile between the lowest-subband and the first-order higher subband coincides with the behavior of semiconductor superlattices. Accordingly, a heterojunction for electrons is confirmed to be formed in the A-GNRs just by changing the transverse width along the length direction.

Furthermore, we examined the role of junction interface between metallic and semiconducting GNRs. Fig. 5 shows the atomic model employed for the purpose, where the two regions are directly connected and thus an unstable atom bonded only to one carbon atom exists at the interfaces as denoted by the red circles. On the other hand, the previous atomic model shown in Fig. 1(a) has a part of GNR configuration with $n = 7$ at the interfaces as shown by the shaded pattern, and all carbon atoms at the edges are bonded to two carbon atoms. Therefore, the interfaces are considered to be more stably constructed in Fig. 1(a). We computed the band structure and probability density distributions for the atomic model of Fig. 5, and clarified the role of the junction interface as presented below. First, Fig. 6 shows the band structure, which indicates that the band gap energy and the effective mass are significantly decreased, as compared with those of Fig. 2(a). This is understood by looking at the electron distributions shown in Fig. 7. Namely, the lowest-subband electrons are not sufficiently confined in the quantum well even for N_{3m} = 5, which results in the reduction of the band gap energy. Due to such an increased electron conductivity through the semiconducting GNR region, the effective mass also decreases as mentioned above. In addition, the first-order higher subband exhibits quite uniform electron distribution as shown in Fig. 7. The above results mean that an ideal electrical heterojunction is not formed with the interface model of Fig. 5.

In conclusion, we have demonstrated that a heterojunction for electrons can be formed in A-GNRs just by changing the transverse width along the length direction, which suggests that a band structure engineering within the atomistic scale is feasible only by using single element, carbon. However, we have pointed out that atomic structure at the junction interface plays a significant role in determining the electronic states in multi-junction A-GNRs.

Acknowledgements This work was supported by a Grant-in-Aid for Scientific Research from the Japan Society for the Promotion of Science (JSPS), and the Japan Science and Technology Agency (JST)/CREST.

References [1] H. Tsuchiya et al., *IEEE-TED* **57** (2010) 406. [2] R. Sako et al., *IEEE-TED* **58** (2011) 3300. [3] K.-T. Lam et al., *IEEE-EDL* **31** (2010) 555. [4] H. Min et al., *PRB* **75** (2007) 155115. [5] Y.-W. Son et al., *PRL* **97** (2006) 216803. [6] R. Sako et al., *IEEE-EDL* **32** (2011) 6.

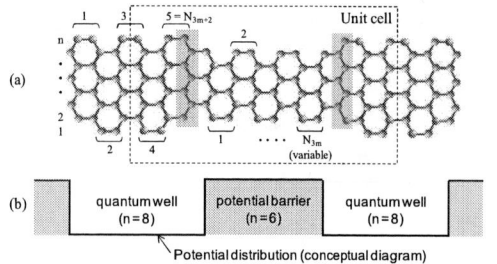

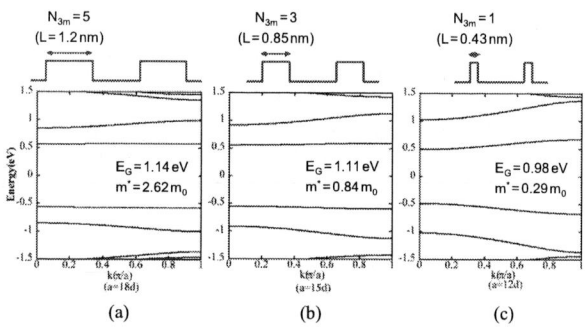

Fig. 1. (a) Atomic model for multi-junction GNR (MJGNR) with armchair-edged structure. Wide and narrow GNR regions consist of $n = 3m+2$ (=8; metallic) and $3m$ (=6; semiconducting) configurations, respectively. Therefore, a periodic potential distribution is considered to be formed as shown in (b). Numbers of six-membered rings in the length direction of wide and narrow GNRs are denoted as N_{3m+2} and N_{3m}, respectively. As for the shaded pattern in (a), refer to the caption in Fig. 5.

Fig. 2. Band structures of MJGNRs computed for various barrier lengths: N_{3m} = (a) 5, (b) 3 and (c) 1. Note that the length of quantum well region is fixed (N_{3m+2} = 5) and thus the band gap only slightly decreases with N_{3m}. In contrast, the effective mass of conduction band minimum significantly decreases with N_{3m}.

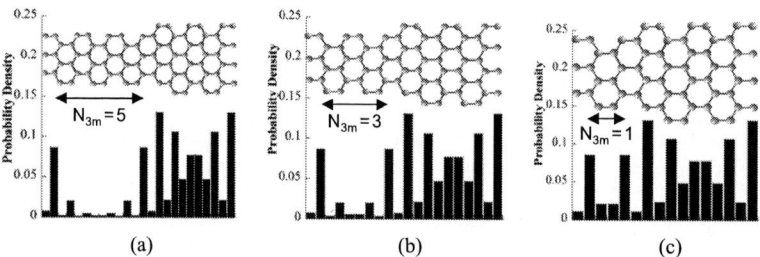

Fig. 3. Probability density distributions of the lowest-subband electrons in MJGNRs computed for N_{3m} = (a) 5, (b) 3 and (c) 1. Wave number k is taken to be 0.5 π/a.

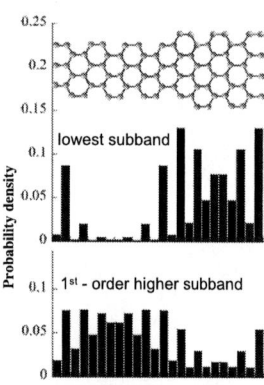

Fig. 4. Probability density distributions of the lowest-subband and the first-order higher subband for $N_{3m} = N_{3m+2} = 5$ at $k = 0.5$ π/a.

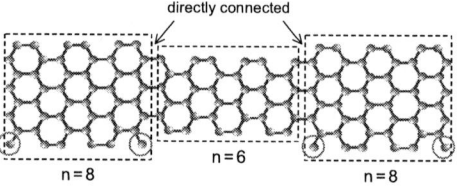

Fig. 5. Atomic model for MJGNR used to investigate the role of junction interface between metallic ($n = 8$) and semiconducting ($n = 6$) GNRs. Note that the two regions are directly connected in this model, and hence an unstable atom bonded only to one carbon atom exists at the interfaces as denoted by red circles. On the other hand, the previous atomic model (Fig. 1) has a part of GNR configuration with $n = 7$ at the interfaces as shown by shaded pattern, and the interfaces may be more stably constructed.

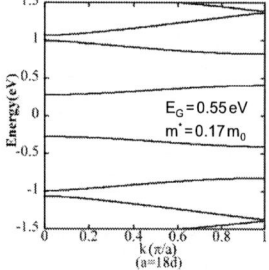

Fig. 6. Band structure of MJGNR with the directly-connected interfaces shown in Fig. 5, where $N_{3m} = N_{3m+2} = 5$.

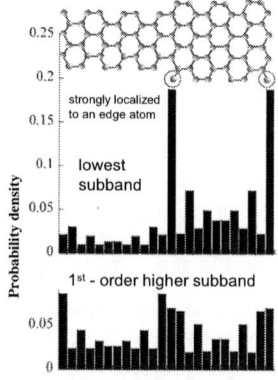

Fig. 7. Probability density distributions of the lowest-subband and the first-order higher subband of Fig. 6 at $k = 0.5$ π/a.

Graphene-Diamond-Silicon Devices with Increased Current-Carrying Capacity: sp^2-Carbon-sp^3-Carbon-on-Silicon Technology

Jie Yu[1], Guanxiong Liu[1], Anirudha V. Sumant[2] and Alexander A. Balandin[1]

[1]Nano-Device Laboratory, Department of Electrical Engineering and Materials Science and Engineering Program, Bourns College of Engineering, University of California – Riverside, Riverside, California 92521 USA

[2]Center for Nanoscale Materials, Argonne National Laboratory, Illinois, 60439 USA

E-mail: balandin@ee.ucr.edu ; web: http://ndl.ee.ucr.edu/

Graphene demonstrated potential for practical applications owing to its excellent electronic and thermal properties. Typical graphene field-effect transistors (FETs) and interconnects built on conventional SiO_2/Si substrates reveal the breakdown current density on the order of 10^8 A/cm^2, which is ~100× larger than the fundamental limit for the metals but still smaller than the maximum achieved in carbon nanotubes. It was discovered by some of us that graphene has excellent thermal conduction properties with the thermal conductivity K exceeding 2000 W/mK at room temperature [1]. Few-layer graphene largely preserves the heat conduction properties [2]. However, the thermally resistive SiO_2, with the thermal conductivity in the range from 0.5 to 1.4 W/mK, creates a bottleneck for heat removal. The latter does not allow graphene to demonstrate its true current-carrying potential. We show that by replacing SiO_2 with synthetic diamond one can substantially increase the current-carrying capacity of graphene to as high as ~ 20×10^8 A/cm^2 under ambient conditions. The two-terminal and three-terminal top-gated graphene devices (see Figure 1) were fabricated on synthetic single-crystal diamond (SCD) and ultrananocrystalline diamond (UNCD). To ensure Si integration, the UNCD layers were grown at low temperatures compatible with Si CMOS technology [3]. Our results indicate that graphene's current-induced breakdown is thermally activated. It was found that the current carrying capacity of graphene can be improved not only on SCD but also on an inexpensive UNCD. The latter was attributed to the decreased thermal resistance of UNCD at elevated temperatures (see Figure 2). The obtained results are important for graphene's hetero-integration on Si substrates. The enhanced current-carrying capacity is beneficial for the proposed applications of graphene in interconnects and high-frequency transistors.

The work in Balandin Group was supported, in part, by the Semiconductor Research Corporation (SRC) and Defense Advanced Research Project Agency (DARPA) through FCRP Center on Functional Engineered Nano Architectonics (FENA), U.S. National Science Foundation (NSF) and the U.S. Office of Naval Research (ONR). The use of the Center for Nanoscale Materials at ANL was supported by the U. S. Department of Energy under Contract DE-AC02-06CH11357.

[1] A. A. Balandin, Thermal properties of graphene and nanostructured carbon materials, *Nature Materials*, **10**, 569 (2011).

[2] S. Ghosh, D.L. Nika, S. Subrina, E.P. Pokatilov, C.N. Lau and A.A. Balandin, Dimensional crossover of thermal transport in few-layer graphene, *Nature Materials*, **9**, 555 (2010).

[3] J. Yu, G. Liu, A.V. Sumant, V. Goyal and A.A. Balandin, Graphene-on-diamond devices with increased current-carrying capacity, *Nano Letters*, **12**, 1603 (2012).

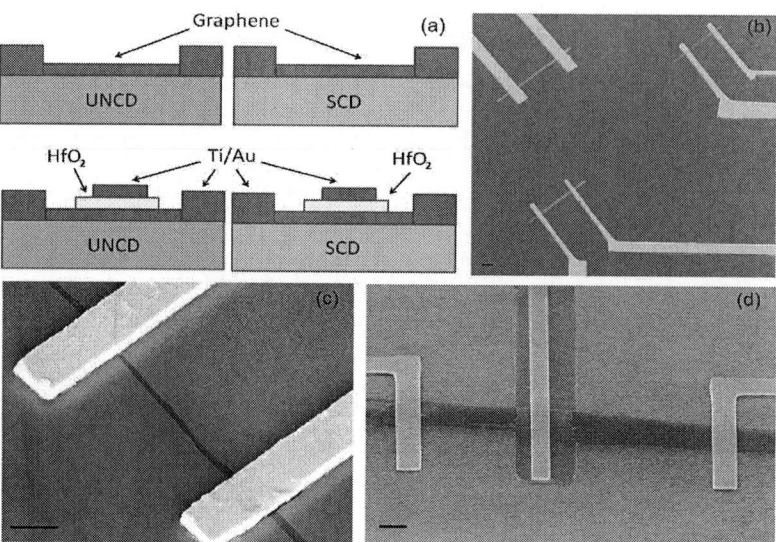

Figure 1: Schematic (a), optical (b) and scanning electron microscopy images (c-d) of graphene-on-diamond devices fabricated for testing of the current density.

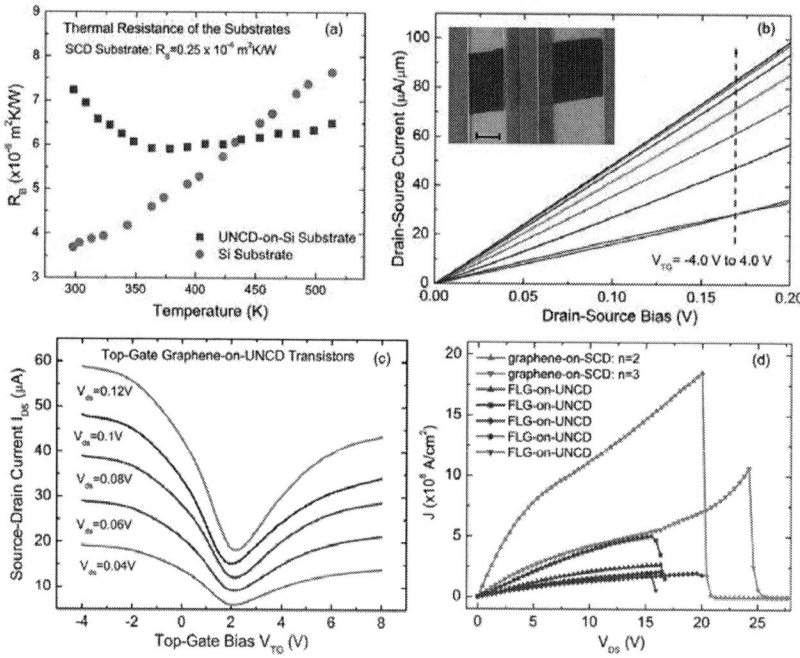

Figure 2: Thermal resistance of UNCD/Si substrate and reference Si wafer (a); current-voltage characteristics of top-gate graphene-on-SCD devices (b); source-drain current in the three-terminal graphene-on-UNCD devices as a function of the top-gate bias (c); breakdown current density in the two-terminal graphene-on-UNCD and graphene-on-SCD devices (d). Note an order of magnitude improvement in the current-carrying ability of graphene devices fabricated on synthetic diamond.

978-1-4673-0996-7/12 $31.00 © 2012 IEEE

Selective Gas Sensing with a *Single* Graphene-on-Silicon Transistor

A.A. Balandin[1], S. Rumyantsev[2,3], G. Liu,[1] M.S. Shur[2] and R.A. Potyrailo[4]

[1]Nano-Device Laboratory, Department of Electrical Engineering and Materials Science and Engineering Program, Bourns College of Engineering, University of California – Riverside, Riverside, California 92521 USA

[2]Center for Integrated Electronics and Department of Electrical, Computer and Systems Engineering, Rensselaer Polytechnic Institute, Troy, New York 12180 USA

[3]Ioffe Physical-Technical Institute, The Russian Academy of Sciences, St. Petersburg, Russia

[4]Chemistry and Chemical Engineering, GE Global Research, Niskayuna, NY 12309 USA

The low-frequency $1/f$ noise in graphene transistors has been studied extensively owing to the proposed graphene applications in analog devices and communication systems [1-5]. The studies were motivated by the fact that the low-frequency noise can be up-converted by device non-linearity and contribute to the phase noise of the system. Similarly, the sensor sensitivity is often limited by the electronic low-frequency noise. Therefore, noise is usually considered as one of the main limiting factors for the device or overall system operation. However, the electronic noise spectrum itself can be used as a sensing parameter increasing the sensor sensitivity and selectivity. Here, we show that vapors of different chemicals produce distinguishably different effects on the low-frequency noise spectra of the graphene-on-Si transistor. Our study showed that some gases change the electrical resistance of pristine graphene devices without changing their low-frequency noise spectra while other gases modify the noise spectra by inducing Lorentzian components with distinctive features. The characteristic corner frequency f_C of the Lorentzian noise bulges in graphene devices is different for different chemicals and varies from f_C=10 – 20 Hz for tetrahydrofuran to f_C=1300 – 1600 Hz for chloroform. We tested the selected set of chemicals vapors on different graphene device samples and alternated different vapors for the same samples. The obtained results indicate that $1/f$ noise in combination with other sensing parameters can allow one to achieve the *selective* gas sensing with a *single* pristine graphene transistor. Our method of gas sensing with graphene does not require graphene surface functionalization or fabrication of an array of the devices with each tuned to a certain chemical. The observation of the Lorentzian components in the vapor-exposed graphene can also help in developing an accurate theoretical description of the noise mechanism in graphene.

[1] Q. Shao, G. Liu, D. Teweldebrhan, A. A. Balandin, S. Rumyantsev, M. Shur and D. Yan, "Flicker noise in bilayer graphene transistors," IEEE Electron Device Letters, 30, 288 (2009).

[2] G. Liu, W. Stillman, S. Rumyantsev, Q. Shao, M. Shur and A.A. Balandin, "Low-frequency electronic noise in the double-gate single-layer graphene transistors," Applied Physics Letters, 95, 033103 (2009).

[3] S. Rumyantsev, G. Liu, W. Stillman, M. Shur and A.A. Balandin, "Electrical and noise characteristics of graphene field-effect transistors: Ambient effects and noise sources," J. Physics: Condensed Matter, 22, 395302 (2010).

[4] S.L. Rumyantsev, G. Liu, M. Shur and A.A. Balandin, "Observation of the "memory steps" in graphene at elevated temperatures," Applied Physics Letters, 98, 222107 (2011).

[5] G. Liu, S. Rumyantsev, M. Shur and A.A. Balandin, "Graphene thickness-graded transistors with reduced electronic noise," Applied Physics Letters, 100, 033103 (2012).

Figure 1: Scanning electron microscopy images of back-gated graphene devices.

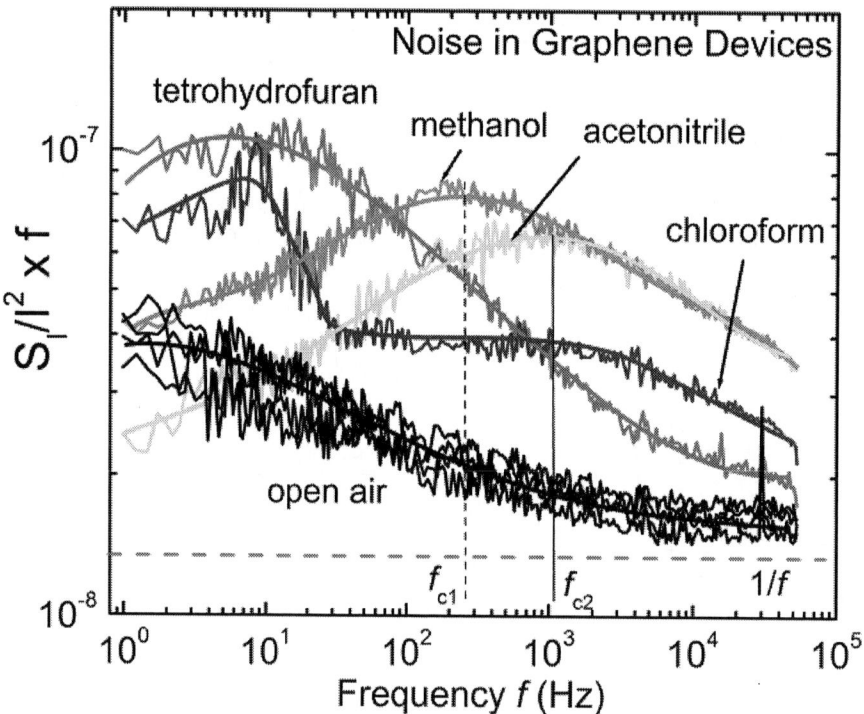

Figure 2: Noise spectral density S_I/I^2 multiplied by frequency f vs. f for the device in open air and under the influence of different vapors. Different vapors induce noise with different characteristic frequencies f_c. The frequencies, f_c, are shown explicitly for two different gases. The difference in the frequency f_c is sufficient for reliable identification of different gases with the same graphene transistor. For comparison the pure $1/f$ noise dependence is also indicated.

978-1-4673-0996-7/12 $31.00 © 2012 IEEE   114

Graphene Fillers for Ultra-Efficient Thermal Interface Materials

K. M. F. Shahil[1,2], V. Goyal[1,3], R. Gulotty[1] and A. A. Balandin[1,*]

[1]Nano-Device Laboratory, Department of Electrical Engineering and Materials Science and Engineering Program, University of California, Riverside, CA 92521 USA
[2]Intel Corporation, Hillsboro, Oregon 97124, USA
[3]Texas Instruments, Dallas, Texas 75243, USA
*corresponding author: balandin@ee.ucr.edu ; web: http://ndl.ee.ucr.edu

Continuous scaling of Si CMOS devices and circuits, increased speed and integration densities resulted in problems with thermal management of nanoscale device and computer chips [1]. Further progress in information, communication and energy storage technologies requires more efficient heat removal methods and stimulates the search for thermal interface material (TIMs) with enhanced thermal conductivity. The commonly used TIMs are filled with the particles such as silver or silica. The conventional TIMs require high volume fractions of the filler (~70%) to achieve thermal conductivity of ~1–5 W/mK. Recently, some of us discovered that graphene has extremely high intrinsic thermal conductivity, which exceeds that of carbon nanotubes [2-4]. To use this property for thermal management of nanoscale electronic devices, we utilized the inexpensive liquid-phase exfoliated graphene and multi-layer graphene (MLG) as filler materials in TIMs (Figure 1). The thermal properties of the obtained graphene-epoxy composites were measured using the "laser flash" technique (Figure 2). It was found that the thermal conductivity enhancement factor exceeded a factor of 23 at 10% of the graphene volume loading fraction [5]. This enhancement is larger than anything that has been achieved using other fillers. We have also tested graphene flakes in the electrically-conductive hybrid graphene-metal particle TIMs. The thermal conductivity of resulting composites was increased by a factor of ~5 in a temperature range from 300 K to 400 K at a small graphene loading fraction of 5-vol.-% [6]. The unusually strong enhancement of thermal properties was attributed to the high thermal conductivity of graphene, strong graphene coupling to matrix materials and the large range of the length-scale – from nanometers to micrometers – of the graphene and silver particle fillers. Graphene-based TIMs have a number of other advantages related to their viscosity and adhesion, which meet the industry requirements. Our results suggest that graphene can become excellent filler materials in the next generation of TIMs for the electronic, optoelectronic and photovoltaic solar cell applications.

The work in Balandin Group was supported, in part, by the Office of Naval Research (ONR), and by the Semiconductor Research Corporation (SRC) and Defense Advanced Research Project Agency (DARPA) through FCRP Center on Functional Engineered Nano Architectonics (FENA).

[1] A. A. Balandin, Chill Out: New Materials Can Keep Chips Cool, IEEE Spectrum, **29**, 35 (2009).
[2] A. A. Balandin, S. Ghosh, W. Bao, I. Calizo, D. Teweldebrhan, F. Miao & C. N. Lau, Superior Thermal Conductivity of Single-Layer Graphene, Nano Lett. **8**, 3, 902 (2008).
[3] S. Ghosh, W. Bao, D. L. Nika, S. Subrina, E. P. Pokatilov, C. N. Lau, & A. A. Balandin, Dimensional crossover of thermal transport in few-layer graphene, Nature Mat., **9**, 555 (2010).
[4] A. A. Balandin, Thermal properties of graphene and nanostructured carbon materials, Nature Mat., **10**, 569 (2011).
[5] K.M.F. Shahil and A.A. Balandin, Graphene-multilayer graphene nanocomposites as highly efficient thermal interface materials, Nano Lett., **12**, 861 (2012).
[6] V. Goyal and A.A. Balandin, Thermal properties of the hybrid graphene-metal nano-micro-composites: Applications in thermal interface materials, Appl. Phys. Lett., **100**, 073113 (2012).

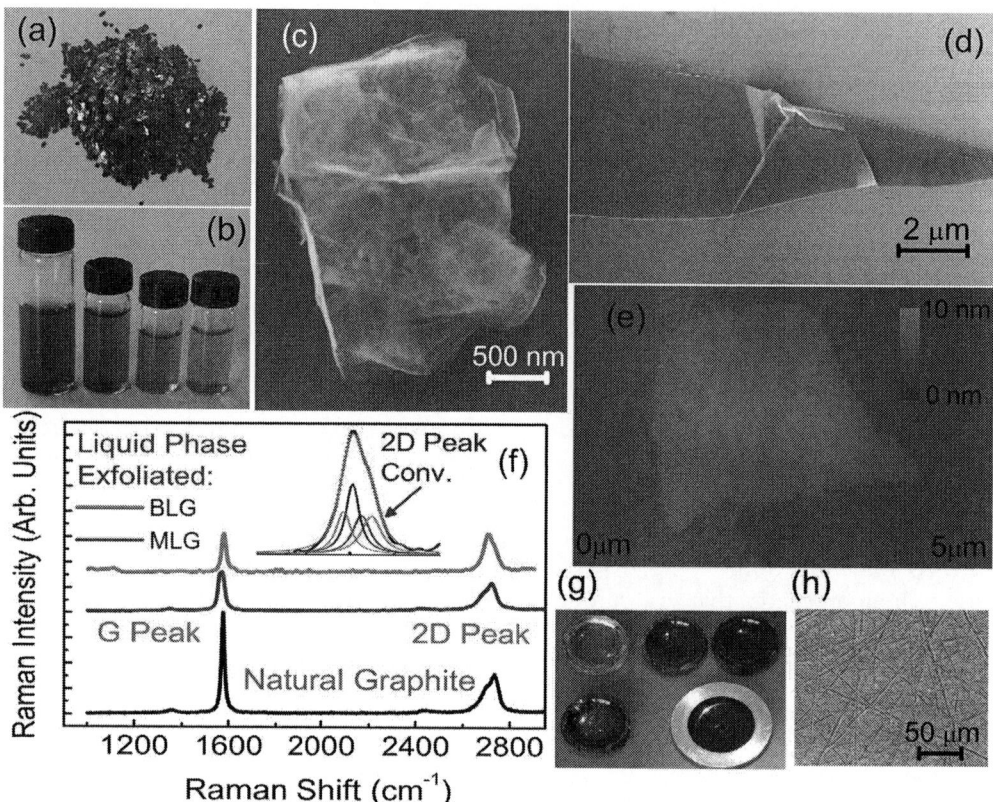

Figure 1: Synthesis and characterization of the graphene-based nanocomposite TIMs. (a) graphite source material; (b) liquid-phase exfoliated graphene in solution; (c) SEM image of graphene flake; (d) SEM image of a large multi-layer graphene (*n*<5) flake extracted from the solution; (e) AFM image of the flake with varying *n*; (f) Raman spectroscopy image of bilayer graphene flakes extracted from the solution; (g) optical image of graphene-based composite prepared for thermal measurements; (d) SEM image of the surface of the resulting graphene based TIMs indicating.

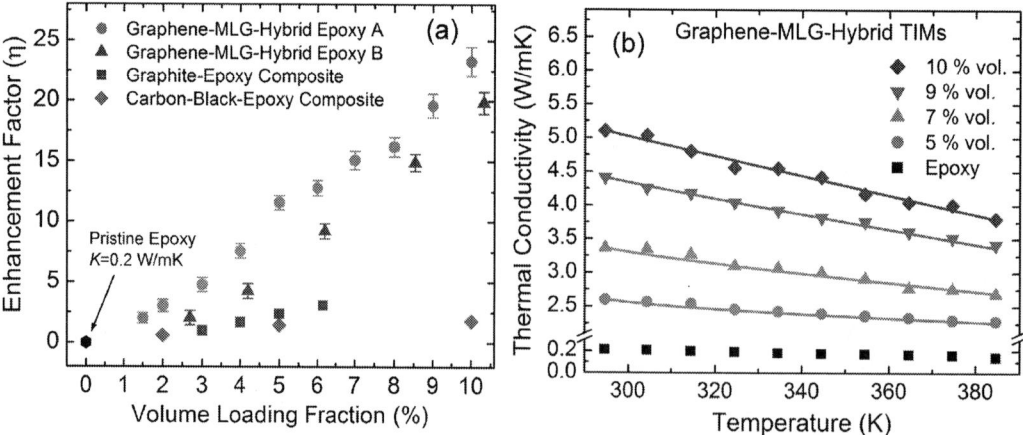

Figure 2: (a) Measured thermal conductivity enhancement factor as a fraction of the graphene filler volume loading fraction; (b) experimentally determined dependence of thermal conductivity of TIMs on temperature for different loading fractions.

978-1-4673-0996-7/12 $31.00 © 2012 IEEE

Silicon Microfabrication Technologies for THz applications

C. Jung-Kubiak, J. Gill, T. Reck, C. Lee , J. Siles, G. Chattopadhyay, R. Lin,
K. Cooper and I. Mehdi

Jet Propulsion Laboratory, California of Technology
4800 Oak Grove Drive, Pasadena, CA 91109

Abstract: Silicon micromachining technology is naturally suited for making THz components, where precision and accuracy are essentials. We report here the development of robust micromachining techniques to enable novel active and passive components in the submillimeter-wave region. These features will enable large format submillimeter-wave heterodyne arrays and 3-D integration in the THz region, where fabricating circuits and structures becomes difficult with conventional machining.

Submillimeter heterodyne instruments are important for a number of applications, from providing quantitative molecular abundance profiles in atmospheres[1] to detection of contra-band and IEDs[2]. To make these instruments compatible on small platforms for the study of the outer planets, it is essential to make them low-mass and low-volume. One way to make highly integrated and compact receiver front-end is to make the waveguide in silicon where the power amplifiers, multipliers, and mixer chips can be integrated in a single silicon micromachined block. By using silicon rather than conventional precision machining tools, we have reduced the volume of the whole receiver by an order of 50, and the mass and cost are divided by a factor greater than 10. Previous studies show that losses associated with waveguides made in silicon are consistent with theoretical predictions and are comparable to the losses in metal waveguides [3].

Using silicon fabrication techniques for THz applications require having very smooth border/bottom surfaces inside the waveguides as well as precision in alignment and patterning. Optimization of the DRIE process is necessary to respect these requirements in order to minimize ohmic losses and to assure good impedance matching across vertical wafer-to-wafer waveguide transitions.

Silicon wafers are processed with a combination of conventional UV lithography and Deep Reactive Ion Etching (DRIE) techniques using thick AZ9260 resist as etching mask. They are initially covered with a thick thermal oxide layer on both sides, which will be used as an additional mask during the DRIE etching step. The DRIE technique used is the well-known Bosch process based on the alternative exposures to SF6 and C4F8 gases. With optimized plasma power and etching gas ratios, we have achieved a 90° etched vertical angle and an 18nm etched surface roughness (figure 1) [4].

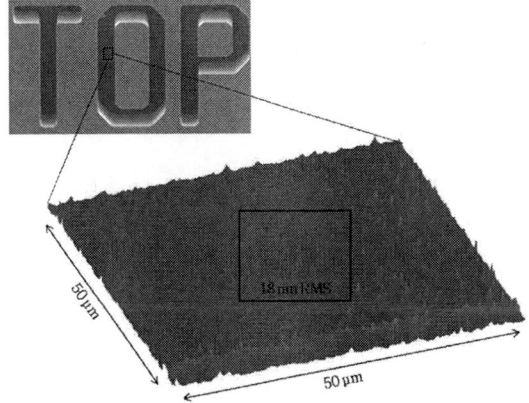

Fig. 1: AFM measurement of silicon micromachined waveguides indicates a 18 nm rms surface roughness.

We are using these micromachining techniques to design, fabricate, and test a novel architecture called Radiometer-On-A-Chip (ROC), that allows for the 3-D integration of the power amplifier, multipliers and mixers in an extremely compact package (figure 2).
Preliminary results at 560GHz give a DSB mixer noise temperature of 4860 K and DSB mixer conversion losses of 12.15 dB at 542 GHz [5].

Fig. 2: Optic picture of the silicon-based 560GHz ROC, including the amplifier stage the fixture size is 25x20x10mm.

All these silicon-based methodologies are very suitable for high frequency circuits and we recently demonstrated a 2.55 THz waveguide HEB mixer block using this approach (figure 3), with a DSB receiver noise temperature of T_{rec}^{DSB} of 2000 ± 100 K (Y-factor of 1.09 ± 0.005) [6].

978-1-4673-0996-7/12 $31.00 © 2012 IEEE

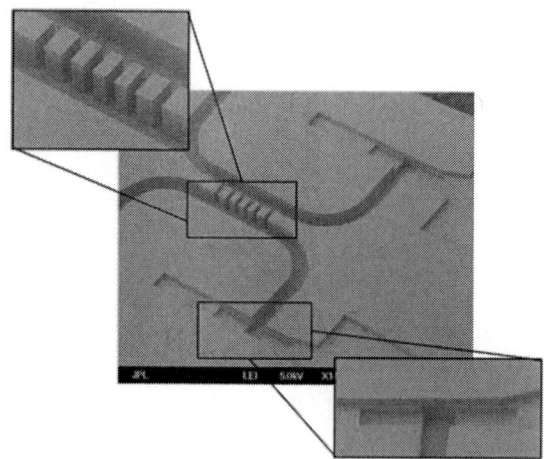

Fig. 3: SEM pictures of 2.55THz silicon-fabricated parts using DRIE techniques.

In addition to these active parts, we are fabricating passive components such as hybrids, twists and OMTs (figure 4) for an integrated radiometer and spectrometer for improved planetary science (PASEO). These parts are currently being tested and results will be published soon [7].

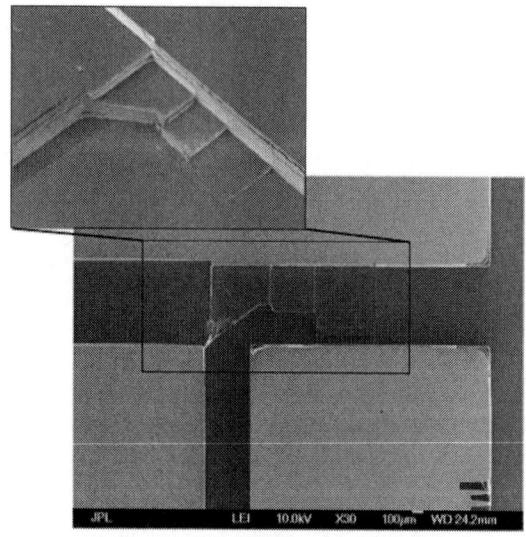

Fig 4: SEM pictures of a silicon-based OMT for a wideband radiometer and spectrometer operating at 520-600 GHz.

ACKNOWLEDGMENT

The research was carried out by the Jet Propulsion Laboratory, California Institute of Technology, under a contract with the National Aeronautics and Space Administration.
Copyright 2012 California Institute of Technology, government sponsorship acknowledged.

REFERENCES

[1] P. Hartogh, *Sub-millimeter Wave Instrument for EJSM*, EJSM Instrument Workshop, MD, USA, Jul. 2009.

[2] P.H. Siegel, *THz Technology*, IEEE Trans. Microwave Theory Tech., 50th Anniversary Issue, vol. 50, no. 3, pp. 910-928, Mar. 2002.

[3] G. Chattopadhyay, J. Ward, H. Manohara, R. Toda, and R. Lin, *Silicon Micromachined Components at Terahertz Frequencies for Astrophysics and Planetary Applications*, International Space Symposium on THz Technologies, Groningen, Netherlands, Apr. 2008.

[4] C. Jung, B Thomas, C. Lee, A. Peralta, G. Chattopadhyay, J. Gill, R. Lin, E. Schlecht, K. Cooper and I. Mehdi, *Compact Submillimeter-wave Receivers made with Semiconductor Nano-Fabrication Technologies*, IEEE-MTT International Microwave Symposium, Baltimore, MD, USA, Jun. 2011.

[5] B. Thomas, C. Lee, A. Peralta, J. Gill, G. Chattopadhyay, S. Sin, R. Lin and I. Mehdi, *600GHz Silicon-based integrated receiver using GaAs MMIC membrane planar Schottky diodes*, International Space Symposium on THz Technologies, Oxford, UK, Mar. 2010.

[6] F. Boussaha, J. Kawamura, J. Stern, C. Jung-Kubiak, A. Skalare and V. White, *2.55 THz Silicon Micro-Machined Waveguide HEB Mixer*, to be published in IEEE Microwave and Wireless Components Letters.

[7] T. Reck, C. Jung, J. Siles, B. Thomas, R. Lin, J. Ward, I. Mehdi and G. Chattopadhyay, *PASEO – An integrated Radiometer and Spectrometer for Improved Planetary Science*, International Space Symposium on THz Technologies, Tokyo, Japan, Apr. 2012.

Contact:
Dr. Cecile Jung-Kubiak
Section 389 - RF Microwave Engineer
JET PROPULSION LABORATORY
4800 Oak Grove Drive - M/S T1721
PASADENA, CA 91109
Phone: (818) 354-1658
Fax: (818) 393-4683
Mail: Cecile.D.Jung@jpl.nasa.gov

Simulation Study on Process Conditions for High-Speed Silicon Photodetector and Quantum-Well Structuring for Increased Number of Wavelength Discriminations

Seongjae Cho[1], Hyungjin Kim[2], Min-Chul Sun[2], Theodore I. Kamins[1], Byung-Gook Park[2], and James S. Harris, Jr[1].

[1]Department of Electrical Engineering, Stanford University, USA

[2]Inter-university Semiconductor Research Center (ISRC) and Department of Electrical Engineering and Computer Science, Seoul National University, Republic of Korea

Email: felixcho@stanford.edu

Abstract — **In this work, process conditions and geometric parameters for high-speed *p-i-n* silicon photodetector are optimized by device simulation. Efforts were made to build up criteria for device fabrication based on silicon epitaxy. For an optimized silicon photodetector, a bandwidth as wide as 80 GHz was obtained at 1 V. Furthermore, a way of increasing wavelength discriminations by introducing silicon-germanium quantum wells for multiple-wavelength signal processing is exploited.**

I. INTRODUCTION

Recently, optical interconnect as a next-generation interconnect technology as well as a collective building blocks for optical signal processing (Fig. 1), is under extensive researches [1-3]. Silicon (Si) photonics is strongly pursued by virtues of cost-effectiveness and compatibility with complementary metal-oxide-semiconductor (CMOS) process. In this work, a *p-i-n* Si photodetector (PD) is optimally designed for high-speed operation aiming 100-GHz bandwidth (BW) with considerations for epitaxy-based process by device simulation [4]. The effect of silicon-germanium (SiGe) quantum wells (QWs) in wavelength modulation is also investigated.

II. DEVICE STRUCTURE AND METHODS

Fig. 2 shows the simulated two-dimensional (2-D) plane of the PD vertically coupled with Si waveguide. Normally incident beam with a power of 1 W/cm^2 was used for direct-current (DC) characterization and small signal with an amplitude of 2 mW/cm^2 was mounted for evaluating the high-frequency (HF) performances. Total thickness of the PD active layer, anode junction depth (X_{aj}), and cathode junction depth (X_{cj}) were the design variables. Multiple models related with recombination, mobility, tunneling, and quantum effect were equipped in the simulation for higher accuracy.

III. SIMULATION RESULTS

Fig. 3 shows the effect of anode thickness (X_{aj} + doping gradient length) on photogeneration. For thicker anode, photocurrent increased due to a larger number of optically generated electrons in the thickened layer, but the amount was not significant. Fig. 4 shows the effect of doping gradient length (distance from peak to reference concentration, 10^{12} cm^{-3}) on cathode current (I_C). For a device with abrupt junctions, great deal of band-to-band tunneling leakage was observed but it was prominently suppressed by even a very short gradient. I_C slightly decreased as the gradient was widened due to increased resistance. The 3-dB roll-off frequency (f_{3dB}), or equivalently bandwidth (BW) (PD shows the frequency response of a low-pass filter (LPF)), increased monotonically as the intrinsic layer got thinner, as shown in Fig. 5. BW from an optimized Si PD with anode layer thickness of 50 nm, intrinsic layer thickness of 200 nm, and doping gradient lengths of 20 nm, for a given width of 5 μm, was 80.1 GHz at an operating voltage of 1 V. The minimum thickness of intrinsic region to achieve a BW wider than 10 GHz was 1.5 μm. For a wavelength within the sensible range, 1 μm, the quantum efficiency was calculated to be 0.81. Fig. 6 shows the effect of introducing multiple SiGe QWs into the intrinsic Si region. Wavelength-discrimination boundaries are made at wavelengths where $dI_C/d\lambda$ is large (steep photogeneration slope) and a certain wavelength for processing can be detected in a region with small $dI_C/d\lambda$. With no QW, there are three regions separated by two discriminating points. However, introducing SiGe QW opens an additional window of which wavelength can be detected with distinction.

IV. CONCLUSIONS

In this work, processing and geometric parameters have been optimized for high-speed and low-power *p-i-n* Si PD. Moreover, it was confirmed that SiGe QWs had an effect of generating a new detection window.

ACKNOWLEDGEMENTS

This work was supported by the Smart IT Convergence System Research Center funded by the Korean Ministry of Education, Science and Technology as Global Frontier Project. Dr. S. Cho is supported by the National Research Foundation of Korea Grant funded by the Korean Government [NRF-2011-357-D00155].

REFERENCES

[1] G. T. Reed and A. P. Knights, *Silicon Photonics*, John Wiley & Sons, Ltd., 2005, pp. 132-149.
[2] Intel's official website for photonics research, http://techresearch.intel.com/ResearchAreaDetails.aspx?Id=26.
[3] Y. Kang, et al., "Monolithic germanium/silicon avalanche photodiodes with 340 GHz gain-bandwidth product," *Nature Photonics*, vol. 3, pp. 59-63, Dec. 2008.
[4] *ATLAS User's Manual*, SILVACO International, Oct. 2011.

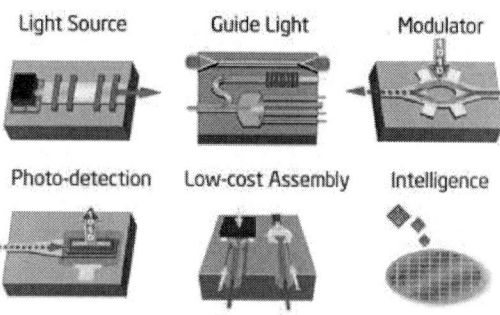

Fig. 1. Building blocks for optical interconnect [2].

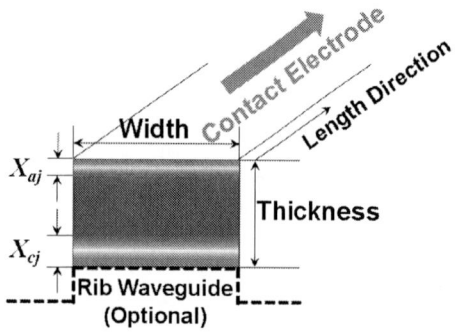

Fig. 2. Si PD vertically coupled with Si waveguide. Colored region indicates the simulated 2-D domain.

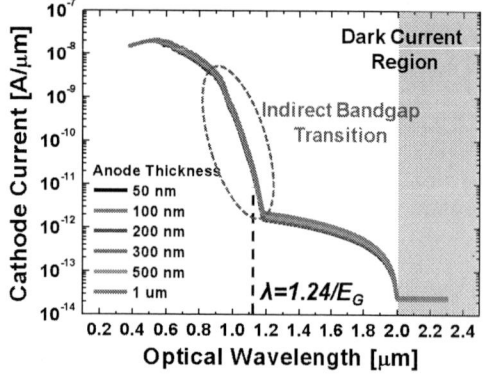

Fig. 3. Cathode current (I_C) vs. optical wavelength (λ) at different anode thicknesses (gradient length = 100 nm).

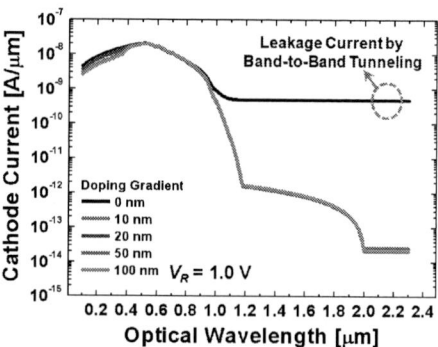

Fig. 4. I_C-λ curves at different doping gradient lengths (Si active thickness = 1 μm and operating voltage = 1 V).

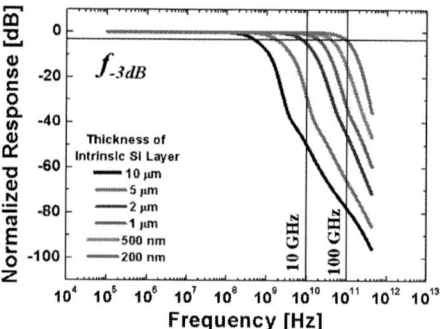

Fig. 5. Normalized frequency response of the Si PD.

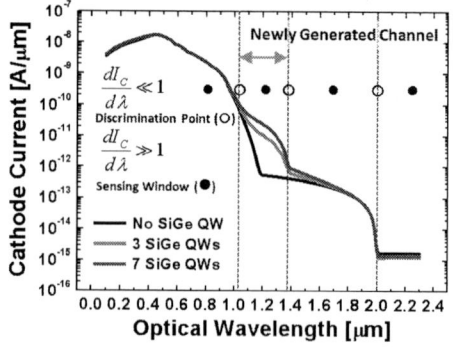

Fig. 6. Opening a new channel by introducing SiGe multiple QWs (thickness = 5 nm, Si active = 500 nm).

Nano-Transfer Printing of Functioning MIM Tunnel Diodes

Mario Bareiß[1], Benedikt Weiler[1], Daniel Kälblein[2], Ute Zschieschang[2], Hagen Klauk[2], Giuseppe Scarpa[1],
Bernhard Fabel[1], Paolo Lugli[1], and Wolfgang Porod[3]

[1]Institute for Nanoelectronics, Technische Universität München, Theresienstraße 90, 80333 München, Germany
[2]Max Planck Institute for Solid State Research, Heisenbergstr. 1, 70569 Stuttgart, Germany
[3]Department of Electrical Engineering University of Notre Dame, IN 46556 USA
Email: porod@nd.edu

Abstract — Nano diodes show great potential for applications in detectors, communications and energy harvesting. In this work, we focus on nano transfer printing (nTP) to fabricate nm-scale diodes over extensive areas. Using a temperature-enhanced process, several millions of diodes were transfer-printed in one single step. We show the reliable transfer of functioning MIM diodes, which were electrically characterized by conductive Atomic Force Microscopy (c-AFM) measurements. Quantum-mechanical tunneling was determined to be the main conduction mechanism across the metal-oxide-metal junction.

I. INTRODUCTION

Nano diodes are important devices for various electronic and optoelectronic applications, such as rectifiers for energy harvesting [1, 2] or infrared detection [3], field-emission cathodes [4], and switching memories [5]. Key challenges for large-scale manufacturing include increasing fabrication reliability and optimizing throughput, while minimizing cost.

In this work, we concentrate on nano transfer printing (nTP) as a scalable, purely mechanical, fabrication technique to manufacture semiconductor devices suitable for various applications[6]. More specifically, we improved conventional protocols for nTP by a temperature-enhanced process step to transfer highly-ordered large-scale arrays of Au and metal-oxide-metal nanostructures on flat substrates, such as p-type Si and SiO2[7]. Scanning electron microscope (SEM) images showed that the transferred devices were structurally intact after transfer-printing, with a transfer yield of almost 90%, we demonstrated that the transferred structures were functional diodes using electrical characterization by c-AFM measurements.

II. FABRICATION

The stamps consist of flat Si wafers structured with nm-scale pillars with height of 80 nm. To provide a hydrophobic anti-sticking surface on the stamps, a self-assembled-monolayer (SAM) was applied on the stamps' surface prior to metal evaporation. The subsequent evaporation of the MIM stack was accomplished using either thermal vapor evaporation or e-beam evaporation. Since Au or AuPd are noble metals and thus provide weak adhesion to any surfaces, the first layer deposited on the stamp consisted of one of these metals. To promote a good adhesion of the metal stack from the stamp onto the target surface, a 4 nm thick Ti layer was evaporated as the last layer on the metal stacks on the stamp. Before transferring, the hydrophilicity of both the metal stack on the stamp and the substrate Si/SiO2 surface was enhanced by a brief plasma treatment. In the subsequent transfer step, stamp and substrate were placed on top of each other and inserted into a NIL 2.5 Nanoimprinter (Obducat). A pressure of 50 bar for 5 min was applied by the machine. Performing this transfer step at a temperature of 200 °C repetitively provided a transfer yield of about 90% of the pillar structures on the stamp.

II. ELECTRICAL CHARACTERIZATION

The IV characteristic of the MIM device comprising the oxide layer fabricated by the oxygen plasma was measured by conductive MFP-3D atomic force microscopy (Asylum Research). The sample was clamped via the conductive substrate to a gold electrode, which was also connected to the cantilever holder. In this way, a closed electrical circuit was built (see Figure 2). The Si tips of the conductive AFM setup were coated with a layer of Ti/Pt (5/20) in order to provide physical strength and low resistivity. In AC tapping mode, the topography of the sample was measured. Then, after a target MIM diode structure was identified, the AFM cantilever tip was brought into direct contact with the surface. After dithering the tip on top of the MIM surface, it was held motionless as an electrical bias was applied through a cyclic, triangle wave pattern to the sample while the current was measured at the tip. The resulting I-V-characteristic is shown in Figure 3. In the voltage range between 0 V and 5 V, the slope for previously fabricated micron-scale MIM diodes are shown [6]. In the range from 5 V to 10 V, the slope for the nano scale diodes is shown. Since they match perfectly, the reliability of this fabrication method is proven. Further, the regime when direct tunneling and Fowler-Nordheim tunneling occurs were determined, as well as static device parameters were extracted, which will be presented.

978-1-4673-0996-7/12 $31.00 © 2012 IEEE

ACKNOWLEDGMENT

The authors acknowledge financial support from the German Research Funding (DFG), the International Graduate School of Science and Engineering (IGSSE) and the Institute for Advanced Studies (IAS), Focus group "Nanoimprint and Nanotransfer," and the German Excellence Cluster 'Nanosystems Initiative Munich' (NIM). The authors thank Dr. Edward M. Nelson and Prof. Gregory Timp for technical support concerning c-AFM measurements.

REFERENCES

[1] L. Novotny and N. van Hulst, "Antennas for light," *Nature Photon.,* vol. 5, pp. 83-90, 2011.

[2] S. Grover and G. Moddel, "Engineering the current–voltage characteristics of metal–insulator–metal diodes using double-insulator tunnel barriers," *Solid-State Electron.,* vol. 67, pp. 94-99, 2012.

[3] J. A. Gómez-Pedrero, J. Ginn, J. Alda, and G. Boreman, "Modulation transfer function for infrared reflectarrays," *Appl. Opt.,* vol. 50, pp. 5344-5350, 2011.

[4] L. Hongzhong, C. Bangdao, L. Xin, L. Weihua, D. Yucheng, and L. Bingheng, "A metal/insulator/metal field-emission cannon," *Nanotechnol.,* vol. 22, p. 455302, 2011.

[5] R. Waser and M. Aono, "Nanoionics-based resistive switching memories," *Nature Mater.,* vol. 6, pp. 833-840, 2007.

[6] M. Bareiß, F. Ante, D. Kälblein, G. Jegert, C. Jirauschek, G. Scarpa, B. Fabel, E. M. Nelson, G. Timp, U. Zschieschang, H. Klauk, W. Porod, and P. Lugli, "High-Yield Transfer Printing of Metal–Insulator–Metal Nanodiodes," *ACS Nano,* vol. 6, pp. 2853-2859, 2012/03/27 2012.

[7] M. Bareiß, M. A. Imtaar, B. Fabel, G. Scarpa, and P. Lugli, "Temperature Enhanced Large Area Nano Transfer Printing on Si/SiO2 Substrates Using Si Wafer Stamps," *J. Adhes.,* vol. 87, pp. 893-901, 2011.

Fig. 1 (a) AFM measurements of MIM nano diodes in (b) c-AFM measurements in contact mode.

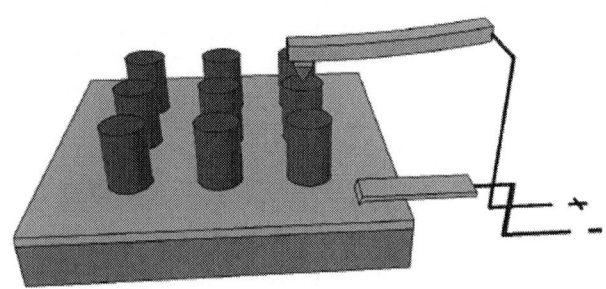

Fig. 2: Schematic of c-AFM setup

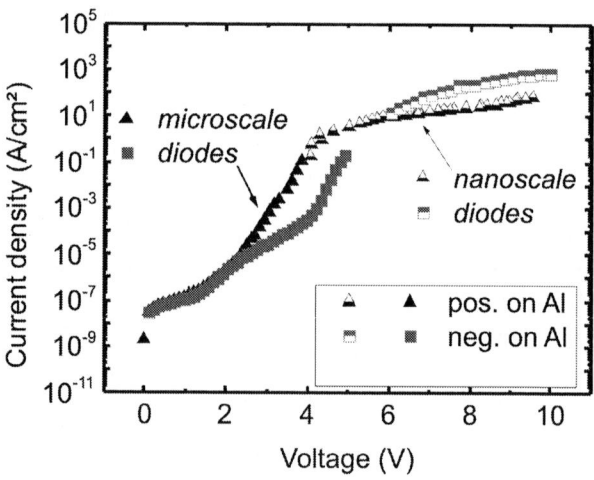

Fig. 3: I-V-characteristics of transfer-printed microscale [6] and nanoscale MIM diodes of this work.

Fabrication and evaluation of heavily P-doped Si quantum dot and back-gate induced Si quantum dot

J.Kamioka[1], T.Kodera[1,2,3], K.Horibe[1], Y.Kawano[1] and S.Oda[1]

[1] Quantum Nanoelectronics Research Center, Tokyo Institute of Technology, 2-12-1, Ookayama, Meguro-ku, Tokyo, Japan
[2] Institute for Nano Quantum Information Electronics, the University of Tokyo, 4-6-1, Komaba, Meguro-ku, Tokyo, Japan
[3] PRESTO, Japan Science and Technology Agency (JST), 4-1-8 Honcho Kawaguchi, Saitama, 332-0012, Japan
Email: kamioka.j.aa@m.titech.ac.jp

Abstract — We realized lithographically-defined electrically-tunable silicon quantum dot (QD) and charge sensor. Two types of device were fabricated and measured. One is heavily P-doped, and the other is back gate (BG)-induced undoped QD device. I-V characteristic of QD and charge sensor was clearly observed in both devices. Then, we estimate capacitance between the charge sensor and QD or two side gates (SGs) from the measurement and the simulation, and compared two devices in terms of their charging energy.

I. INTRODUCTION

The electron spin in Si quantum dot have a high potential for quantum information process. Spin-related tunneling phenomena are successfully observed in lithographically-defined Si double QD device in previous studies. Top-gate (TG) had been used to induce electron in the QD, but in case of TG device, it is difficult to attach magnetic structure just above the quantum dot in order to obtain electron spin resonance. In case of heavily P-doped Si QD device, attaching magnetic structure is possible in simple structure (schematically shown in Fig1. (C)). However, P-doped QD device has inevitable fluctuation of dopant. Here we improved device fabrication techniques, and made P-doped QD with less fluctuation. Also, we introduced BG-induced QD device. This device had a problem of less effectivity because of the micro-order thickness of Si substrate of Silicon-on-insulator (SOI). To solve this problem, we employed ion-implantation process to the Si substrate to make resistance lower. Then, we achieved charge sensing of both heavily P-doped and BG-induced undoped QD devices.

II. DEVICE FABRICATION

Both devices are defined by using electron beam lithography (EBL), as schematically shown in Fig. 1. (a). The same pattern is formed by Cl_2 dry-etching on undoped (Thickness of 20 nm) and heavily P-doped (Thickness of 45 nm) SOI wafer, with 100 nm buried oxide (BOX). Dose amount of heavily P-doped SOI is 1×10^{15} /cm^2. Fig. 1. (b) is the image of the real P-

doped QD device taken by scanning electron microscope (SEM). The lower QD is the charge sensor of the upper QD. QDs are connected to source and drain, and two SGs besides each dot. The BG voltage is applied directly to the Si substrate.

III. MEASUREMENT AND DISCUSSION

In order to estimate the each capacitance and charging energies, we observed Coulomb diamond [1]. Fig. 2. shows the contour plots of I_{CS} measured in P-doped device as a function of the voltages applied to the SG2, V_{sg2}, and the drain voltage of the charge sensor, V_{d2}, at 4.5 K. In a diamond regime shown as dashed line, source-drain current I_{CS} is suppressed by Coulomb blockade. Fig. 3(a) and (b) shows the contour plots of I_{CS} measured in P-doped device and BG-induced device respectively. In both results, Coulomb peak current shows parallel lines with some shifts shown in red dashed lines. The shifts of Coulomb peak lines are due to the changes in the number of electrons confined in upper QD. Change in the number of electron cause the change of electrical field, then I-V characteristic of the charge sensor shifts. These parallel Coulomb peak and their shifts shown in parallel red dashed lines are the signature that the both QDs are capacitively coupled to two SGs. In both devices, we successfully achieved clear charge sensing. It is notable that in BG-induced QD device, charge sensing was observed where BG bias, V_{bg}, was 6.8 V, low enough that there was almost no leakage current. From these result, we roughly estimate capacitances of the equivalent circuit shown in Fig. 4(a). For confirmation, we simulated charge sensing with the estimated capacitances [2]. The result is shown in Fig. 4(b) and (c), for P-doped and BG-induced devices respectively. The experimental result (Fig. 3) and simulation (Fig. 4) seems to be quite similar in each device. It ensures that the estimation is reasonable. Then we estimated charging energy of the charge sensor, 8 meV for P-doped device and 9 meV for BG-induced device. From SEM image, size of the charge sensor of the P-doped device was a bit smaller. In this point of view, charging energy of P-doped charge sensor should be larger. The reason of this discrepancy is supposed to be the difference of effective dot thickness. In BG-induced device, thickness of the QD should be around 10 nm and not correspond to SOI thickness, because electron in QD is induced only by field effect. P-doped QD should have effective thickness same as the SOI due to the existence of dopant. This should cause larger charging energy of BG-induced charge sensor.

SUMMARY

We fabricated heavily P-doped Si QD devices and BG-induced undoped Si QD devices defined by lithography. In P-doped QD devices, we suppressed the fluctuation of dopant by fabricating smaller QD. In BG-induced devices, low-effectiveness of BG was improved by applying ion-implantation to the Si substrate. Finally, we succeed in charge sensing of the QD in both devices, and larger charging energy was estimated in BG-induced device by experiment and simulation.

ACKNOWLEDGEMENT

This work was financially supported by JSPS KAKENHI (22246040), JST-PRESTO, Yazaki Memorial Foundation for Science and Technology, and Project for Developing Innovation Systems of the Ministry of Education, Culture, Sports, Science and Technology (MEXT).

REFERENCES

[1] L P Kouwenhoven, D G Austing and S Tarucha. *Rep. Prog. Phys.* **64** (2001) 701–736

[2] H. Grabert and M. H. Devoret. *Single Charge Tunneling*. Plenum Press, 1991.

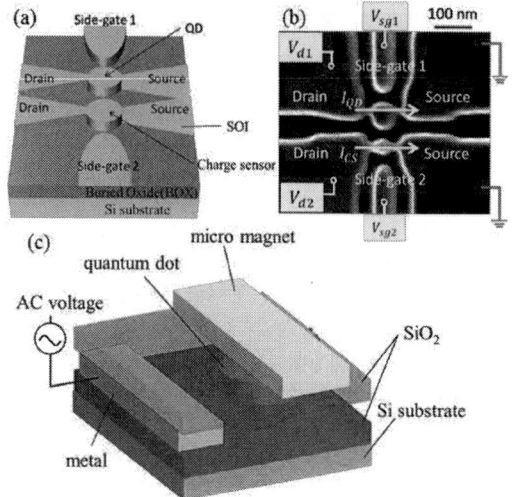

Fig. 1. (a) Schematic image of the QD device. Two side gates (SG1 and SG2), sources, and drains are made of P-doped or undoped SOI. (b) Scanning electron

microscope (SEM) image of quantum dot device. (c) Schematic image of the device integrated with a micro magnet. The micro magnet just above the QD induce position-dependent magnetic field gradient.

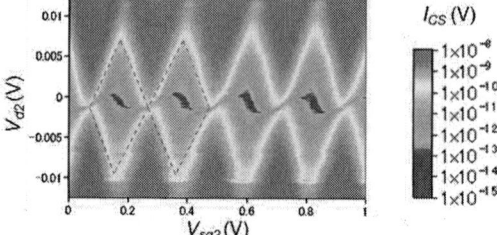

Fig. 2. Coulomb diamond, observed in the charge sensor current I_{CS} of P-doped device.

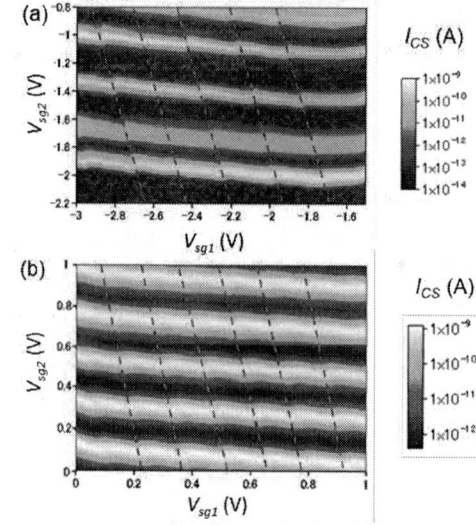

Fig. 3. Contour plot of I_{CS} (a) Observed in P-doped device, where V_{d1} =1 mV, V_{d2} =1 mV, V_{bg} =4.5 V, at 4.5 K (b) Observed in BG-induced device, where V_{d1} =0.5 mV, V_{d2} =1 mV, V_{bg} =6.8 V, at 4.2 K.

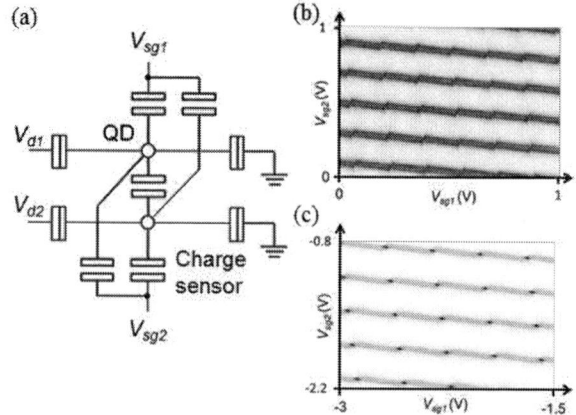

Fig. 4 (a) The equivalent circuit of the device. (b), (c) Simulated current counter for P-doped and BG-induced device, respectively. Current flows in the dark regime.

Microwave manipulation of electrons in silicon quantum dots

T. Ferrus[1], A. Rossi[1], T. Kodera[2,3,4], T. Kambara[2], W. Lin[2], S. Oda[2] and D. A. Williams[1]

[1]Hitachi Cambridge Laboratory, J. J. Thomson Avenue, CB3 0HE, Cambridge, United Kingdom
[2]Quantum Nanoelectronics Research Centre, Tokyo Institute of Technology, 2-12-1
Ookayama, Meguro-ku, Tokyo, 152-8552 Japan
[3]Institute for Nano Quantum Information Electronics, University of Tokyo, 4-6-1, Komaba, Meguro, Tokyo,
Japan
[4]PRESTO, Japan Science and Technology Agency (JST), Kawaguchi, Saitama 332-0012, Japan

In quantum computation, the choice for the qubit implementation drives the method to be used to manipulate qubits and read out the computed state. For this, either electro-magnetic pulses or photons can be used. In semiconductors, qubit states in GaAs may be manipulated using a combination of surface acoustic waves in the GHz range while applying radiofrequency pulses [1] or static magnetic fields [2] to access the spin states but exciton states in InAs/GaAs quantum dots necessitate photons of energy about 1.3 eV to be manipulated. Optical access is also possible for accessing Rydberg states of implanted atoms in silicon as well as for realizing quantum operations [3]. However, in the case of the Kane-related proposals, electrical pulses are generally preferred and privileged.

In our approach technical choices have oriented research towards the use of an isolated double quantum dot structure (IDQD), with the aim of minimizing the number of connected leads that may carry unwanted noise. In this system, two possible charge qubits states could be implemented depending on the nature of the coupling between the states in each individual dot: a molecular state where electrons of each dot are coupled and formed a bonding/anti bonding state or localized states where an excess of charge in one of the dot determines the state of the qubit. In both cases the energy necessary to do a swap operation is of the order of few meV so that, only THz photons are expected to be able to operate the qubit. To this end, most previous investigations have been carried out using electrical pulses on the gates [4].

However, recent results have indicated the possibility of inducing spatial Rabi oscillation with microwave photons [5].

Here we present the results of an investigation on microwave-induced effects that we have observed in silicon devices, including phosphorous doped and Metal-Oxide-Semiconductor Single Electron Transistors (SET) as well as IDQD [6]. Continuous pulsed microwave and single shot measurements are used to demonstrate that photons in the range of 10-15 GHz allow manipulation of the electron number in the island of a doped SET, despite the high value for the charging energy and in a regime where photon assisted tunnelling is not observable. The method is applied to a device made of a SET with a capacitively coupled IDQD. Partial control of the qubit is obtained and results in the possibility of manipulating charge states in an isolated structure with GHz photons.

This work was supported by Project for Developing Innovation Systems of the Ministry of Education, Culture, Sports, Science and Technology (MEXT), Japan.

[1] S. Furuta et al, Phys. Rev. B 70, 205320 (2004)
[2] C. H. W. Barnes et al, Phys. Rev. B 62, 8410 (2000)
[3] P. T. Greenland et al, Nature 465, 1057 (2010)
[4] J. Gorman et al, Phys. Rev. Lett. 95, 090502 (2005)
[5] A. Rossi et al, J. Appl. Phys. 108, 034509 (2010)
[6] T. Ferrus et al, New J. Phys. 13, 103012 (2011)

978-1-4673-0996-7/12 $31.00 © 2012 IEEE

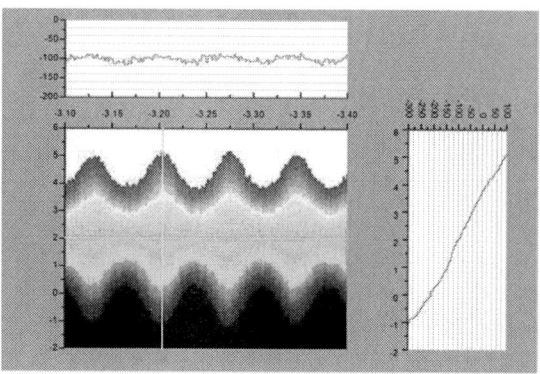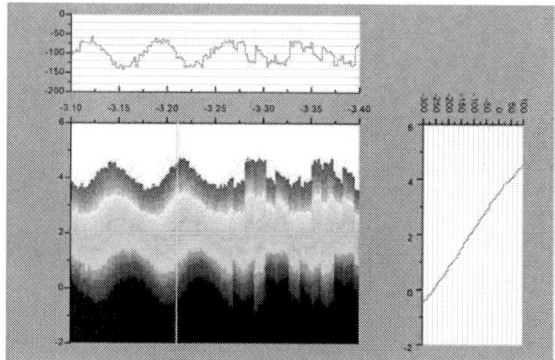

Fig.1 : Evolution of Coulomb diamonds at fixed frequency, without microwaves (left) and at medium power (right). Coulomb diamonds appear distorted (sawtooth like) before instabilities appear as a superposition of two diamonds at two different positions. At high power, heating destroys the Coulomb blockade oscillations.

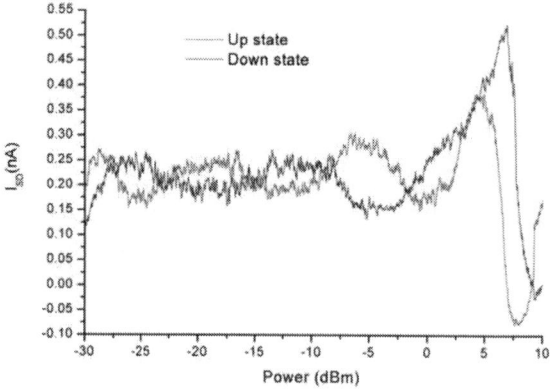

Fig. 2 : Microwave power dependence of the source-drain current of the SET at 12.83GHz. The 'up state' refers to an initial accumulation of charge in the upper IDQD dot and the 'down state' to an accumulation in the lower dot, at the beginning of the measurement. There is a clear dephasing by π between the two curves which is expected in case of oscillations of charge between the two IDQD dots.

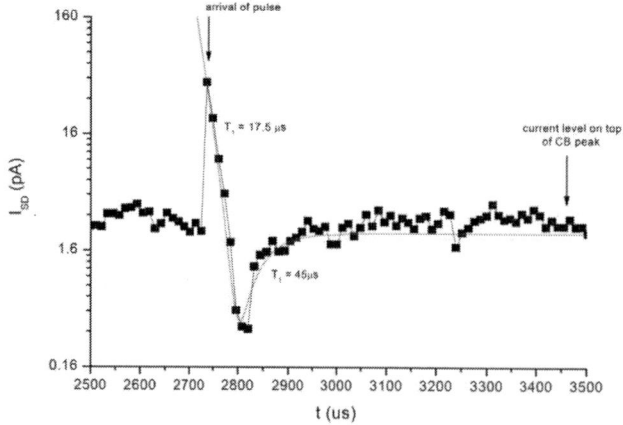

Fig.3 : Single shot measurement with electron feedback action (dip) at 4.2 K in a SET

978-1-4673-0996-7/12 $31.00 © 2012 IEEE   126

Charge sensing of a Si triple quantum dot system using single electron transistors

R. Mizokuchi[1*], T. Kodera[1,2,3], K. Horibe[1], Y. Kawano[1], and S. Oda[1]

[1]Quantum Nanoelectronics Research Center, Tokyo Institute of Technology, 2-12-1 , Ookayama, Megruro-ku, Tokyo, Japan

[2]Institute for Nano Quantum Information Electronics, the University of Tokyo, 4-6-1, Komaba, Meguro-ku, Tokyo, Japan

[3]PRESTO, Japan Science and Technology Agency (JST), 4-1-8 Honcho Kawaguchi, Saitama, 332-0012, Japan

[*]Tel/Fax: +81-3-5734-2542, E-mail: mizokuchi.r.aa@m.titech.ac.jp

Abstract — We fabricate a serial triple quantum dot (TQD) system, which is made on a silicon-on-insulator (SOI) wafer by dry etching and integrated with single electron transistors (SETs) as charge sensors. We observe charge transitions of a dot in the TQD in the characteristic of the charge sensor which is the furthest to the dot. It implies a SET charge sensor has a capability of sensing of all the charge transitions in TQD.

Introduction

Recently, electron spin quantum bit using quantum dot (QD) attracts a great deal of attention because of the long coherence time. A three-spin quantum bit in a TQD can be controlled only by using nearest-neighbor exchange interaction [1], which enables simple quantum device structure. Using a III-V heterostructure, a few-electron TQD has been realized and the spin manipulation has been achieved [2]. While III-V semiconductor has nuclear spins which cause spin decoherence, Si has few nuclear spin. Then Si is expected to make long electron spin coherence time.

We measure TQD characteristics by using SET charge sensors. SET has high sensitivity to electrons. In few electron regime, even though electron transitions cannot be observed by the current of a TQD, a SET will be able to sense of the transitions.

In this work, we fabricated a TQD system integrated with SETs as charge sensors and measured the property of Si TQD by using SETs.

Device fabrication

Fig.1 (a) and (b) show the schematic structure and a scanning electron microscope (SEM) image of the device, respectively. A TQD, two SETs and side gates (SGs) are fabricated on SOI by using electron beam lithography and Cl_2 dry etching. The SGs control the energy levels of the device. The area around the TQD is undoped. Ionized impurities prevent the formation of few electron regime. In the TQD area electrons are generated by using back gate voltage as two dimensional electron gas.

Measurement results and discussion

The results of the measurement are shown in Fig.2 (a) and (b), where the derivatives of the TQD current and the right SET current with SG_L voltage are depicted respectively as a function of SG_L voltage V_{SGL} and SG_R voltage V_{SGR}. Because the left SET can't work, the only right SET is used. The measurement is carried out in a condition where the temperature is 4.2 K, the TQD drain-source voltage V_{DS1} is 5 mV, the SET drain-source voltage V_{DS2} is 1.5 mV, the back gate voltage V_{BG} is 6 V, SET SG voltage V_{α} is -4 V and the other SG voltages are 0 V.

A TQD characteristic has different three negative gradients each of which corresponds to each dot. The three slopes can be read out partly in Fig.2 (a) and one of them is obvious in Fig.2 (b). A slope represented by a light green line corresponds to a slope of a dot of TQD. The dot is thought in opposite side to the charge sensor because the line has dependence on V_{SGL} mainly. Lines of dots closer to the charge sensor are not visible in the SET characteristic. This is because the charge sensor does not have an appropriate property as a single quantum dot due to parasitic dots near the constriction of SET. If the charge sensor SET had an ideal characteristic, current lines from other dots should be visible, which suggest a SET can observe charge transitions in every TQD because of the visibility of the transition of the furthest dot.

Conclusions

We fabricated Si TQD integrated with SETs as charge sensors. One of the SET works as a charge sensor and it reads out electron transitions of the furthest dot of TQD. The improvement of charge sensor characteristic by redesign of SET will achieve charge sensing of all the charge transitions in TQD.

Acknowledgement

This work was financially supported by JSPS KAKENHI (22246040), JST-PRESTO, Yazaki Memorial Foundation for Science and Technology, and Project for Developing Innovation Systems of the Ministry of Education, Culture, Sports, Science and Technology (MEXT).

References

[1] D. P. DiVincenzo, D. Bacon, J. Kempe, G. Burkard, and K. B. Whaley, Nature 408, 339-342 (2000)
[2] L. Gaudreau, G. Granger, A. Kam, G. C. Aers, S. A. Studenikin, P. Zawadzki, M. Pioro-Ladrière, Z. R.Wasilewski and A. S. Sachrajda, Nature Phys. 8, 54-58 (2012)

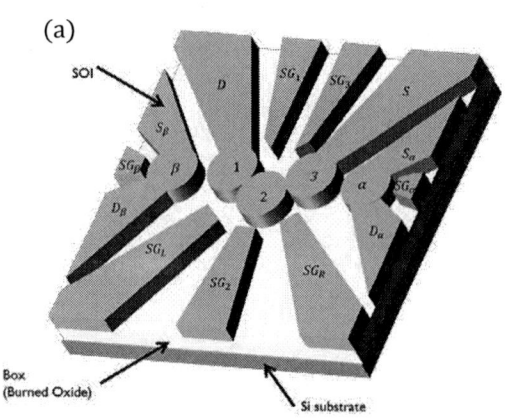

(a)

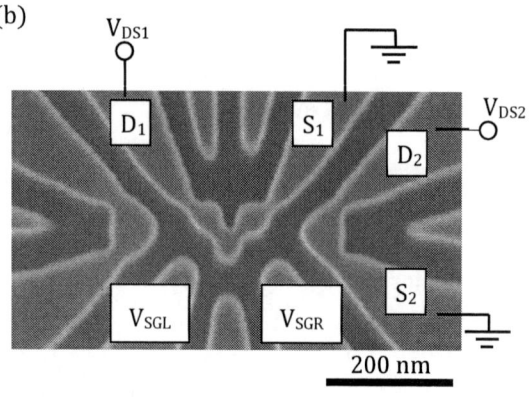

(b)

Fig.1. (a) Schematic image of the device used in this experiment. Back gate (BG) induces carriers. Side gates, source and drain are made of SOI. (b) SEM image of the device. Dark color area is burned oxide (BOX), and light color area is SOI. In the measurement,

the TQD current and the right SET current are measured and side gate voltages V_{GL} and V_{SGR} are swept when the TQD and SET source-drain voltages V_{DS1} and V_{DS2} and back gate voltage V_{BG} are applied.

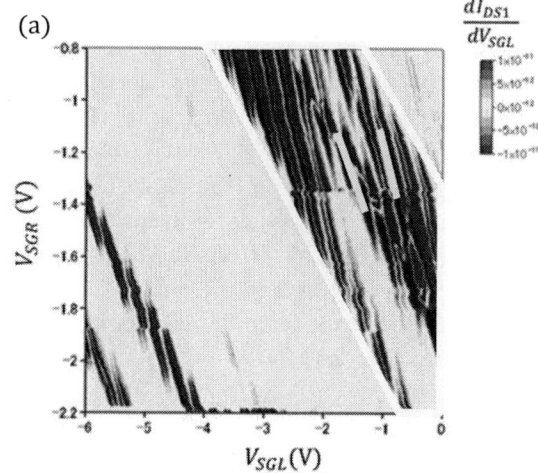

(a)

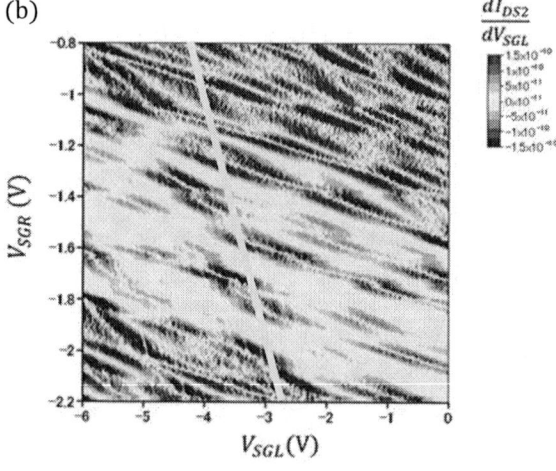

(b)

Fig.2. The derivatives of the TQD current and the right SET current with SG_L voltage where V_{DS1} is 5 mV, V_{DS2} is 1.5 mV and V_{BG} is 6 V (T = 4.2 K). (a) Light green line, light blue line and yellow area indicate each transition slope of the TQD. (b) Light green line represents vague lines. It agrees with the same color line in Fig.2 (a).

Fabrication and characterization of Si/SiGe quantum dots with capping gate

Tetsuo Kodera [1,2,3*], Yuji Fukuoka [1], Kenta Takeda [4], Toshiaki Obata [4], Katsuharu Yoshida [4], Kentaro Sawano [5], Ken Uchida [1], Yasuhiro Shiraki [5], Seigo Tarucha [2,4], and Shunri Oda [1]

[1] QNERC, Department of Physical Electronics, Tokyo Institute of Technology, 2-12-1-S9, Ookayama, Meguro, Tokyo, Japan

[2] Institute for Nano Quantum Information Electronics, The University of Tokyo, 4-6-1, Komaba, Meguro, Tokyo, Japan

[3] PRESTO, Japan Science and Technology Agency (JST), 4-1-8 Honcho Kawaguchi, Saitama, Japan.

[4] Department of Applied Physics, The University of Tokyo, 7-3-1, Hongo, Bunkyo, Tokyo, Japan

[5] Advanced Research Laboratories, Tokyo City University, 8-15-1, Todoroki, Setagaya, Tokyo, Japan

* Tel: +81-3-5734-3350, Fax: +81-3-5734-3350, Email: kodera.t.ac@m.titech.ac.jp

Abstract —We study transport properties of quantum point contacts (QPCs) and quantum dots (QDs) with a global capping gate, fabricated on a Si/SiGe high electron mobility transistor (HEMT) wafer. By biasing the capping gate negatively, we succeed in making QPC operation point of surface Schottky gate negatively smaller and then reducing noise. We also observe Coulomb oscillations using a QD structure by suppressing charging noise with negative capping gate voltage.

I. INTRODUCTION

Electron spin quantum bit (qubit) using Si quantum dots (QDs) attracts a great deal of attention because it is expected to have long electron spin coherence time. Si is nominally nuclear free material so that the electron spin in Si QDs is only weakly influenced by spin-orbit interaction. Recent studies on QDs fabricated in Si/SiGe heterostructure have shown extremely long relaxation time T_1 exceeding seconds [1] and dephasing time T_2 of 360 ns [2] much longer than that obtained in GaAs QDs. Our final research goal is manipulating electron spin and making electron spin qubit in Si/SiGe QDs experimentally. In order to realize stable qubit operation, it is important to reduce the noise coming from device itself. In this work, we fabricate and characterize quantum point contacts (QPCs) and QDs formed by Pd Schottky gates and a global capping gate on Si/SiGe HEMT wafer, and suppress charging noise with capping gate voltage.

II. DEVICES

Fig. 1 shows schematic cross section of our device on Si/SiGe HEMT wafer. Two dimensional electron gas (2DEG) is located approximately 60 nm below the surface. Pd Schottky gates are fabricated on the HEMT wafer using electron beam lithography and lift off process. Above the Pd Schottky gates, insulator of Al_2O_3 is deposited by atomic layer deposition and then global capping gate metal of Ti/Pd/Au is evaporated using an electron beam evaporator.

Scanning electron microscope (SEM) images of a QPC and QDs are shown in Fig. 2 (a) and (b), respectively. By applying negative voltages to the Schottky gates, electrons in 2DEG beneath the gates can be depleted so QPC and QDs can be formed.

III. MEASUREMENT RESULTS AND DISCUSSIONS

All the measurements described below are performed at a temperature of 60 mK. Figure 3 (a) shows the current through the QPC as a function of Schottky gate voltage with capping gate voltage V_{CG} changed as parameters using the device shown in Fig. 2 (a). With increasing negative voltage applied to Schottky gates, electrons beneath the gates are completely rejected and finally the current through the QPC does not flow at a pinch-off voltage. The pinch-off voltage of QPC is controlled from -2.3 V to -1.7 V as V_{CG} is changed from 0 V to -1.5 V because of lowering electron densities of 2DEG. By making QPC operation point of surface Schottky gate negatively smaller, we can reduce noise caused by leakage current and charge trapping from the Schottky gate. Figure 3 (b) and (c) show QPC current measured for another device with narrower gap between two gates as a function of Schottky gate voltage with V_{CG} =0 V and -1.5 V, respectively. When V_{CG} =0 V, we observe the first conductance plateau at $4e^2/h$ as shown in Fig. 3 (b). On the other hand, when V_{CG} =-1.5 V, we observe both $2e^2/h$ and $4e^2/h$ plateaus as shown in Fig. 3 (c). These results indicate that doubly-degenerated valleys of Si are split by applying V_{CG}. The valley degeneracy can be lifted by making quantum confinement perpendicular to 2DEG stronger with electrical field, which may suppress decoherence of electron spins caused by the valley degeneracy.

Figure 4 shows the current through the QD as a function of T gate voltage at V_{GC} =-2V using the device shown in Fig. 2 (b). We succeed in observing Coulomb oscillation by suppressing charging noise with negative capping gate voltage. This is the first observation of forming Si/SiGe QDs with the global capping gate on the Schottky gates.

978-1-4673-0996-7/12 $31.00 © 2012 IEEE

IV. SUMMARY

We fabricated and characterized both the QPCs and the QDs with the global capping gate, using the Si/SiGe HEMT wafer. With negative capping gate voltage, we succeeded in making more positive QPC operation point of Pd Schottky gates. We also observed Coulomb oscillations using the QD structure by suppressing charging noise with negative capping gate voltage.

ACKNOWLEDGEMENTS

This work was financially supported by JSPS KAKENHI (22246040), JST-PRESTO, Yazaki Memorial Foundation for Science and Technology, Project for Developing Innovation Systems of the Ministry of Education, Culture, Sports, Science and Technology (MEXT), and Funding Program for World-Leading Innovative R&D on Science and Technology (FIRST), Japan.

REFERENCES

[1] C. B. Simmons, J. R. Prance, B. J. Van Bael, Teck Seng Koh, Zhan Shi, D. E. Savage, M. G. Lagally, R. Joynt, Mark Friesen, S. N. Coppersmith, and M. A. Eriksson, Phys. Rev. Lett. 106, 156804 (2011).

[2] B. M. Maune, M. G. Borselli, B. Huang, T. D. Ladd, P. W. Deelman, K. S. Holabird, A. A. Kiselev, I. Alvarado-Rodriguez, R. S. Ross, A. E. Schmitz, M. Sokolich, C. A. Watson, M. F. Gyure, and A. T. Hunter, Nature 481, 344-347, (2012)

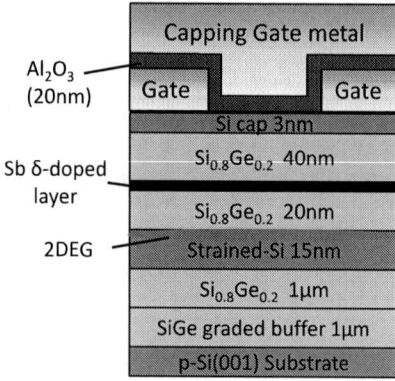

Fig. 1 Schematic cross section of our device, composed of Si/SiGe heterostructure, insulator and capping gate metal. The Sb δ-doped layer and 2DEG are located approximately 40 nm and 60 nm below the surface, respectively. The carrier density and mobility of the 2DEG are 2.0-2.3×10^{11}cm^{-2} and 150000 cm^2/Vs, respectively. Pd local fine gates are fabricated on the wafer using electron beam lithography and lift off process. We can form QPCs and QDs by applying negative voltage to the local gates and depleting electrons beneath the gates. Capping gate structure is also fabricated in order to tune carrier density independent with local gates.

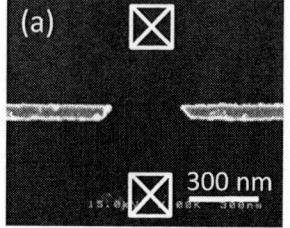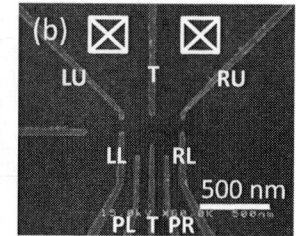

Fig. 2 (a) Scanning electron microscope (SEM) image of our QPC device before deposition of insulator and capping gate metal. Ohmic contacts are schematically indicated by squares. (b) SEM image of our QD device. All voltages applied to Pd Schottky gates (LU, LL, PL, T, PR, RL, and RU) can be controlled independently.

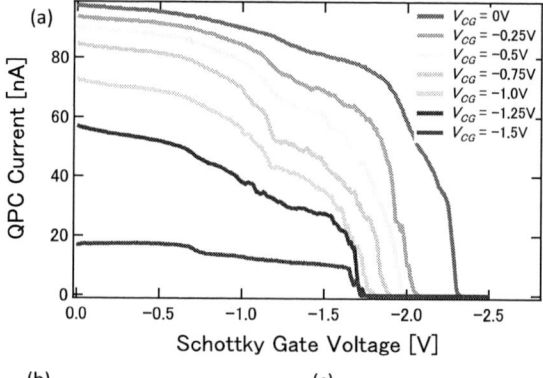

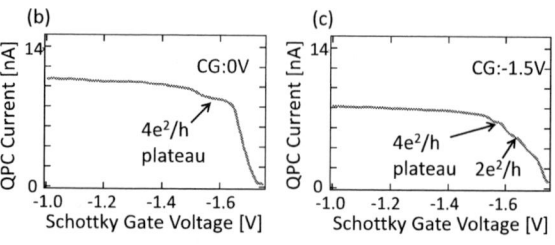

Fig. 3 (a) QPC current as a function of Schottky gate voltage measured for QPC device at capping gate voltage $V_{GC} = 0$ V to -1.5 V. The gap between two gates is 210 nm. (b,c) QPC currents as a function of Schottky gate voltage measured for another QPC device at $V_{GC} = 0$ V (b), and -1.5 V (c), respectively.

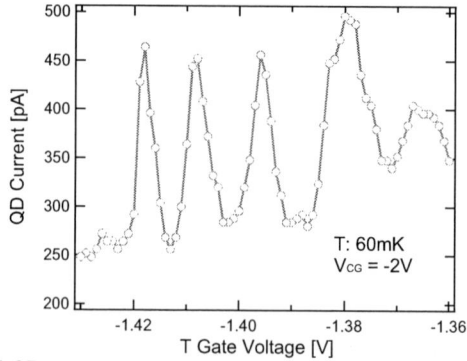

Fig. 4 QD current as a function of T gate voltage measured for QD device at a temperature of 60 mK. Here $V_{CG} = -2$ V. Several Coulomb oscillation peaks are observed.

Single Ge quantum dot placement along with self-aligned electrodes for effective management of single electron tunneling

I. H. Chen, K. H. Chen, and P. W. Li

Department of Electrical Engineering, National Central University, ChungLi, Taiwan, ROC, 32001
E-mail: pwli@ee.ncu.edu.tw

Abstract-We demonstrate controlled number and placement of the Ge quantum dot (QD) along with tunnel junction engineering through a self-organized approach for effective management of single electron tunneling. In this approach, a single Ge QD (~11 nm) self-aligning with nickel-silicide electrodes is realized by thermally oxidizing a SiGe nanorod bridging a 15-nm-wide nanotrench in close proximity to electrodes via a spacer bi-layer of Si_3N_4/SiO_2. The fabricated Ge QD single electron transistor exhibits clear Coulomb staircase and Coulomb diamond behaviors at $T = 120$–300 K.

Single-electron transistors (SETs), consisting of nanometer-sized QD weakly coupled to source, drain and gate electrodes through a few nanometer-thick tunneling barriers, are an ultimate electronic device for controlling charge transport in a single electron precision based on the so-called Coulomb blockade effect. For an effective charge manipulation in SETs, it requires precise not only control over the QD shape, size, and crystallinity relating to the QD electronic structure, but also the QD number and its lateral location (tunnel junctions) between electrodes. Additionally to enhance the signal-to-noise ratio for SET operation, device structure engineering is indispensable for reducing the coupling between gate and source/drain electrodes, thus minimizing background leakage emanating from gate induced tunnel barrier lowering (GIBL) and parasitic capacitances between electrodes. However, making electrical contacts to a specific nanoscale QD presents a major challenge because of the limits on the minimum feature size and the accurate overlay alignment of lithography.

The authors have demonstrated precise placement of a single Ge QD in center of a 30-nm-wide oxided SiGe nanotrench, which is separated from electrodes by SiO_2 or Si_3N_4 spacers. Notably the thickness of deposited spacers directly determines the tunnel paths between electrodes and QDs. These promising results build up solid cornerstones for effective SETs with controllable QD number/position. In this paper, the fabrication of Ge QD SETs began with a 80 nm-thick SiO_2, a 30 nm-thick Si_3N_4 and a 250 nm-thick poly-Si layers deposition on top of a (100) p-Si substrate. A trench with a width of 250 nm and a depth of 220 nm was then defined using electron-beam lithography (EBL) and plasma etching. The deposition of a bi-layer (15 nm-thick Si_3N_4 and 90 nm-thick SiO_2) sidewall spacer further shrinks the trench width to 35 nm. After etching away the superfluous poly-Si layer below the trench, a 10 nm-thick Si_3N_4 spacer was deposited again to form the second sidewall spacers, reducing the trench width further to 15 nm and acting as tunnel barriers between adjacent poly-Si and Ge QD, which would be generated by subsequent processes. Next, the trench was refilled with a bi-layer of poly-$Si_{0.85}Ge_{0.15}$/poly-Si (15 nm/150 nm), whose thickness was reduced to 15 nm by a direct-etch-back process, followed by the delineation of a 50 nm-wide SiGe nanorod therein. After completely oxidizing the SiGe rod, a single Ge QD is generated in the center of the trench. Subsequently, a 150 nm-thick poly-Si film was deposited into the trench as the gate material. EBL and plasma etching were performed to delineate poly-Si electrode pads, followed by nickel deposition and 500 °C annealing processes for forming nickel-silicide gate, source, and drain electrodes. The schematic diagram of designed device structure and key process flow are summarized in Fig. 1, in which the cross-sectional scanning electron microscopy (SEM) and plane-view transmission electron microscopy (TEM) images clearly shows a single, spherical, single-crystalline 13 nm Ge QD self-aligns with NiSi electrodes via Si_3N_4/SiO_2 (**10 nm/ 3 nm**) tunnel barriers.

Charge tunneling through energy levels of the Ge QD appears to be well directed by applied gate and drain voltages (V_G and V_D). Figure 2(a) illustrates temperature-dependent current-voltage (I_D-V_D) and differential conductance ($G_D \equiv \partial I_D/\partial V_D$)-voltage ($G_D$-$V_D$) characteristics for a Ge QD SET with gate grounded. Under drain bias modulation we observed distinct current plateaus and differential conductance peaks once V_{D+} (V_{D-}) exceeds a threshold value of +26 (-24) mV, which is required for the line-up of the ground state E_1 of the QD with fermi energies of source and drain electrodes. The staircase and oscillatory patterns retain all the way from $T = 120$ to 300K, while increasing temperature significantly broadens the full-width at half-maximum (FHWM) of each G_D peak. The Coulomb gap ($|V_{D+} - V_{D-}|$) and the G_D peak positions are well preserved with indiscernible shift in V_D (Fig. 2(b)), indicating that thermal stable current staircase and G_D oscillatory behaviors are governed by charges tunneling through energy levels of the Ge QD instead of charge random hopping across trap states. This also manifests the effectiveness of the proposed QD placement technique on producing symmetrical tunnel paths for charges into and out of the QD as well as suggests an equal of drain-QD capacitance (C_d) and source-QD (C_s) capacitance. A Coulomb gap of 50 mV corresponds to an effective capacitance ($C_d + C_s$) of 3.2 aF and an estimated QD diameter of 12 nm, very close to the statistic value (~11 nm) obtained from TEM observations. Decreasing temperature from 300 to 120 K gives rise to a significant reduction in the magnitude of tunneling current and within the Coulomb gap, the temperature dependence of tunneling current follows the power law of $\ln(I_D/T^2) \propto 1/T$ (Fig. 2(c)) together with an extracted barrier of $\phi_B \sim 70$–80 meV if a typical Schottky emission equation of $I \sim T^2\exp[(a\sqrt{V_D}/T - q\phi_B/kT]$ is fitted. This suggests that the initial charge injection from metal electrode to the QD is dominated by thermionic emission across a Schottky barrier across the NiSi/poly-Si, probably because some portion of the thin poly-Si layer below the first "spacer" of Si_3N_4/SiO_2 is not completely silicided.

Increasing V_G from 0 to 150 mV progressively washes out the feature of current staircases and makes the G_D peaks shift toward to smaller V_D (Fig. 3). The diminishing amplitude of G_D peaks together with a broadening FWHM suggests the applied gate voltage gradually fine-tuning energy levels of the QD away from resonance conditions. Increasing V_G does not induce a dramatic increase in tunneling current, manifesting a self-aligned gate electrode for a SET effectively suppressing tunnel resistance decrease and tunnel capacitance increase caused by GIBL. Figures 4(a) illustrates the gate modulated tunneling current characteristics at T =120 K under various drain bias. When a small V_D of 10 mV is applied, the first and the second Coulomb oscillatory peaks occur at V_G = 1.53 and 3.28 V, respectively. These peaks shift steadily to lower V_G with an increase of V_D from 10 to 20 mV or -10 to -20 mV (Fig. 4(b)), because of the energy level getting close alignment with the chemical potentials of the electrode with the aid of V_D. The experimental temperature-dependence of I_D-V_D and the drain-voltage dependence of I_D-V_G evidence strong Coulomb blockade effects on charges transport within the QD, making it possible to resolve the electronic structures of the Ge QD via the fabricated Ge QD SET. The single-electron charging energy for the ground state is estimated to be 12.7 meV by multiplying the peak gate voltage spacing and the gate modulation factor.

This work was supported by National Science Council of R. O. C. under Contract Nos. 100-2120-M-008-004 and 99-2221-E-008-095-MY3.

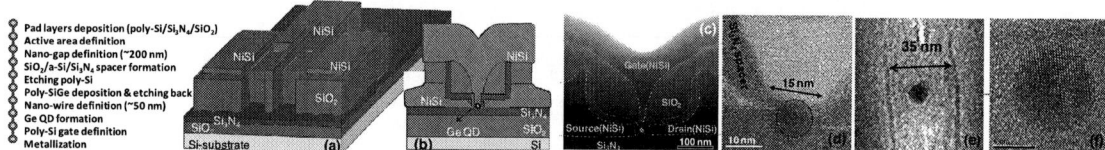

Fig. 1. (a)/(b) schematic device structure of designed SET, (c)/(d)/(f) cross-sectional and (e) plane-view SEM/TEM images of fabricated Ge QD SETs, in which a single 11 nm Ge QD self-aligns with NiSi source/drain electrodes via 10–12 nm-thick tunnel barriers of Si_3N_4 and with gate electrodes through gate oxide of SiO_2.

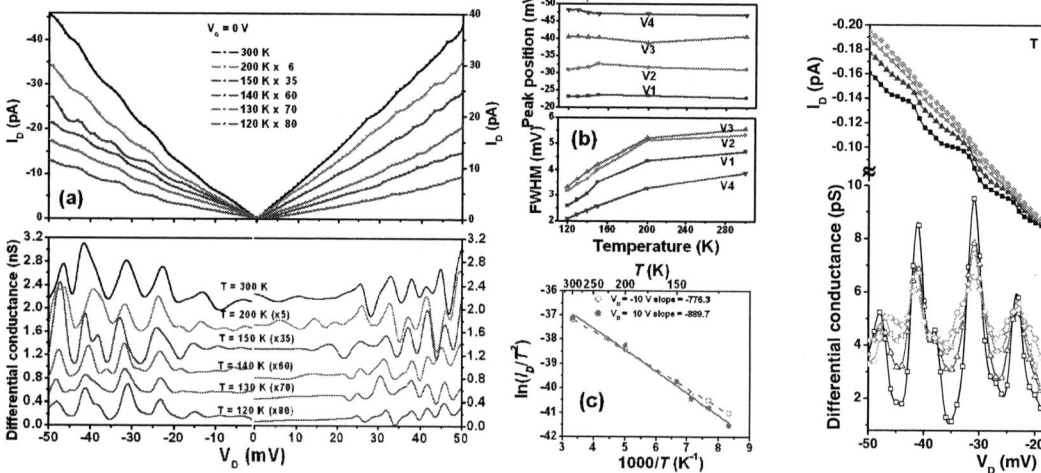

Fig. 2 (a) I_D-V_D and G_D-V_D characteristics of fabricated Ge QDs SET at $T = 120$-300 K. Each G_D curve is vertically offset -0.2 nS for clarity. (b) temperature dependence of the G_D peak voltage and FWHM, and (c) temperature dependence of the background current arising from charge thermionic emission across the metal-poly-Si interface.

Figure 3 Gate-voltage dependent I_D–V_D and G_D-V_D for Ge QD SETs at $T = 120$ K.

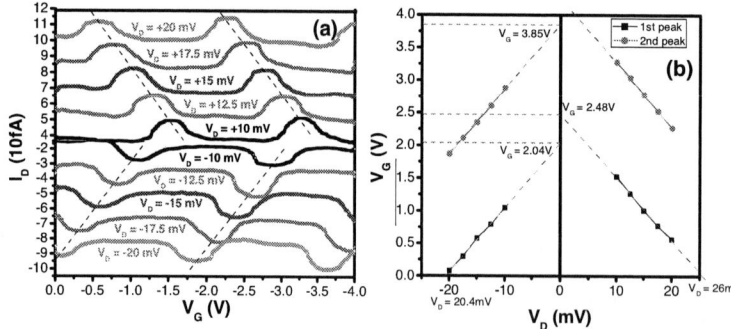

Fig. 4 (a) Drain-voltage dependent I_D–V_G for Ge QD SETs at $T = 120$ K. (b) oscillatory peaks as functions of drain voltage and gate voltage.

978-1-4673-0996-7/12 $31.00 © 2012 IEEE

Single-electron transport through a single donor at elevated temperatures

Earfan Hamid, Daniel Moraru, Takeshi Mizuno and Michiharu Tabe

Research Institute of Electronics, Shizuoka University, 3-5-1 Johoku, Hamamatsu 432-8011, Japan

Tel/Fax: +81-53-478-1335 e-mail: earfan@rie.shizuoka.ac.jp

Introduction

In recent years, single-electron transport mediated by individual dopants in silicon transistors was observed, in particular at low temperatures, below 20 K.[1-5] When the dopants are embedded in nanostructured channels, it has been predicted that their ionization energy is enhanced compared to the bulk case due to dielectric and quantum confinement.[6]

Here, we present experimental results for SOI-MOSFETs with nanoscale channels doped with phosphorus. By measuring the temperature (T) dependence of their electrical characteristics, we observed that, by increasing T, new current peaks successively emerge at lower gate voltages (V_G). This result may allow us to develop single-dopant tunneling devices working even at room temperature.

Electrical characteristics of doped SOI-FETs

The devices under study, schematically shown in Figs. 1(a) and 1(b), are SOI-FETs with thin channels (< 5 nm, as observed from the TEM image shown in Fig. 1(c)), having lateral dimensions below 100 nm. The SOI layer was uniformly doped with phosphorus (P) with a concentration $N_D \cong 1\times10^{18}$ cm^{-3}. It is expected that, in these devices, discrete donors in the channel form quantum dots (QDs) that control single-electron tunneling transport at low temperatures. Fig. 1(d) shows the I_D-V_G characteristics measured for one device as a function of T. At low T (20 K), isolated current peaks can be observed, due to transport through individual donors.[4,5]

From our measurements on a large number of devices, we observed the emergence of one or more new current peaks as T is increased. These new peaks, such as those indicated by arrows in the zoom-in view in Fig. 1(e) for $T = 25$ and 30 K, appear at lower V_G's, suggesting that they are due to donors with relatively deeper energies. Donor energy may be modified due to the dimensions of the silicon nanostructures and their geometry. An analysis of the effects of channel size and channel pattern on the temperature evolution of the electrical characteristics was also performed. We found that, for 1-disk devices, the number of emerging current peaks is larger than for nanowire devices, i.e., 1-disk pattern is favorable for finding deep donors in the channel. We also found that the highest temperature at which new peaks appear depends on the channel width, as shown in Fig. 2(a). For the smallest 1-disk devices ($W_J = 25$ nm), a final new peak is observed at $T = 100$ K, as shown in Fig. 2(b). With increasing channel width, the highest temperature of the final new peak decreased. On the

other hand, nanowire devices did not show strong dependence on channel width. For the small nanowire devices (W = 5~35 nm), the average temperature of the final new peak appearance is $\cong$ *50-60 K*, as shown in Fig. 2(c). We found that smallest 1-disk devices are preferable in terms of donor deepening and high temperature tunneling operation due to dielectric and quantum size effect. In the smallest nanowire devices a similar effect should be present, but, as the channel is directly coupled to electrodes, screening effect of the donor potential is more significant.

Stability diagrams of nanoscale devices

Measurements of source-drain current (I_D) in the source-drain bias vs gate voltage (V_D-V_G) space, i.e., stability diagrams, were taken at 20 K and 65 K, as shown in Fig. 3. The first current peak corresponds to the deep donor in the channel, as indicated by the dashed line. At 20 K, the first peak was not clearly observed in I_D-V_G characteristics at small $V_D = 5$ mV due to the low current level, because of high barrier and low tunneling rate. At 65 K, current is enhanced due to thermal activation and the new peak becomes prominent. For calculating the ionization energy, a lever-arm parameter, $\alpha = 0.8$, was extracted from the slopes of the Coulomb diamonds. The deep donor peak appears at $V_{GI} = 0.33$ V (dashed line in Fig. 3) and the threshold voltage is estimated to be $V_{TH} = 0.6$ V (not shown). According to the equation, $E_I = \alpha \times (V_{TH} - V_{GI})$, the calculated ionization energy for the first donor is 216 meV.

Summary

We showed that, in nanoscale doped SOI-FETs, new current peaks become observable as temperature is increased. For smallest 1-disk devices, a final new tunneling current peak has been observed even at $T = 100$ K, indicating that such patterned-channel devices are suitable for high temperature tunneling operation. Ionization energy was estimated to be about 5 times larger than for bulk Si, due to dielectric and confinement effect.

References

[1] H. Sellier *et al.*, Phys. Rev. Lett. **97**, 206805 (2006).
[2] Y. Ono *et al.*, Appl. Phys. Lett. **90**, 102106 (2007).
[3] M. Pierre *et al.*, Nat. Nanotechnol. **5**, 133 (2010).
[4] M. Tabe *et al.*, Phys. Rev. Lett. **105**, 016803 (2010).
[5] E. Hamid *et al.*, Appl. Phys. Lett. **97**, 262101 (2010).
[6] M. Diarra *et al.*, Phys. Rev. B **75**, 045301 (2007).

Acknowledgements

This work was partly supported by KAKENHI 20246060, 22656082, and 23226009.

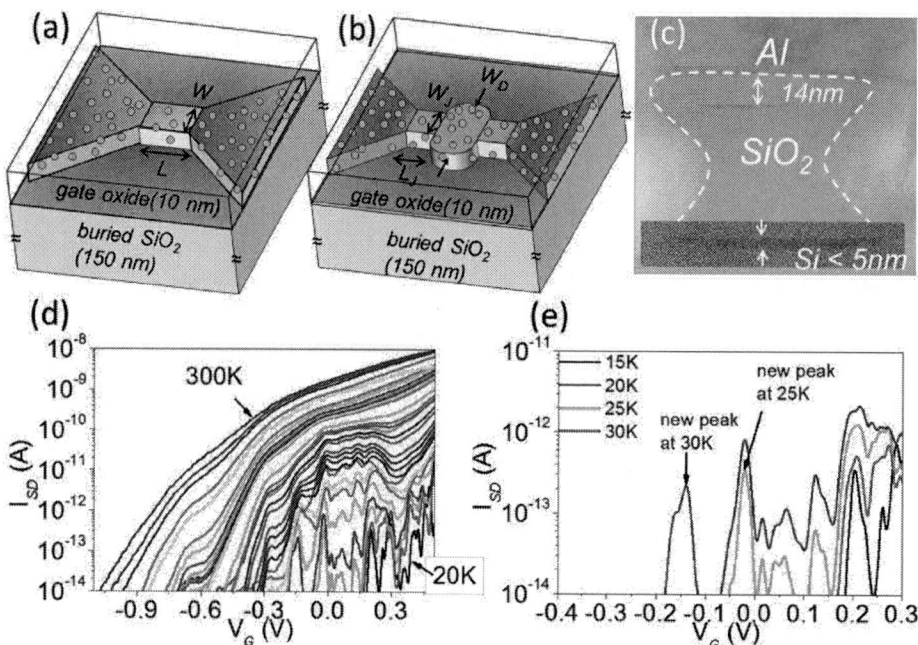

Fig. 1. Schematic view of SOI-FET nanowire device [(a)] and 1-disk device ([b]) (c) Cross-section TEM image of 1-disk device, with a zoom-in view that shows the real thickness of the Si channel (<5 nm). (d) I_D-V_G characteristics for a nanoscale SOI-FET as a function of temperature, from 20 to 300 K. (e) Zoom-in view of Fig. 1(d), showing two new peaks, indicated by arrows, emerging at T = 25 K and 30 K, respectively.

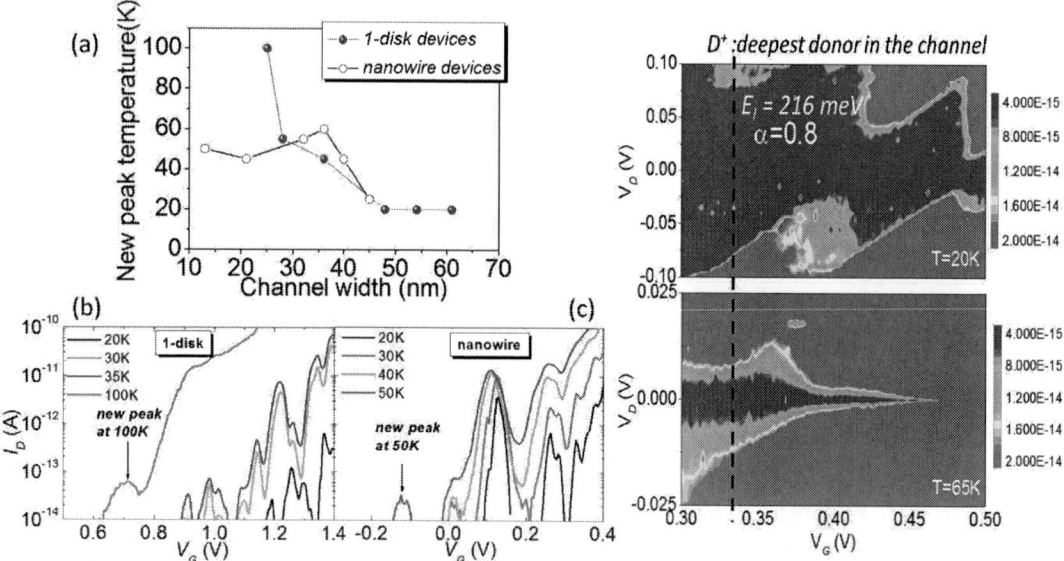

Fig. 2. (a) Final new peak appearance temperature as a function of channel width. (b) I_D-V_G characteristics (V_D = 5 mV) of smallest 1-disk device, showing a final new peak appearing at 100 K. (c) I_D-V_G characteristics (V_D = 5 mV) of smallest nanowire device, showing a final new peak appea

Fig. 3. Stability diagrams of 1-disk device measured at 20 K (top) and at 65 K (bottom). At V_G = 0.33 V, the current peak region was not observable, but it becomes prominent at 65 K, as indicated by the dashed line.

978-1-4673-0996-7/12 $31.00 © 2012 IEEE   134

The Interplay of Self-Heating Effects and Static RTF in Nanowire Transistors

D. Vasileska, A. Hossain[*] and S. M. Goodnick
School of Electrical, Computer and Energy Engineering, Arizona State University, Tempe, AZ
[*]Intel Corp., Chandler, AZ
e-mail: vasileska@asu.edu, arif.hosain@intel.com, goodnick@asu.edu

Random Telegraph Noise Fluctuations (RTF) manifest themselves as fluctuations in transistor threshold voltage and drive (on) current. RTF is caused by random trapping and detrapping of charges lying at the inversion channel of the device under consideration close to the oxide-semiconductor interface [1]. Traditionally RTF were important only in analog design at low frequencies [2]. However, as CMOS is scaling into sub-100 nm regime, the effect of RTF as well as its variability is not negligible any more even in digital design [3]. In fact, we have illustrated in past work that the presence of a single negatively charged trap at the source end of the channel in a nanowire transistor can have very degrading effect on the on-current [4]. In these simulations we have utilized 3D Monte Carlo device simulator in which the short-range portion of the Coulomb interaction was accounted for by our real-space molecular dynamics model, details of which can be found in Ref. [5]. The model properly accounts for both the short-range and the long-range component of the Coulomb interaction and has been well recognized in the literature and applied in many other studies [6].

The purpose of this work is to present the results of our current investigations of the influence of the negatively charged trap on the magnitude of the on-current for the case when in addition to the short-range Coulomb interactions, self-heating effects are incorporated in the theoretical model. The nanowire FET being simulated in this work has gate oxide 0.8 nm thick and the BOX is 10 nm thick (Fig. 1). The dimensions of the silicon nanowire are: 10 nm channel length, 7 nm channel thickness and 10 nm channel width. For the thermal conductivity, that appears in the acoustic phonons energy balance solvers, we have taken the value from Li Shi measurements [7] that correspond to wire with cross-section of 7×10 nm.

The incorporation of self-heating effects in the existing model is achieved by self-consistently solving the 3D Monte Carlo/molecular dynamics solver coupled to a 3D Poisson equation solver self-consistently with 3D energy balance solver for the acoustic and optical phonon temperatures [8,9]. In all the simulations presented in this work, the bottom (substrate) electrode is taken to be at lattice temperature T=300 K and the top gate is also assumed to be at temperature T=300K. For all the other boundaries, Neumann boundary conditions are assumed. At the device plane under the gate oxide, we have assumed that we have copper metal gates. Copper has thermal conductivity of 400 W/m-K. Neumann boundary conditions are assumed at the edges of the metal boundaries.

Simulation results for the current degradation for a trap located right in the middle of the source end of the channel and a trap located 1 nm towards the drain are presented in Table 1 and Fig. 2. From the results presented we may conclude that for a trap located at the source edge of the channel (Fig. 1), the degradation in current is 0.47%. Such small degradation is attributed to the source electrons screening the potential of the negatively charged trap. For the case of a trap located 1 nm towards the drain (Fig. 1) the current degradation is 2.12 %, much higher than in the previous case. This behavior is attributed to the reduced screening of the source charges. The larger impact of the second trap is also reflected on the magnitude of the peak lattice temperature which is 2 K lower due to a larger reduction of the current and therefore smaller impact of self-heating effects. In summary, this simulation example illustrates that charge trapping and self-heating effects must be simultaneously considered to get proper values for the current degradation.

REFERENCES

[1] M. J. Kirton and M. J. Uren, "Noise in solid-state microstructures: A new perspective on individual defects, interface states and low frequency (1/f) noise", Advances in Physics, Vol. 38, pp. 367-468, (1989).

[2] K. S. Ralls, W. J. Skocpol, L. D. Jackel, R. E. Howard, L. A. Fetter, R.W. Epworth and D. M. Tennat, "Discrete Resistance Switching in Submicrometer

Silicon Inversion Layers: Individual Interface Traps and Low-Frequency (1/f ?) Noise", Physical Review Letters, Vol. 52, No. 3, pp. 228-231 (1984).

[3] H. Kurata, K. Otsuga, A. Kotabe, S. Kajiyama, T. Osabe, Y. Sasago, S Narumi, K. Tokami, S. Kamohara and O. Tsuchiya, "The impact of Random Telegraph Signals on the Scaling of Multilevel Flash Memories,", in 2006 Symposium on VLSI Circuits, pp. 125-126.

[4] D. Vasileska and S. S. Ahmed, "Narrow-Width SOI Devices: The Role of Quantum Mechanical Size Quantization Effect and the Unintentional Doping on the Device Operation", IEEE Trans. Electron Devices, Volume 52, pp. 227 – 236, Feb. 2005.

[5] W. J. Gross, D. Vasileska and D. K. Ferry, "A Novel Approach for Introducing the Electron-Electron and Electron-Impurity Interactions in Particle-Based Simulations," IEEE Electron Device Letters Vol. 2 No. 9, pp. 463-465, (1999).

[6] Z. Aksamija and I. Knezevic, "Anisotropy and boundary scattering in the lattice thermal conductivity of silicon nanomembranes", Phys. Rev. B., Vol. 82, 045319, 2010.

[7] D. Li, Y. Wu, P. Kim, L. Shi, P. Yang, and A. Majumdar, "Thermal conductivity of individual silicon nanowires," Appl. Phys. Lett., vol. 83, no. 14, pp. 2934–2936, 2003.

[8] D. Vasileska, K. Raleva and S. M. Goodnick, Self-Heating Effects in Nano-Scale FD SOI Devices: The Role of the Substrate, Boundary Conditions at Various Interfaces and the Dielectric Material Type for the BOX, IEEE Trans. Electron Devices, Vol. 56, No. 12, pp. 3064-3071 (2009).

[9] D. Vasileska, K. Raleva, S.M. Goodnick, "Modeling heating effects in nanoscale devices: the present and the future", Journal of Comp. Electronics, DOI 10.1007/s10825-008-0254-y (2008).

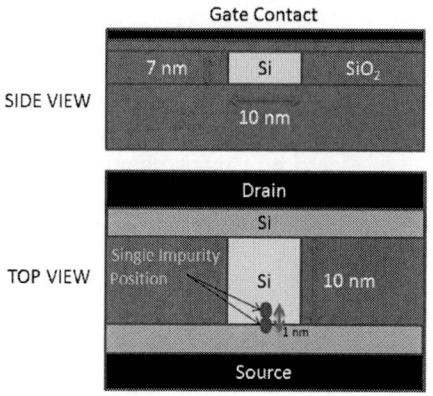

Fig. 1. Side and Top view of the structure being simulated.

Table 1. Single Negative Impurity Impact.

GUMMEL CYCLE	No impurity case [µA/µA]	Source Edge [µA/µA]	1 nm Towards Drain [µA/µA]
1	4154	4122	4075
3	4068	4030	3974
5	4052	4035	3968
Degradation	N/A	0.47%	2.12%

Screening of the source charges
Reduces the impact of the negative trap.

Maximal impact

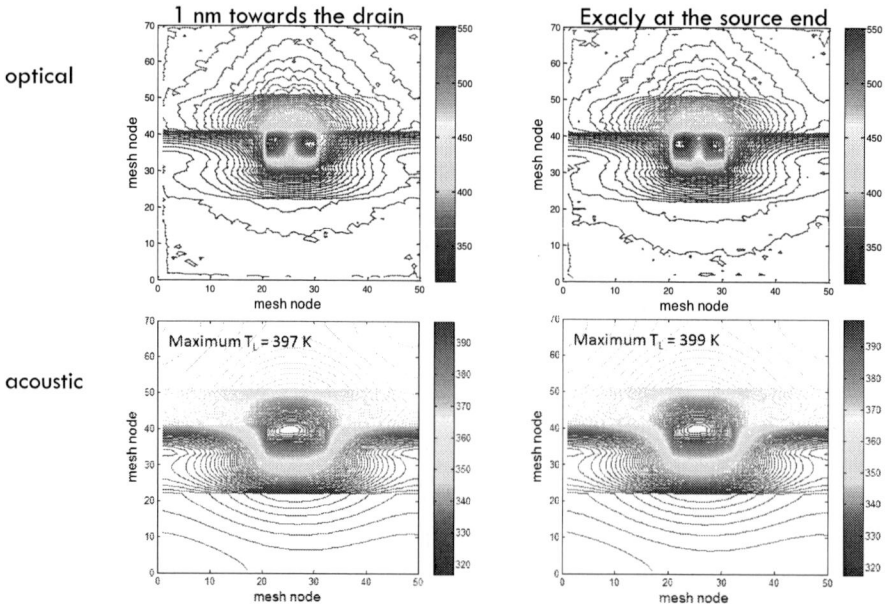

Fig. 2. Top panel - Optical phonon temperature profile which shows large degree of localization of the heat because of small group velocity of optical phonons. Bottom panel – Acoustic (lattice) phonon temperature. Notice the shift of the hot-spot towards the drain end of the channel.

Effect of Interfacial States on the technological variability of Trigate MOSFETs

E. González-Marín*, F. G. Ruiz, A. Godoy, I. M. Tienda-Luna, F. Gámiz

Dpto. Electrónica, Facultad de Ciencias, Universidad de Granada. Av. Fuentenueva, S/N, 18071 – Granada (Spain)
*Email: egmarin@ugr.es

Abstract

This work studies the influence of the interfacial states on the performance of Trigate MOSFETs and, specifically, on the Subthreshold Swing (SS) and threshold voltage (V_T) variability. To do so, a solver for the 2D Schrödinger-Poisson coupled equation system has been developed, including the effects of interfacial states (D_{it}). Different $D_{it}(E)$ profiles have been considered, analyzing their influence for several device geometries.

Introduction

Multigate (MuG) FETs are good candidates for future technology nodes [1], due to their higher electrostatic control on the channel, limiting the Short Channel Effects, and then reducing the SS to nearly ideal values [2]. Moreover, since the channel doping can be reduced to nearly intrinsic values, V_T variability may be neglected. However, both V_T and SS can be modified by the presence of D_{it} [3], increasing the technological variability. D_{it} are associated to the interruption of the periodic structure of the silicon crystal at the Si-SiO$_2$ interface, creating the P_b centers [4]. These P_b centers are amphoteric, i.e., they behave as donor-like (acceptor-like) in the lower (upper) half of the bandgap (Fig. 1). Recent studies have shown behaviors in the energy profile of D_{it} that differ from those expected for P_b centers on usual crystal orientations [5]. As the energy profile determines the surface potential energy needed to capture or release an electron, this fact can be relevant in the device performance. Moreover, when MuG devices with 2D confinement are considered, the potential profile is not simple, as it varies along the Si-SiO$_2$ interface [6]. Thus, the D_{it} activation along the interface occurs at different V_g values depending on the specific geometry of the device, complicating the analysis.

Numerical Method

The 2D Schrödinger and Poisson equations have been numerically solved in the cross section of a long-channel device. The details of the simulator can be found elsewhere [6,7]. The main improvement of this work is the self-consistent simulation of Q_{it} (Fig. 2). The Fermi occupation function is approximated by a step function for simplicity, and the relation with the surface potential (φ_s) is included implicitly through $E_i(\vec{r})$. $D_{it}(E)$ has been modeled as a Gaussian function (Fig. 2), which is shown in Fig. 3 along with the experimental one from [5].

Results

In our work, a 10x10 nm^2 Trigate SOI MOSFET has been simulated. Undoped body, corner rounding (R_{curv}=1nm), a midgap metal gate and 1nm thick SiO$_2$ as gate-insulator were considered. Unless otherwise stated, (100)/[011]-oriented devices are used (see Fig. 4). Fig. 5 shows the SS calculated as a function of V_g in the presence of the D_{it} at different temperatures. Our simulations reproduce the behavior shown in [5]: for small V_g, the SS values are close to the ideal limit, $\ln(10)\cdot k_B T/q$. However, there is a kink close to V_T related to the presence of the D_{it}. We have compared the SS values achieved for different A and σ values of the analytical D_{it} curve (Fig. 3), at T=50K. First, the amplitude value A was modified. The higher A, the higher total Q_{it} and, as expected from the $-Q_{it}/C_{ox}$ term of V_T [8], the larger ΔV_T (Tab. I). The amplitude and broadening in the V_g range of the SS kink (Fig. 6(a)) increase with A. This effect degrades the on-off transition regime, as shown in Fig. 6(b), where the total charge density *vs.* V_g is represented. An alternative explanation of this effect can be found when showing the φ_s behavior along the Si-SiO$_2$ interface, Fig. 6(c). The linear relation between φ_s and V_g is broken when D_{it} begin to fill: at that point, the increase of Q_{it} screens the potential and reduces $|d\varphi_s/dV_g|$. As a consequence, the subband energy levels, E_n, show a plateau region which extends through the V_g range where D_{it} keep filling up (see Fig. 6(d)). The E_n rise is

responsible for a lower level occupation and thus the charge reduction shown in Fig. 6(b). A variation of σ also modifies the behavior of the SS kink, as shown in Fig. 7, where the high correlation between the SS kink and the shape of the D_{it} curve is evidenced. In this case, the D_{it} curves have been adjusted to have the same total integrated charge by modifying also the curve amplitude, and thus ΔV_T is very similar for all the σ values (Tab II).

A. D_{it} variations along the device interface

Experimental results have shown that the D_{it} associated to lateral and top/bottom regions of MuG devices can be significantly different [4,5,9,10], due to the fabrication processes and the interface orientation [9]. In Fig. 8 we compare the SS achieved when the D_{it} are placed in all the Si-SiO$_2$ interfaces and in each of them separately: corners, top, lateral and Si-BOX (bottom) regions are considered. The maximum ΔV_T and wider SS kink are produced by charges located at the Si-BOX interface. This result is related to the reduction of the gate control on such interface, which may be modeled as a reduction of the C_{ox} associated to that region, which would increase V_T. The stronger effect of the lateral regions compared to the top one is due to their higher total charge ($2H_{Si} > W_{Si}$). The potential decrease due to D_{it} screening in each region is depicted in Fig. 9: lower values are found near the regions where the D_{it} are placed. Again, the stronger effect of D_{it} at the Si-BOX interface is perceived. Moreover, we have considered two alternative transport orientations: [001] and [011] (both on (100)) wafers, corresponding to the top surface of the Trigate): the results show that the orientation does not significantly impact neither the SS nor the ΔV_T, due to the fairly noticeable orientation dependence of the charge distribution on the subthreshold regime, Fig. 10 (the different peak value of the charge is related to a V_T variation, shown in Fig. 8). Experimental results [5,9] show a higher charge density in the lateral regions compared to the top ones. This effect has also been studied as can be seen in Fig. 11: higher lateral interface charge densities increase the impact on both, the SS kink and in ΔV_T, degrading the device performance.

B. Gate insulator variability

The relation between the SS degradation and ΔV_T has been analyzed in [3], without successful results. One of the reasons may be the influence of the T_{ox} thickness variability, also perceived in the TEM scans of SiNWs in [5]. We have modified the gate insulator geometry as shown in the insets of Fig. 12, where T_{ox} at the center of the top/lateral regions is increased or decreased to better fit fabricated devices. As shown in Fig. 12, the thicker insulator thickness provokes a degradation on the SS and a higher ΔV_T, as a consequence of the lower electrostatic control, being the variations of the lateral regions more important due to the larger interface length.

Conclusions

From all the aforementioned results, we conclude that: i) the reduction of the gate control increases the negative effects of the D_{it}; ii) as a consequence, the increase of the gate insulator thickness or the presence of a noticeable D_{it} at the Si-BOX interface of SOI devices degrades the device performance; iii) the previous conclusions indicate that the improvement of the electrostatic control on the gate insulator interface reduces the technological variability associated to D_{it}. Hence, the increase of the number of gates achieves the highest reduction of SS and ΔV_T (see Fig. 13).

Acknowledgment-Work supported by the Junta de Andalucia (P09-TIC-4873) and the Spanish Government (FIS-2008- 05805 and FIS-2011 -26005). E. González-Marín also acknowledges the FPU program.

978-1-4673-0996-7/12 $31.00 © 2012 IEEE

References

[1]http://www.intel.com/content/www/us/en/silicon-innovations/intel-22nm-technology.html, [2] Colinge, IEEE-EDL, 24, 2003, [3] Hong, IEEE-EDL, 32, 9, 2011, [4] Y. Nishi, JJAP, 11, p. 85, 1972, [5] Sato, APL, 98, 233506, 2011. [6] F. G. Ruiz, IEEE-TED, 54, 12, 2007. [7] I. M. Tienda-Luna, IEEE-TED, 58, 2011. [8] Y. Tsividis, Oxford Univ. Press, 2008 [9] Kapila, IEEE-EDL, 28, 3, 2007. [10] Cassé, APL, 96, 123506, 2010.

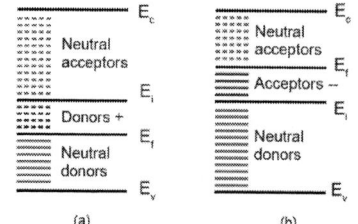

(a)   (b)

Fig. 1. Silicon and D_{it} band diagram. Fermi level E_F below (a) and above (b) intrinsic level E_i. Occupied states are plotted as continuous lines.

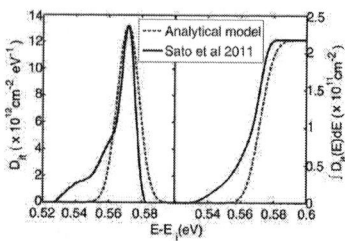

Fig. 3. Analytical D_{it}: A_0 and E_{C0} have been chosen to fit the curves in [5] (left). σ_0 was chosen to fit the charge when all states are occupied.

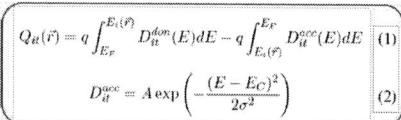

$$Q_{it}(\vec{r}) = q \int_{E_F}^{E_i(\vec{r})} D_{it}^{don}(E)dE - q \int_{E_i(\vec{r})}^{E_F} D_{it}^{acc}(E)dE \quad (1)$$

$$D_{it}^{acc} = A \exp\left(-\frac{(E-E_C)^2}{2\sigma^2}\right) \quad (2)$$

Fig. 2. Equations for Q_{it} and D_{it}. $\vec{r} \in$ Si-SiO$_2$ interface, E_i and E_F are intrinsic and Fermi energies, $D_{it}^{don/acc}$ is the donor / acceptor D_{it}.

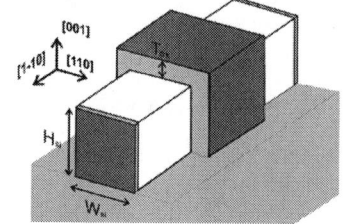

Fig. 4. Geometry of a W_{Si} x H_{Si} Trigate SOI FET.

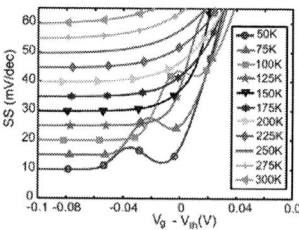

Fig. 5. SS *vs.* V_g-V_T as a function of temperature.

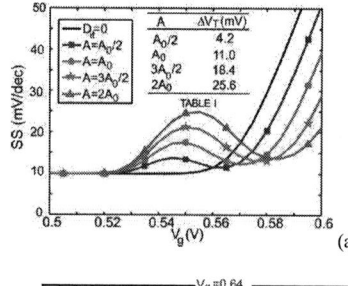

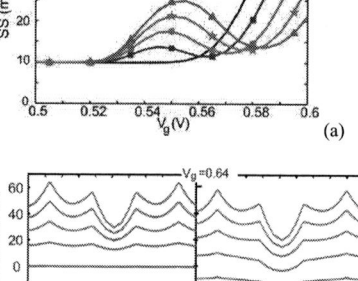

(a)

(c)

Fig. 6. Study of the D_{it} amplitude (A) variation, from $A_0/2$ to $2A_0$: (a) SS *vs.* V_g, (b) Charge density *vs.* V_g, (c) longitudinal φ_S *vs.* V_g (left without D_{it}, right with D_{it}), (d) First energy level (E_1) *vs.* V_g.

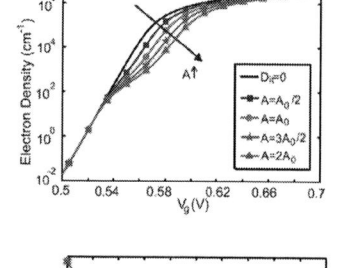

(b)

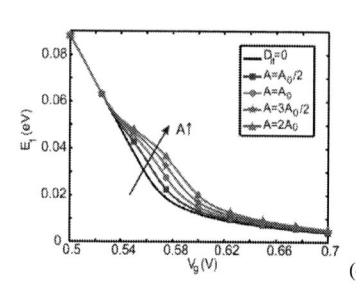

(d)

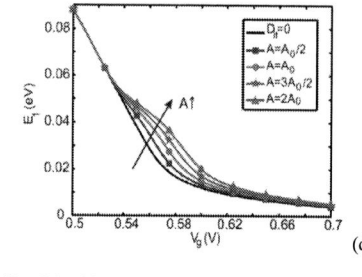

Fig. 7. SS *vs.* V_g for different σ values of D_{it}.

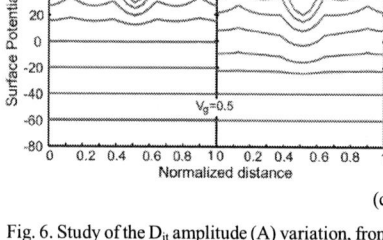

Fig. 8. SS *vs.* V_g for D_{it} placed at different device regions.

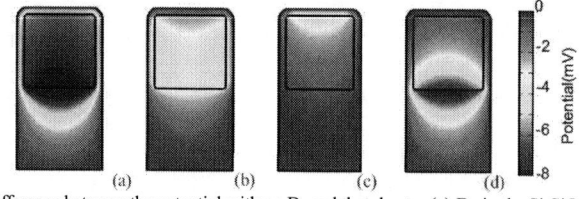

(a)   (b)   (c)   (d)

Fig. 9. Difference between the potential with no D_{it} and that due to: (a) D_{it} in the Si-SiO$_2$ interface, (b) D_{it} only in the lateral regions, (c) D_{it} only in the top region, and (d) D_{it} in the Si-BOX region.

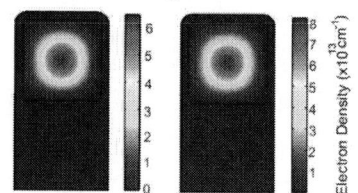

Fig. 10. Electron density for (100)/[011] (left) and (100)/[001] (right) Trigate FETs at V_g=0.53V (D_{it}=0).

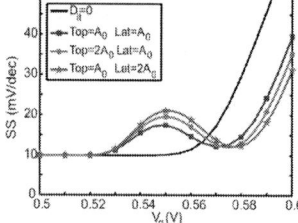

Fig. 11. SS *vs.* V_g for different charge densities (D_{it} curve amplitude variations).

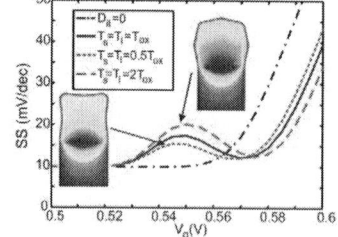

Fig. 12. SS *vs.* V_g for varying T_{ox} at the center of the lateral and top regions (see inset figures).

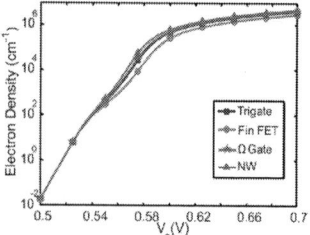

Fig. 13. Electron density *vs.* V_g for different device architectures.

978-1-4673-0996-7/12 $31.00 © 2012 IEEE

Physical Model for Random Telegraph Noise Amplitudes and Implications

Richard G. Southwick III[1], K.P. Cheung*[1], J.P. Campbell[1], S.A. Drozdov[2], J.T. Ryan[1], J.S. Suehle[1], A.S. Oates[3]

[1]National Institute of Standards & Technology, Gaithersburg, MD, USA; [2]ECE Dept. Univ. of Maryland, College Park, MD, USA;
[3]TSMC, Hsinchu, Taiwan *kin.cheung@nist.gov

Abstract: Random Telegraph Noise (RTN) has been shown to surpass random dopant fluctuations as a cause for decananometer device variability, through the measurement of a large number of ultra-scaled devices [1]. The most worrisome aspect of RTN is the tail of the amplitude distribution – the limiting cases that are rare but nevertheless wreak havoc on circuit yield and reliability. Since one cannot realistically measure enough devices to imitate a large circuit, a physics-based quantitative model is urgently needed to replace the brute force approach. Recently we introduced a physical model for RTN [2-3] but it contains a serious error. In this paper, we developed and experimentally verified a new model that provides a physical understanding of RTN amplitude. By providing a quantitative link to device parameters, it points the way to control RTN in decananometer devices.

Introduction: RTN amplitude has been a challenging topic for decades. Among the large body of literature on RTN amplitude modeling, Yau et al. [4] was the first to consider the simple picture of a trapped charge creating a "hole" (reduced carrier density) in the inversion layer. This idea was extended by Reimbold [5] to include the carrier density reduced to zero (i.e. cored-out "hole"). Ohata et al. [6] developed a simple equation to link the "hole" size to the RTN amplitude, eq. (1). They concluded that the hole size, deduced from measured RTN data, was too large (compared to the expected screening length) to be correct. , Asenov et al. [7] came to the same conclusion via atomistic simulations. This simple model therefore appears dead. However, is it reasonable to assume the "hole" size is equal to the screening length? Screening length is defined as electric field decay to 1/e (36.8%) level, certainly not the same as "inversion charge density drops to threshold" level. Thus, what is the proper "hole" size that makes physical sense?

Model: A charge located between the gate electrode and the inversion layer has a local field that affects the local inversion layer charge density. The "hole" size is largest when the charge is buried within the inversion layer and zero when right at the gate electrode. Thus, determining the maximum "hole" size gives a calibration point and provides the foundation for modeling the hole at any location in the gate stack. From a physics standpoint, trapped charge in the inversion layer differs from the rest of the inversion charges only by being localized. Thus, intuitively, the "hole" created by the trapped charge is equal to the area it consumes with relation to other carriers in the channel, giving it a radius r equal to the average distance between carriers in the inversion layer, eq. (2). This simple conceptual breakthrough on the size of the "hole" is the foundation of our model. For large gate overdrives (V_{OD}), the 2D inversion charge density Q is written as (3), thus relating the maximum "hole" size to V_{OD}. For a charge trapped anywhere between the gate electrode (d_1) and the inversion layer (d_2), the hole is reduce from (2) by the ratio $d_1/(d_1+d_2)$, see Fig. 1. We therefore write the "hole" size as (4).

Using (1), (3), and (4), we can quantitatively calculate the worst-case RTN amplitude (one defect) for a planar MOSFET with an undoped channel. Intuitively, the worst-case occurs when the charge is trapped at the oxide semiconductor interface located in the center of the channel.

When the gate overdrive is small, additional considerations are needed. First, (3) is no longer valid because the decrease in inversion charge density slows down for a given decrease in V_{OD}. This correction can be handled using the measured CV curve. Second, when the "hole" is larger than the channel length, (1) simplifies to (5). When the "hole" is larger than the channel width, $\Delta I_{Drain}/I_{Drain}$ reaches 1 and cannot grow larger. All of these factors are included in our model calculations.

Experimental: To verify the model, we compare the maximum observed RTN amplitude (single trap) from many devices to the worst case prediction of the model (an upper bound). As the model only considers undoped channels, measuring RTN at large V_{OD} is necessary to minimize random dopant effects in our devices. RTN was measured from threshold (V_{TH}) to V_{DD} ($V_D = 50mV$) on five different (W/L) sized nMOSFETs using a time window of $50~\mu s < t < 20$ s. The total sample size was 280. When multi-level RTN was encountered, only the largest two-level amplitude was used. The channel charge density was extracted from measured CV on large area devices using [8]. An important subtlety is that the appropriate channel length used in the modeling is not the physical gate length or the metallurgical channel length. The length of the inversion layer is significantly shorter than these quantities as demonstrated by Asenov et al. [7]. Since all the devices are from the same 40nm technology, we choose a constant 20 nm ΔL.

Results and Discussion: Fig. 2 compares our model and the measured RTN amplitudes as a function of V_{OD} for the five different device dimensions (dimensions provided in each figure). The model curve predicting an upper bound of RTN amplitude is clearly verified.

Note that there are a few points lying slightly outside the model's prediction. This is expected as doped channel devices were measured (while the model assumes an undoped channel). RTN can be exacerbated by random dopant effects as demonstrated by Asenov et al. [7]. This interaction can be large at low V_{OD} but eventually disappears at high V_{OD}. For future technology nodes, undoped channels are necessary to overcome random dopant effects and the model will agree more with experiment.

With the model successfully verified, it can now be used to perform quantitative predictions for future technology nodes. Fig. 3a shows the maximum "hole" size for silicon channel devices for various EOTs. The minimum size device which meets the criteria of a 10% maximum $\Delta I_{Drain}/I_{Drain}$ RTN is extracted from the hole radius (Fig. 3a) using (1) and shown in Fig. 3b. Fig. 3b, shows the link between RTN, V_{OD}, and device scaling. As device dimensions decrease, a larger V_{OD} is necessary to limit RTN in I_{Drain}. RTN therefore places a new constraint on device and V_{OD} scaling. A possible way out of this predicament is to use very small EOT – an extremely challenging proposition. For example, the ITRS projection for L_{min} at the 10nm node (dashed line in Fig. 3b) should have an EOT less than 0.3nm to limit RTN to 10% at large V_{OD}. However, ITRS projects the EOT at the 10nm node to be 0.8nm and no known solution exists for sub 0.65nm.

For SRAM, threshold voltage variability (ΔV_{th}) is a huge issue. We can estimate the maximum RTN induced ΔV_{th} by equating ΔV_{th} to the V_{OD} necessary to reduce the "hole" to 90% of the device width. The result is shown in Fig. 4. The maximum ΔV_{th} increases rapidly below the 22 nm node, pointing to a serious challenge for SRAM scaling.

We can also examine the case of alternative channel materials. Fig. 5 compares the minimum size device for a maximum $\Delta I_{Drain}/I_{Drain}$ of 10% due to a single trap for Si, Ge, GaAs, InGaAs, and InAs. Little difference in RTN amplitude is estimated for the different channel substrates, a surprising result given the dark space (distance between oxide interface and channel inversion layer) varies considerably [8]. On the other hand, dark space and density of states are a coupled quantity. The larger dark space is offset by the smaller inversion charge density, leading to similar RTN amplitude.

Recently, it is reported that low frequency noise in high-k/metal gate devices are lower than the SiO_2 technology [9-10]. Since low frequency noise results from the superposition of RTNs, it is consistent with our model that lower EOT leads to lower RTN amplitude. Thus, for the first time, a predictable way to control 1/f noise is made possible.

Conclusions: We developed a simple physical model for RTN amplitudes. The model allows for a quantitative prediction of RTN amplitude (worst case) of a single trap in future devices. We simulated the worst-case RTN for a single defect only (most likely case for highly scaled devices). However, more than one defect could exist and this model can serve as the foundation for multi-trap cases.

978-1-4673-0996-7/12 $31.00 © 2012 IEEE

References: [1] N. Tega, *et al.*, *VLSI Symp.*, pp. 50-51, 2009.
[2] K. P. Cheung, *et al.*, VLSI-TSA, 2011.
[3] K. P. Chenug, *et al.*, IRPS to be published, 2012.
[4] L. D. Yau, *et al.*, *IEEE TED*, vol. 16, pp. 170-177, 1969.
[5] G. Reimbold, *IEEE TED*, vol. 31, pp. 1190-1198, 1984.
[6] A. Ohata, *et al.*, *JAP*, vol. 68, pp. 200-204, 1990.
[7] A. Asenov, *et al.*, *IEEE TED*, vol. 50, pp. 839-845, 2003.
[8] Southwick, et al., *IEEE TDMR*, vol. 11 pp. 236-243, 2010.
[8] A. Lubow, *et al.*, *APL*, vol. 96, pp. 122105-3, 2010.
[9] C. H. Jan, *et al.*, *IEEE IEDM*, pp. 27.2.1-27.2.4, 2010.
[10] E. G. Ioannidis, *et al.*, *IEEE IEDM*, pp. 449-451, 2011.

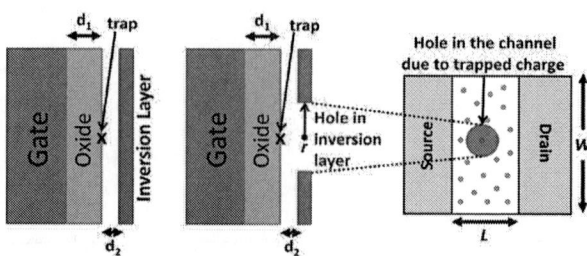

Fig. 1: A hole in the inversion layer due to a trapped charge. If the trapped charge is in the inversion layer, the radius of the hole is the average distance between charge carriers. A distance of 0.4nm (EOT) is used for d_2 in silicon (distance between oxide/semiconductor interface and inversion layer).

Equations:

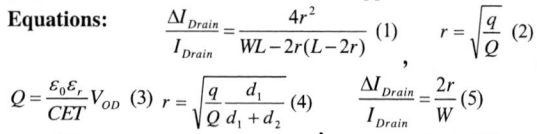

$$\frac{\Delta I_{Drain}}{I_{Drain}} = \frac{4r^2}{WL - 2r(L-2r)} \ (1) \qquad r = \sqrt{\frac{q}{Q}} \ (2)$$

$$Q = \frac{\varepsilon_0 \varepsilon_r}{CET} V_{OD} \ (3) \quad r = \sqrt{\frac{q}{Q} \frac{d_1}{d_1 + d_2}} \ (4) \qquad \frac{\Delta I_{Drain}}{I_{Drain}} = \frac{2r}{W} \ (5)$$

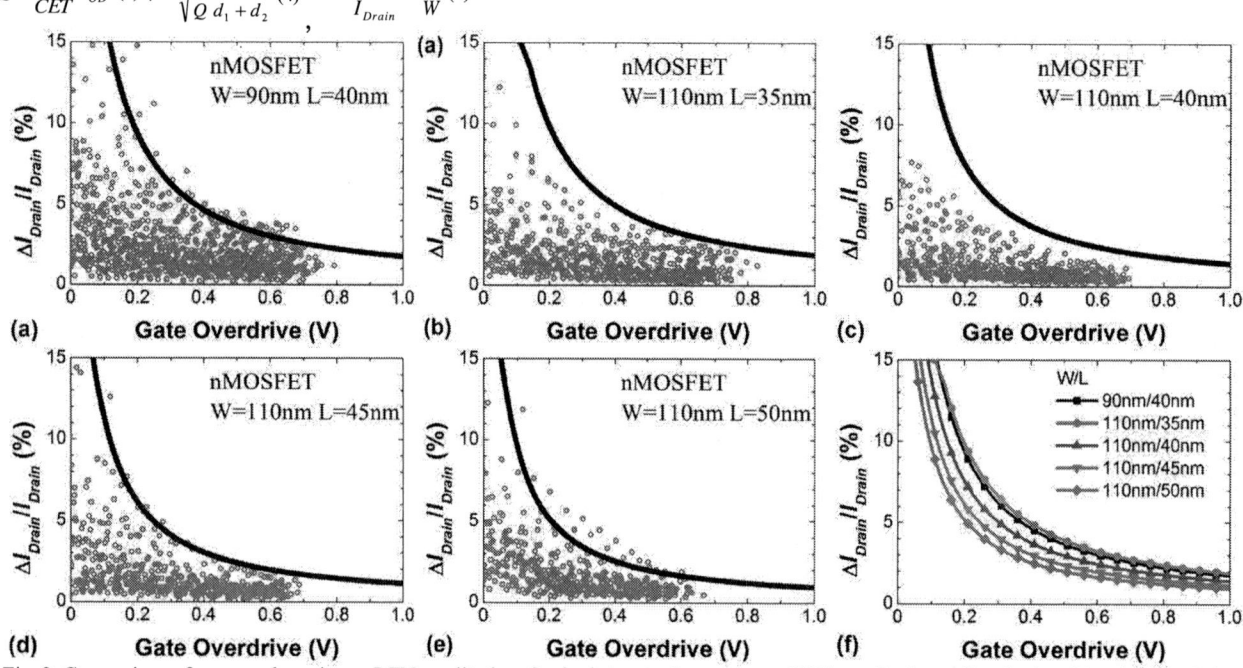

Fig. 2: Comparison of measured maximum RTN amplitudes of a single trap to the maximum RTN amplitude predicted using (1) and (3) for 5 sized devices (a-e). Part (f) shows a comparison of the modeled maximum RTN amplitude for the 5 device sizes. The model serves as an upper bound for the observed RTN amplitudes. At lower gate overdrive voltages measured RTN amplitudes exceed the model due to random dopant fluctuations. The model uses a measured CET of 1.86nm and a dark space (d_2) of 0.4nm (EOT).

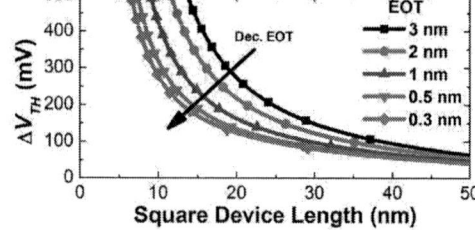

Fig. 4: Estimated maximum threshold voltage change for a square device for various EOTs.

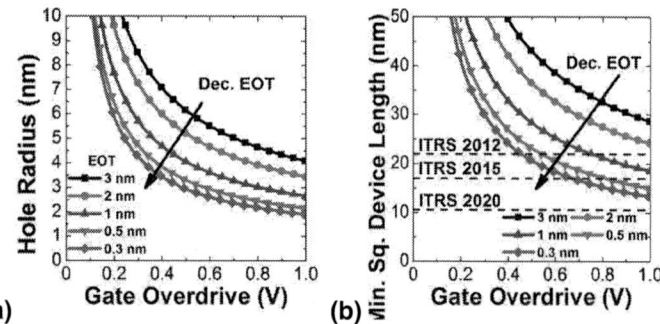

Fig. 3: Inversion layer hole radius (a) and corresponding minimum size square device for a maximum $\Delta I_{Drain}/I_{Drain}$ of 10% (b). The ITRS projections for future technology nodes are given. Extremely thin EOT are necessary to reduce device size and V_{OD}.

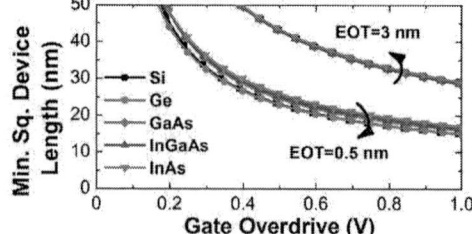

Fig. 5: Estimated minimum square device size for max $\Delta I_{Drain}/I_{Drain}$ of 10% for a single trap for channel substrates of Si, Ge, GaAs, InGaAs, and InAs.

978-1-4673-0996-7/12 $31.00 © 2012 IEEE

Optoelectrical Lifetime Evaluation of Single Holes in SOI MOSFET

Wei Du[1], Dedy Septono Putranto[1,2], Hiroaki Satoh[1], Atsushi Ono[1], Purnomo Sidi Priambodo[2],
Djoko Hartanto[2], and Hiroshi Inokawa[1,*]

[1]Research Institute of Electronics, Shizuoka University, 3-5-1 Johoku, Naka-ku, Hamamatsu, 432-8011 Japan
[2]Electrical Engineering Department, University of Indonesia, Kampus Baru UI, Depok 16424, Indonesia
*Email: inokawa06@rie.shizuoka.ac.jp

Abstract — **Optoelectrical method to evaluate the lifetime of single holes in SOI MOSFET is presented, in which the device is illuminated with a continuous light and the histograms of the digitized drain current is analyzed. It was found that smaller number of holes and the higher transverse electric field greatly enhance the hole lifetime.**

I. INTRODUCTION

Lifetime of the majority carriers in SOI MOSFET is of great importance in relation to the parasitic bipolar effect [1], the retention time of the one-transistor DRAM [2], the maximum count rate of the SOI single-photon detector [3], etc. Especially when the device sizes are scaled down, the behavior of the small number of majority carriers accumulated in the body of SOI MOSFET needs to be analyzed. In this paper, we will report the optoelectrical method to evaluate the lifetime of single holes in the n-channel fully-depleted SOI MOSFET.

II. DEVICE STRUCTURE AND EXPERIMENTAL METHODS

Figure 1 shows the cross-sectional view of the device. A 300-nm length n$^+$ polysilicon gate (LG) is delineated above the 110-nm wide, 50-nm thick silicon channel with p$^-$ dopant concentration less than 10^{15} cm^{-3}. Although an upper gate (UG) exists and hinders the light entrance in this particular design, this can be removed without affecting the measurement. For the detection of the stored holes, both the top and bottom channels can be used in principle, but only the bottom one, enabled by the biasing condition of $V_{LG}<0$ and $V_{SUB}>0$, can be used this time due to the limitation set by the signal-to-noise ratio. In this case, photo-generated holes are stored under the LG whereas bottom electron channel is used as an electrometer to detect the presence of the stored holes. Photo-generation of carriers and their recombination will modulate the electron current, and can be observed as pulses.

In the experiment, we changed V_{LG}, and correspondingly adjust V_{SUB} to keep the baseline drain current at the same level of 1 nA. For different V_{LG}, we evaluated the hole lifetime by the analysis of the drain current histogram.

III. EXPERIMENTAL RESULTS AND DISCUSSION

Figure 2 shows the I_D-V_{LG} curves for different V_{SUB} from −10 to 10 V. As we mentioned, the operating region is located where $V_{LG}<0$ and $V_{SUB}>0$ and electrons flow in the bottom channel, considering the noise level affected by the bias condition. We evaluated a series of devices, and

observed low noise level when $0< V_{SUB} <5$ V, as shown in Fig. 3.

Figure 4(a) shows typical drain current waveforms for different intensity intensities at wavelength of 550 nm. We shift each waveform for clarity. It shows that pulse count increases as the light intensity increases. Moreover, discrete current levels can be seen clearly, corresponding to the different number of holes stored under the lower gate. Figure 4(b) shows an example of histogram of drain current corresponding to the Fig. 4(a). The closed symbols are obtained data and solid lines are fitting curves with Gaussian distribution. The peaks from left to right correspond to the number of stored holes of 0, 1, 2 and 3. When incident light intensity increases, more and more holes are generated, thus possibility of holes being stored under LG increases, resulting in the higher peaks for more stored holes. Hole lifetimes can be extracted, assuming the rate equations, $f_i/\tau_i=f_{i-1}R$ and $\sum f_i=1$, where τ_i is the lifetime for the number of stored holes of i (=0, 1, 2 and 3), f_i is the probability of state obtained from the peak height, and R is the hole generation rate [3].

Figure 5 shows the recombination rate (inverse of the lifetime) as a function of number of stored holes at $V_{LG}=-1$ V. Recombination rate is proportional to the number of stored holes. In other V_{LG} cases ($V_{LG}=-0.97, -1.03$ and -1.06 V), the results show similar tendency, although some data points a little deviate from the proportionality mainly due to statistical error for larger i.

The hole lifetimes at different V_{LG} are depicted in Fig. 6. The hole lifetime increases significantly as V_{LG} decreases. It is estimated that lower V_{LG} (higher transverse electric field) separates the stored hole and electron more effectively, and reduces the probability of recombination, leading to the longer lifetime.

For the application to the single-photon detection [3], quantum efficiency (QE) of the hole generation is important. As shown in Fig. 7, there is proportionality between hole generation rate and incident light intensity, and the nominal QE, assuming the photosensitive area of 300×110 nm^2, is almost independent of the V_{LG}, indicating that the hole lifetime can be controlled without affecting the QE.

IV. CONCLUSIONS

Lifetime of the single holes stored in the body of the n-channel SOI MOSFET was evaluated by the optoelectrical method. It was found that smaller number of holes and the higher transverse electric field applied by the opposite gate and substrate voltages greatly enhance the hole lifetime.

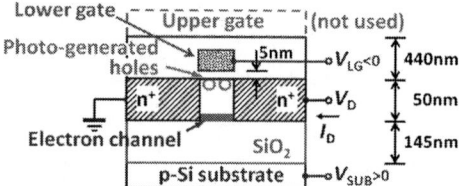

Fig. 1 Device structure (cross-sectional view). The thicknesses of buried oxide, SOI, lower gate (LG) oxide and insulator below the upper gate (UG) are 145, 50, 5 and 440 nm, respectively.

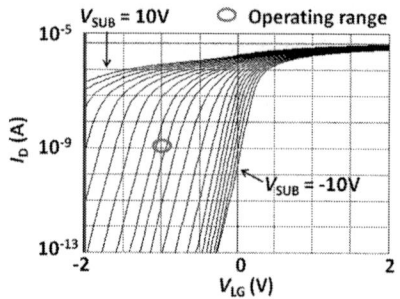

Fig. 2 I_D-V_{LG} curve with V_{SUB} as a parameter. We use bottom electron channel and photo-generated holes are stored below the LG.

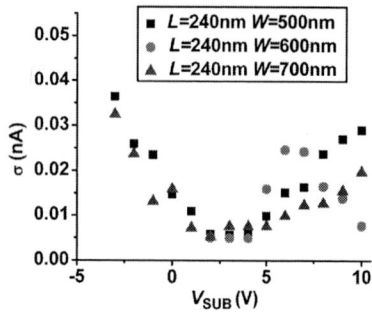

Fig. 3 Noise evaluation in dark condition. σ is the standard deviation of the drain current for the bandwidth of 800 Hz.

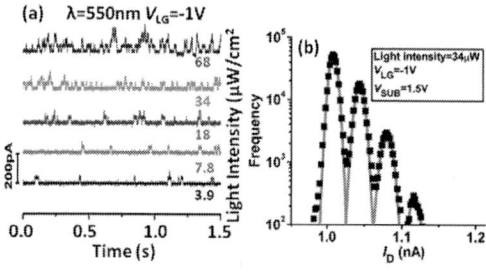

Fig. 4 (a) Drain current waveforms at $V_{LG}= -1$ V. We shifted each waveform for clarity. (b) Histogram of digitized drain current corresponding to (a) with a continuous light intensity of 34 μW/cm^2.

REFERENCES

[1] J.-Y. Choi and J.G. Fossum, *IEEE Trans. Electron Devices*, vol. 38, no. 6, pp. 1384-1391, 1991.

[2] E. Yoshida and T. Tanaka, *IEEE Trans. Electron Devices*, vol. 53, no. 4, pp. 692-697, 2006.

[3] W. Du, H. Inokawa, H. Satoh and A. Ono, *Optics Letters*, vol. 36, no. 15, pp. 2800-2802, 2011.

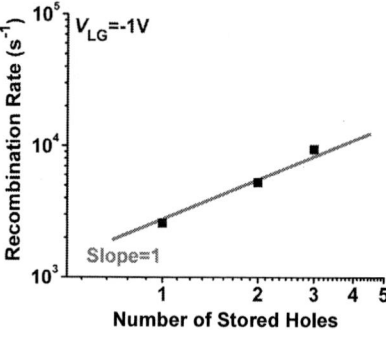

Fig. 5 Hole recombination rate as a function of number of stored holes at $V_{LG}= -1$ V. The data are derived from the analysis of the drain current histograms. Slope of the fitting line is one.

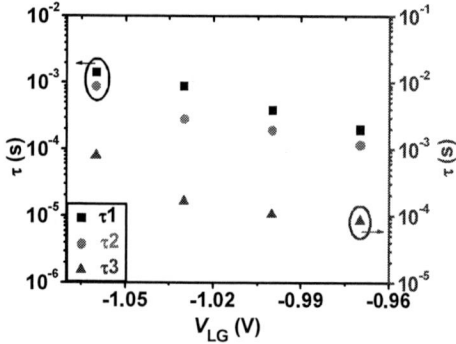

Fig. 6 Hole lifetime as a function of V_{LG}. τ_1, τ_2 and τ_3 are the lifetimes when the number of the stored holes are one, two and three, respectively.

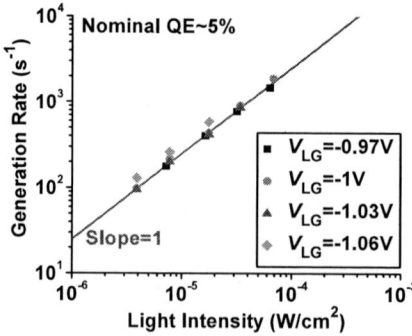

Fig. 7 Hole generation rate as a function of incident light intensity for each bias condition. Slope of the fitting line is one.

Ab initio analysis of donor state deepening in Si nano-channels

D. Moraru[1], Y. Kuzuya[1], E. Hamid[1], T. Mizuno[1], M. Tabe[1], and H. Mizuta[2,3]

[1]Research Institute of Electronics, Shizuoka University, 3-5-1 Johoku, Hamamatsu 432-8011, Japan
Tel: +81-53-478-1335 Fax: +81-53-478-1335 e-mail: daniel@rie.shizuoka.ac.jp
[2]School of Materials Science, Japan Advanced Institute of Science and Technology, Japan
[3]Nano Group, ECS, Faculty of Physical and Applied Sciences, University of Southampton, United Kingdom

1. Introduction

As device dimensions are continuously scaled down, the role of individual dopants in nanoscale silicon devices becomes significantly more active. A new technology field, i.e., *single-dopant devices*, can be developed using one dopant as the active unit for tunneling transport [1-4]. In this frame, it becomes essential to clarify the properties of dopants in Si nanostructures, which are expected to be significantly different from those of bulk Si due to effects such as dielectric and quantum size confinement [5].

In this work, we study by *ab initio* atomistic simulations the energy spectrum of individual dopants in Si nm-size structures. We find that the ionization energy of dopants becomes significantly larger than the value known for bulk Si. Simulation results are supported by the single-electron tunneling characteristics obtained by electrical measurements of nanoscale doped SOI-FETs.

2. Density-of-states spectrum and ionization energy of single donors in nanostructures

In order to obtain the density-of-states spectrum, we used a fully atomistic simulation based on the density functional theory (DFT). Donors were embedded in substitutional positions within Si nano-rods (with typical diameter of 1 nm) or Si nano-plates, coupled to Au electrodes, as illustrated in Fig. 1. Density-of-states (DOS) spectra are also shown in Fig. 1 for nano-rods without dopants [(a)] and with a P donor located near the center [(b)], respectively. Even for the undoped nano-rod, it can be seen that the band gap becomes ~2.4 eV, suggesting a strong quantum size effect. For the case of the nano-rod containing one P donor, near the edge of the conduction band (around $E = 0$ eV, as shown in the zoom-in), a series of discrete DOS peaks are noticed and ascribed to the donor's energy levels. After the identification of the donor ground state (E_{GS}) and LUMO level (E_{LUMO}), as indicated in Fig. 1(b), ionization energy can be calculated as: $E_I = E_{LUMO} - E_{GS}$.

Figure 2 shows the DOS spectra obtained for donors located in a 1-nm-diameter nano-rod, at different positions. The results indicate the s-orbital nature of the donor's ground state and a weak dependence on the donor location. The analysis was extended to nano-rods of different lengths (1~5 nm). Ionization energy (E_I) for all these cases is plotted in Fig. 3. We conclude that E_I is enhanced up to values on the order of 1 eV, about 20 times larger than for shallow donors in bulk Si. E_I is only weakly dependent on the donor position, with

donors at the center of the nano-rod exhibiting the largest E_I, as indicated in Fig. 3.

Our findings are consistent with the data of Diarra *et al.* [5] obtained from a tight-binding approach for nano-rods with radii larger than 1 nm. This leads to donor deactivation, which was experimentally measured as an increase in the resistivity for doped nanowires [6] and as a shift in the threshold voltage in transistors [7].

3. Impact on doped nanoscale-device electrical characteristics

For the experimental study, we fabricated SOI-FETs with nanoscale channels, as shown in Fig. 4(a), doped with P donors. The thickness of the channel is below 5 nm (TEM image in the inset), while the lateral dimensions are ~20 nm for the smallest devices. The Si layer is specifically patterned with the central part of the channel significantly isolated from the source and drain pads, thus limiting the screening effects due to a larger number of donors in the leads. The channel is covered by a 14-nm-thick thermally-grown SiO_2 layer and an Al front gate. In these devices, dielectric and quantum confinement are expected to play a key role in transport.

The I_D-V_G characteristics ($V_D = 5$ mV) are shown in Fig. 4(b) for a temperature $T = 35$ K for one of the smallest disk-pattern FET. A series of several isolated current peaks can be observed at low V_G's, indicating single-electron tunneling transport. The number of P donors in the channel is ~5, and the donors are uncontrollably located at different positions within the channel. It can be expected that the ground-state energy strongly depends on the donor location within the channel. Therefore, we can ascribe consecutive peaks to transport through different donors with different donor ground states. According to our analysis, tunneling barrier height estimated from the experimental characteristics is on the order of 100 meV. These findings may give guiding principles for enhancing the operation temperature of single-dopant devices working in tunneling mode by adjusting the donor's environment.

4. Conclusions

We analyzed by *ab initio* atomistic simulations the energy spectrum of individual donors in Si nanostructures and found significantly enhanced ionization energy (~ 1 eV). By correlating these findings to experimental measurements of doped nanoscale SOI-FETs, design rules can be clarified for tunneling operation of *single-dopant devices* towards room temperature.

Acknowledgements

This work was partly supported by Grants-in-Aid for Scientific Research from MEXT Japan (KAKENHI 22310085, 20246060, 22656082, and 23226009).

References

[1] H. Sellier *et al.*, Phys. Rev. Lett. **97**, 206805 (2006).
[2] M. Tabe *et al.*, Phys. Rev. Lett. **105**, 016803 (2010).
[3] D. Moraru *et al.*, Nanoscale Res. Lett. **6**, 479 (2011).
[4] P. M. Koenraad and M. E. Flatté, Nature Mater. **10**, 91 (2011).
[5] M. Diarra *et al.*, Phys. Rev. B **75**, 045301 (2007).
[6] M. T. Björk *et al.*, Nature Nanotechnol. **4**, 103 (2009).
[7] M. Pierre *et al.*, Nature Nanotechnol. **5**, 133 (2010).

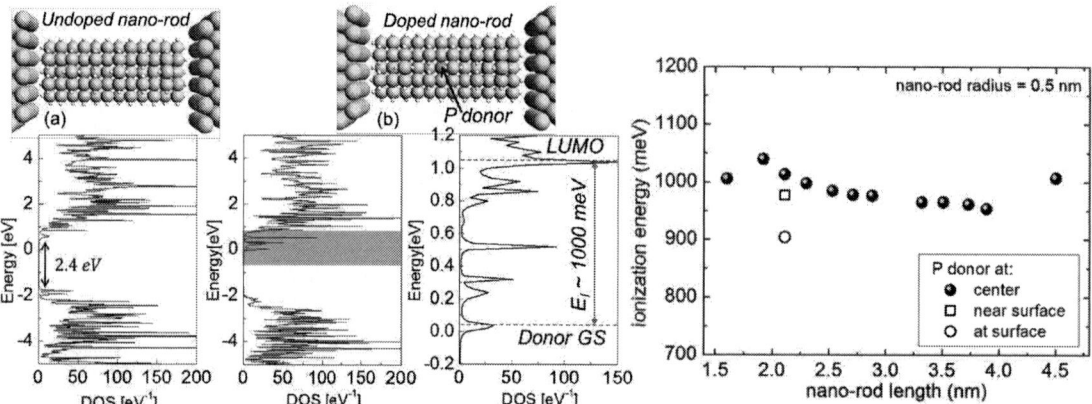

Fig. 1. Atomistic view of Si nano-rods, undoped [(a)] and doped with one phosphorus atom [(b)] and calculated density-of-states spectra for both cases. Right panel shows a zoom-in for energies close to the conduction band edge, indicating the donor ground state, excited states and the LUMO level.

Fig. 3. Ionization energy calculated for a P donor in a nano-rod of different lengths. The small effect due to different radial positions of the donor is also shown.

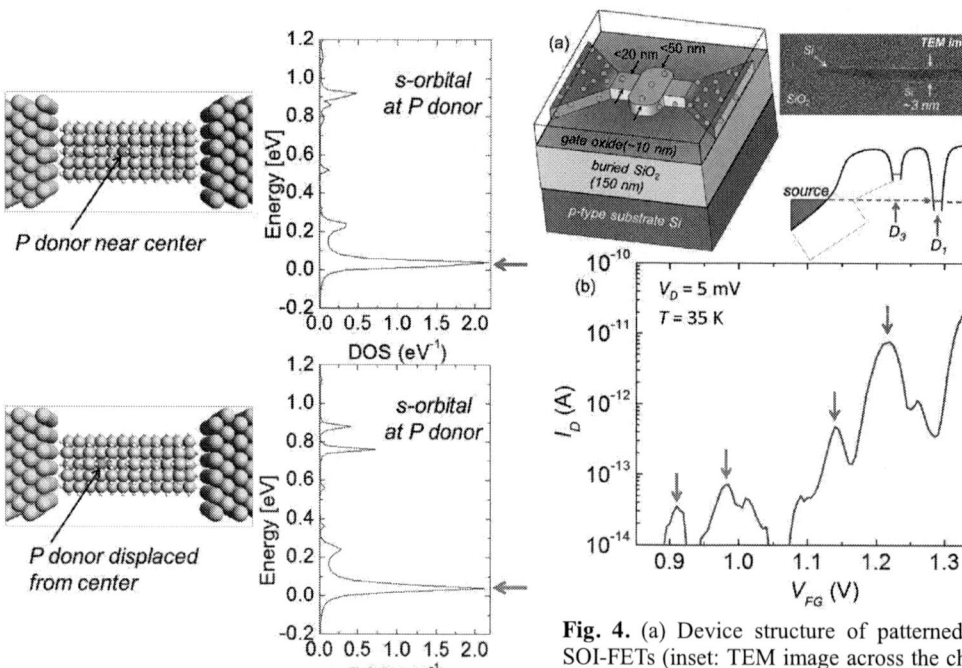

Fig. 2. S-orbital density-of-states spectra projected at the P donor (zoom-in around the conduction band edge) for 2 different Si nano-rods, containing single P donor at different locations along the nanowire axis.

Fig. 4. (a) Device structure of patterned nanoscale SOI-FETs (inset: TEM image across the channel). (b) Experimental results of I_D-V_G characteristics (V_D = 5 mV) measured for a nanoscale disk-patterned SOI-FET for T = 35 K. Current peaks are ascribed to single-electron tunneling via donors with different ground-state energies, as shown in the model.

Channel Length-Dependent Series Resistance?

J.P. Campbell, K.P. Cheung*, S.A. Drozdov**, R.G. Southwick, J.T. Ryan, A.S. Oates[+], J.S. Suehle

Semiconductor and Dimensional Metrology Division, NIST, Gaithersburg, MD 20899, 301-975-8308, *kin.cheung@nist.gov

**Department of Electrical and Computer Engineering, University of Maryland, College Park, MD 20740

[+]TSMC Ltd., Hsin-Chu, Taiwan 300-77, R.O.C

ABSTRACT

A recently developed series resistance (R_{SD}) extraction procedure from a single nanoscale device is shown to be highly robust. Despite these virtues, the technique unexpectedly results in a channel length-dependent R_{SD} which is observed across a wide range of channel lengths and across many different technologies (SiO2, SiON, and high-k) (see Figs. 1a-f). This observation obviously raises some concerning issues and implications as R_{SD} is universally accepted as channel length-independent. However, careful examination of the R_{SD} extraction procedure as well as comparison between R_{SD}-corrected field effect mobility (u_{FE}) and geometric magnetoresistance mobility (u_{MR}) suggests that this unexpected observation may be valid.

RESULTS AND DISCUSSION

As the channel length scales into the nanometer regime, R_{SD} is becoming an increasingly deleterious part of the total resistance (R_{TOT}) [1-3]. This has led to significant processing efforts to reduce the overall R_{SD} [3]. Consequently, proper R_{SD} extraction (especially in short channel devices) is gaining importance. Nearly all R_{SD} extraction techniques rely on I_D-V_G measurements on an array of channel lengths [4]. These "L-array" methods have been shown to be error prone [5]. This issue motivated the effort to develop a new R_{SD} extraction method which requires only a single device [6]. The new method utilizes the ratio of two I_D-V_G measurements taken on the same device at similar (linear) drain biases (V_D). Its appeal lies in no requirement of "difficult to measure" parameters like mobility (μ), effective channel length (L_{eff}), and oxide capacitance (C_{OX}). Recently, the method was benchmarked against "Y-function" based methods and shows strong agreement for short channel devices [7].

In this work we extend this methodology to include the body coefficient term (m) in the linear (low) drain current expression (Eq. 1-2) as well as proper gate leakage current correction (Eq. 3) [8]. These improvements have led to more precise R_{SD} extractions for a large array of linear V_D combinations. This is illustrated in Fig. 2a where $R_{TOT} = V_D/I_D$ and R_{SD} are plotted as a function of gate overdrive for 12 different V_D combinations. In this methodology, R_{SD} extraction is most accurate at higher gate overdrives and is typically taken at operation gate voltages [6]. Notice the tight R_{SD} distribution of 41 Ω at V_G-V_{th} = 0.5V. The method's precision is demonstrated via the insertion of known (symmetric) external series resistances ($R_{external}$), shown in Figs. 2b and 2c. $R_{external}$ values are replicated across all V_D combinations. This correspondence is shown in Fig. 2d.

However, this fairly precise method also results in channel length-dependent R_{SD} (Figs. 1a-f, 2a, 3a-c). While this surprising result counters accepted concepts, it is important to note that this is the first time such an experimental study is possible.

Since the possible error sources associated with the proposed R_{SD} extraction technique are limited, each can be carefully examined. The first of which involves cancellation of the μ, L_{eff}, and C_{OX} terms from Eq. 1. The validity of these cancellations are well justified and have been discussed in detail [6]. The second possible error could involve proper V_{th} extraction (which is error prone). In an attempt to examine the impact of such errors, we artificially shift the extracted V_{th} (linear extrapolation at maximum g_m) by ± 10mV, ± 50mV, and ± 100mV and then re-extract R_{SD}. Fig. 4 shows the percentage change in the extracted R_{SD} as a function of channel length. While V_{th} errors do introduce a small change in the extracted R_{SD}, it affects shorter channel lengths more than longer channel lengths (opposite of our trend in Fig. 1). Thus, it is very unlikely that V_{th} are responsible for the R_{SD} channel length-dependence. Lastly, we examine R_{SD} sensitivity to the body coefficient (m). We extract m from a series of I_D-V_G measurements as a function of body bias [9] and found that in

our case $m \approx 1.2$ for a large number of devices and technologies. Since m is thought to vary between 1.1 and 1.4 [4], we artificially vary m across this range and re-extract R_{SD}. Fig. 5 shows that R_{SD} is relatively insensitive to the choice of m (± 2.5% error across all values of m) and can safely be ignored.

After examining all the potential error sources we reveal a rather robust method. Even though we do not have a solid physical argument for such channel length dependence, we are left to ponder the possibility that the observed R_{SD} channel length-dependence may actually be real.

The implications of a channel length-dependent R_{SD} are undoubtedly copious. Let us examine the effects on the R_{SD} corrected field effect mobility (μ_{FE}) as this parameter receives the majority of attention. As illustrated in Eq. 5, the field effect mobility is easily obtained with knowledge of R_{TOT}, R_{SD}, as well as split-CV measurements to determine C_{OX} and the metallurgical channel length ($L_{MET} = L$ -ΔL). Fig. 6 illustrates (1) μ_{FE} corrected for $R_{SD}(L)$, (2) μ_{FE} corrected for a channel length-independent R_{SD} (taken as the intercept of the R_{TOT} vs. L plot), and (3) μ_{MR} [10] (which is weakly dependent on R_{SD}) at V_G-V_{th} = 0.5V. We note a similar channel length dependence for the μ_{FE} corrected for $R_{SD}(L)$ and the μ_{MR} while $\mu_{FE}(R_{SD}=const)$ appears to be independent of channel length.

In recent years, mobility degradation with decreasing channel length has been reported by many groups [11-14]. As shown in Fig. 6, channel length-dependent mobility results from a length-dependent R_{SD} correction, **not from a length-independent correction**. This channel length-dependent μ_{FE} is also observed for all the other technologies surveyed (Fig. 7a-e) and is not coincidental. Since there is little possibility for R_{TOT} extraction errors, the $\mu_{FE}(R_{SD}=const)$ seems to be in disagreement with the literature.

Accepting our $\mu_{FE}[R_{SD}(L)]$ trend leads to disturbingly high mobility values. Here we turn to the measured μ_{MR} for comparison. μ_{MR} has been found to be channel length-dependent as well [10,15,16]. The relationship between μ_{MR} and other electrically-determined mobilities centers on the geometric magnetoresistance factor and the Hall factor. The observed correspondence between the $\mu_{FE}[R_{SD}(L)]$ and μ_{MR} values is an indication that the Hall factor, and geometric magnetoresistance factor, are very close to 1. The Hall factor is predicted theoretically and demonstrated experimentally to be very close to 1 for the level of channel doping in these advanced devices [17-19]. Thus, the observed high mobility values are in agreement with these literature results.

As for why the extracted mobility values exceed the universal mobility, we note the comment of Cristoloveanu [20] that the universal mobility curve is "constructed based on systematic and careful measurements". Most of these measurements involve the assumption of channel-length independent R_{SD}.

CONCLUSIONS

The channel length-dependent μ_{FE} results are admittedly uncomfortable but seemingly without error. This evidence led to a re-examination of published mobility extraction techniques and resulted in a new awareness that mobility is rarely corrected for R_{SD} and never corrected for a channel length-dependent R_{SD}. We admit that our extracted μ_{FE} values seem quite high, but note that the $\mu_{FE}(R_{SD}=const)$ values are consistent with much of the reported literature (indicating accurate C_{OX} and L_{MET} extractions). Furthermore, since this R_{SD} extraction technique is derived using equations which should be most accurate for longer channels and least accurate for shorter channels, it is very difficult to argue that the long channel results are in error. Thus, we are left with multiple indications which suggest that R_{SD} is channel length-dependent.

References: [1] K. K. Ng and W. T. Lynch, *IEEE Trans. Electron Devices,* **34**, 503 (1987), [2] K. Seong Dong, *et al., IEDM*, 723 (2000) [3] S. Thompson, *et al., Symp. on VLSI Tech.,* 132 (1998), [4] Y. Taur and T.H. Ning, *Fundamentals of Modern VLSI Devices,* NY, NY: Cambridge University Press (1998) [5] S. Biesemans, *et al., IEEE Trans. Electron Dev.,* **45**, 1310 (1998) [6] J.P. Campbell, *et al., IEEE Electron. Dev. Lett.,* **32**, 1047 (2011) [7] L. Pantisano, *et al.,* to be published in *VLSI-TSA,* (2012) [8] P.M. Zeitzoff, *et al., IEEE Electron Dev. Lett.,* **24**, 275 (2003) [9] D.K. Schroder, *Semiconductor Material and Device* Characterization, NY,NY, John Wiley and Sons, (1998) [10] J.P. Campbell, *et al., Symp. on VLS Tech.,* 75 (2011) [11] M.S. Shur, *IEEE Electron Dev. Lett.* **23**, 511 (2002), [12] F. Lime, *et al., ESSDERC,* 525 (2005) [13] A. Cros, *et al., IEDM,* 663 (2006) [14] J. Song, *et al., IEEE Trans. Electron Dev.,* **56**, 533 (2009) [15] Y.M. Meziani, et al., *J. Appl., Phys,* **96**, 5761 (2004) [16] M. Casse, *et al., J. Appl. Phys.,* **105**, 084503 (2009) [17] P. Norton, *et al., Phys. Rev. B,* **8**, 5632 (1973) [18] A. Toriumi, *et al, IEDM* (2006) [19] C. Jungemann, *et al., IEEE Trans. Electron Dev.,* **46**, 1803 (1999) [20] S Cristoloveaunu, *et al, Trans. Electron Dev.,* **57**, 1327 (2010)

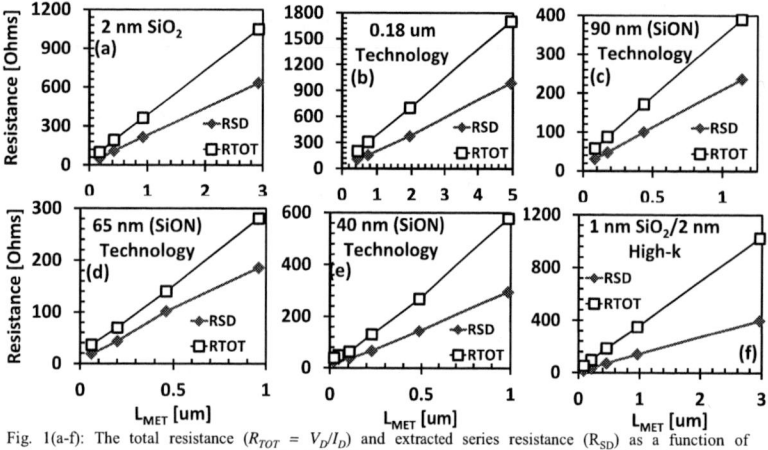

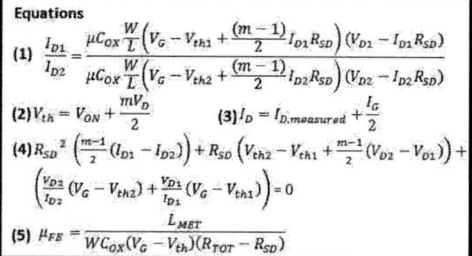

Equations

$$(1) \quad \frac{I_{D1}}{I_{D2}} = \frac{\mu C_{OX} \frac{W}{L}\left(V_G - V_{th1} + \frac{(m-1)}{2} I_{D1} R_{SD}\right)(V_{D1} - I_{D1} R_{SD})}{\mu C_{OX} \frac{W}{L}\left(V_G - V_{th2} + \frac{(m-1)}{2} I_{D2} R_{SD}\right)(V_{D2} - I_{D2} R_{SD})}$$

$$(2)\ V_{th} = V_{ON} + \frac{mV_D}{2} \qquad (3)\ I_D = I_{D,measured} + \frac{I_G}{2}$$

$$(4)\ R_{SD}^2\left(\frac{m-1}{2}(I_{D1}-I_{D2})\right) + R_{SD}\left(V_{th2}-V_{th1}+\frac{m-1}{2}(V_{D2}-V_{D1})\right) + \left(\frac{V_{D2}}{I_{D2}}(V_G-V_{th2})+\frac{V_{D1}}{I_{D1}}(V_G-V_{th1})\right) = 0$$

$$(5)\ \mu_{FE} = \frac{L_{MET}}{WC_{OX}(V_G-V_{th})(R_{TOT}-R_{SD})}$$

Fig. 1(a-f): The total resistance ($R_{TOT} = V_D/I_D$) and extracted series resistance (R_{SD}) as a function of metallurgical channel length ($L_{MET} = L\text{-}\Delta L$) for 6 different technologies spanning SiO$_2$, SiON, and High-k. All exhibit a channel length dependent R_{SD}. All channel widths are 10 um (as drawn).

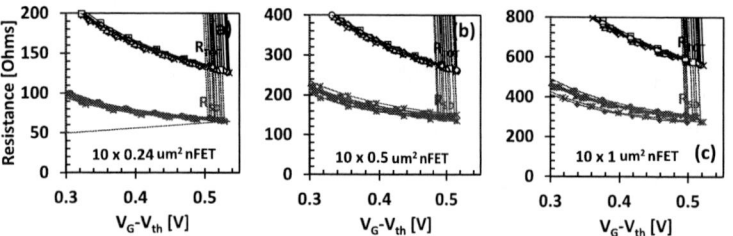

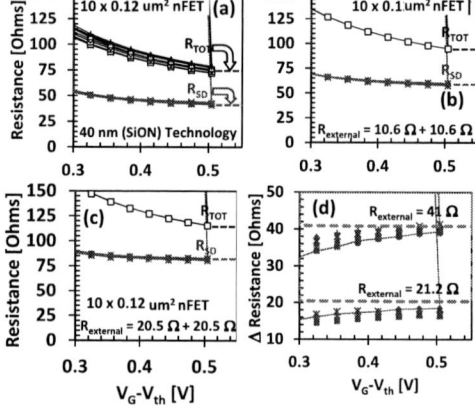

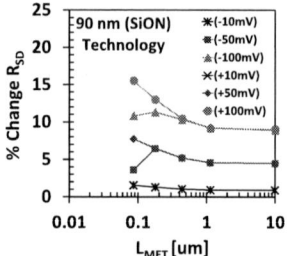

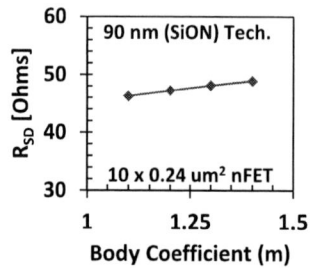

Fig. 3(a-c): R_{TOT} and R_{SD} as a function of gate overdrive for 3 additional channel lengths on the 40 nm (SiON) technology exhibiting a channel length dependent R_{SD}. All channel widths are 10 um (as drawn).

Fig. 2(a) R_{TOT} and R_{SD} as a function of gate overdrive for 12 different combinations of V_D ranging from 5 to 100 mV. R_{TOT} and R_{SD} respond to the introduction of known external resistors ($R_{external}$) for both (b) 21.2 Ω and (c) 41 Ω. (d) Summarizes the $R_{external}$ contributions.

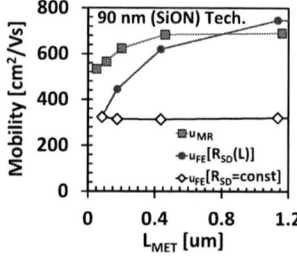

Fig. 4: R_{SD} errors introduced by altering the extracted V_{th} by ±10mV, ±50mV, and ±100mV. All channel widths are 10 um (as drawn).

Fig. 5: R_{SD} extraction is insensitive to errors in body coefficient (m) with errors ± 2.5% across the entire range of m. Similar errors are found for the other devices/technologies examined in this study.

Fig. 6: The geometric magnetoresistance mobility (u_{MR}), field effect mobility corrected for a channel length dependent R_{SD} ($u_{FE}[R_{SD}(L)]$) and the field effect mobility corrected for a constant R_{SD} ($u_{FE}[R_{SD}=\text{const}]$) as a function of L_{MET} (V_G-V_{th} = 0.5V). Note the similar channel length dependent mobility seen in the u_{MR} and the $u_{FE}[R_{SD}(L)]$. All channel widths are 10 um (as drawn).

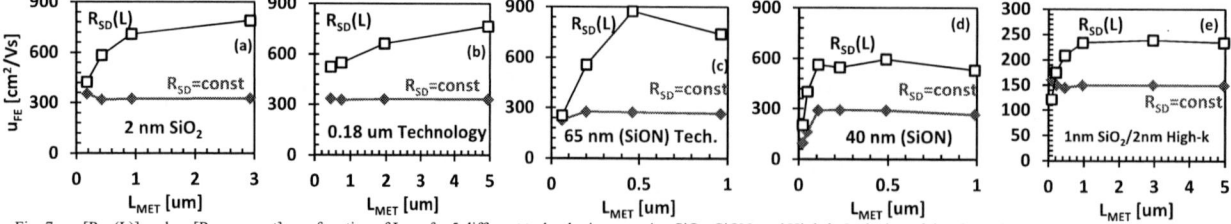

Fig. 7: $u_{FE}[R_{SD}(L)]$ and $u_{FE}[R_{SD} = \text{const}]$ as a function of L_{MET} for 5 different technologies spanning SiO$_2$, SiON, and High-k. Inclusion of the channel length dependent R_{SD} introduces a channel length dependent mobility which has been recently reported in the literature. All channel widths are 10 um (as drawn).

978-1-4673-0996-7/12 $31.00 © 2012 IEEE

Effects of Amorphous Silicon Atomic Density Variation on Series and Contact Resistances in Nanoscale Thin-Film Structures

Min Woo Ryu, Sung-Ho Kim, and Kyung Rok Kim*

School of Electrical and Computer Engineering, Ulsan National Institute of Science and Technology
Ulsan, 689-798 Korea, Tel: +82-52-217-2122, Fax: +82-52-217-2109, *E-mail: krkim@unist.ac.kr

Abstracts – In this study, we investigate the effects of amorphous silicon (a-Si) mass density variations on the electrical series and contact resistance of nanoscale structures for thin-film transistors (TFTs). Impurity distributions according to the variation of a-Si mass density ($\rho_{a\text{-Si}}$) are obtained from Monte-Carlo (MC) method and the resistance extraction are performed by using device simulation based on transfer length method (TLM) with a-Si mobility and Schottky contact model. Under the small variations of ±5% from standard $\rho_{a\text{-Si}}$, electrical resistances are significantly changed with 30% variations from its typical characteristics in nanoscale TFTs.

1. Introduction

Amorphous silicon (a-Si) has been widely used owing to its cost-effective and flexible thin film formation rather than the crystalline silicon (c-Si) epitaxial growth especially in the photovoltaic solar cell and flat panel display industry. The atomic density of a-Si can vary with process conditions [1]. During the intentional ion implantation or indirect plasma-related process, however, impurity distributions in a-Si thin film have not been investigated extensively according to the a-Si atomic density variations yet.

In this work, we report the effects of a-Si atomic density variations on the impurity distributions in a nanoscale a-Si thin film by using Monte-Carlo (MC) ion implantation model [2]. In addition, the resultant electrical characteristics such as series and contact resistance of a nanoscale a-Si thin film have been investigated according to the average atomic density variations of a-Si thin film structures based on device simulation with transfer length method (TLM).

2. Impurity Profiles by Monte-Carlo (MC) Method

Figure 1 shows the a-Si thin film structure with a thickness of $T_{a\text{-Si}}$ in a nanoscale range. A typical 2-dimensional (2D) impurity distribution is presneted in the inset of Fig. 1 by MC implant simulation [3]. MC implant model can predict the impurity profiles in a-Si material considering the average atomic density $N = 1/L^3$, where L is the average path length of projectile bewteen collisions [2]. Figure 2 shows the calculation results of a-Si mass density ($\rho_{a\text{-Si}}$) by considering the average number of a-Si atoms as in case of c-Si unit cell. The variation range of $\rho_{a\text{-Si}}$ has been taken by the average number change from 6 to 10 in a c-Si unit cell dimension, which coresspond to ±25% variations from the standard mass density.

One more advantage of MC simulation is that all kinds of impurity profiles can be predicted such as nitrogen (N) and hydrogen (H) ion as shown in Fig 3 (a) and (b), respectively. These impurities and their composite ions can be implanted during the plasma-related process. Accoding to the variations of $\rho_{a\text{-Si}}$, the significant changes of profile metrics such as projection range (R_p) and junction depth (X_j) have been confirmed.

For the evaluation of electrical resistance variability of a-Si thin film structures according to the $\rho_{a\text{-Si}}$ variations, we also performed MC implant simulations with the conventional impurities such as boron (B), phosporous (P), and arsenic (As) as shown Fig. 4. As $\rho_{a\text{-Si}}$ decreases, the R_p and X_j increases and this tendency become significant in the lighter impurity such as H and B compared with N and P at the same conditions. The values of R_p and X_j are plotted as a function of $\rho_{a\text{-Si}}$ in Fig. 5 for B, P, and As, respectively.

3. Electrical Characteristics and Discussions

To investigate the electrical characteristics from the impurity profile variations, 3-dimensional (3D) device simulations have been performed by applying TLM structure to the nanoscale a-Si thin film with the range of thickness ($T_{a\text{-Si}}$) from 5 nm to 300 nm. The doping-dependent a-Si mobility and Schottky contact model are used in this device simulation [3,4]. Figure 6 (a) and (b) show the series resistance (R_s) and contact resistance (R_c) as a function of $\rho_{a\text{-Si}}$ in P-doped n-type a-Si thin film, respectively. As $\rho_{a\text{-Si}}$ decreases, R_p and X_j increases and thus, R_s increases as $\rho_{a\text{-Si}}$ and $T_{a\text{-Si}}$ decreases. According to the small variation of ±0.1 g/cm^3 (±5%) from standard $\rho_{a\text{-Si}} = 2$ g/cm^3, significant R_s variations of ±11.4 kΩ (27.4%) are observed in a nanoscale sample with $T_{a\text{-Si}} = 5$ nm. In case of Schottky contact, R_c is less sensitive to barrier width determined by doping profile near surface than the barrier height (Φ_B) and thus, R_c changes more significantly according to the $T_{a\text{-Si}}$ variations under 10 nm by the physical confinement of the barriers in nanoscale film.

4. Conclusions

In this work, quantitative analysis of nanoscale a-Si thin film characteristics was performed according to the average atomic density variations. In the nanoscale a-Si thin-film structures, the atomic density variation effects on the electrical device performance should be taken in careful considerations.

Acknowledgements

This work was supported by the National Research Foundation of Korea (NRF) funded by the Ministry of Education, Science and Technology (2011-0005342).

References

[1] D. Gracin et. al., J. Appl. Cryst., 40, s373 (2007)
[2] S. Tian, J. Appl. Phys., 93(10), p.5893 (2003)
[3] M.C. Wang et, al., Appl. Phys. Lett., 90, 192114 (2007)
[4] Sentaurus TCAD Manual v. D-2010.03 (Synopsys, 2010)

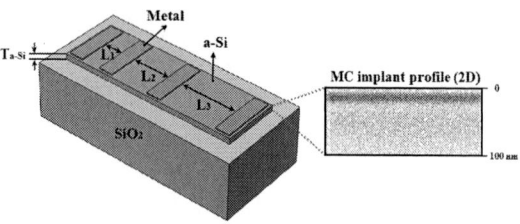

Fig. 1. Schematics of a-Si thin film structure with a thickness of T_{a-Si} in a nanoscale range (L_1, L_2 and L_3 are values of contact distance for TLM structure). A typical 2-dimensional (2D) MC implant profile is presented as inset.

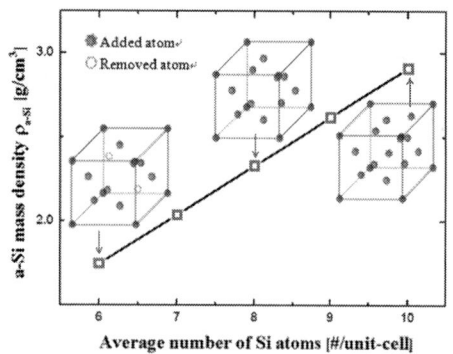

Fig. 2. The calculation results of a-Si mass density by considering the average number of a-Si atoms as in case of c-Si unit cell. The average number change from 6 to 10 in a c-Si unit cell dimension(inset), which coresspond to ±25%.

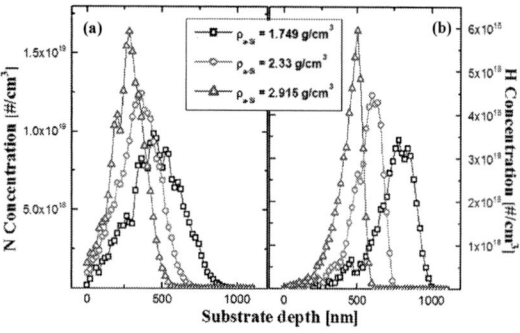

Fig. 3. MC implant simulation results for (a) nirtrogen and (b) hydrogen profiles according to the a-Si mass density (ρ_{a-Si}) variation. (energy= 10 keV, dose= 5e13 cm^{-2})

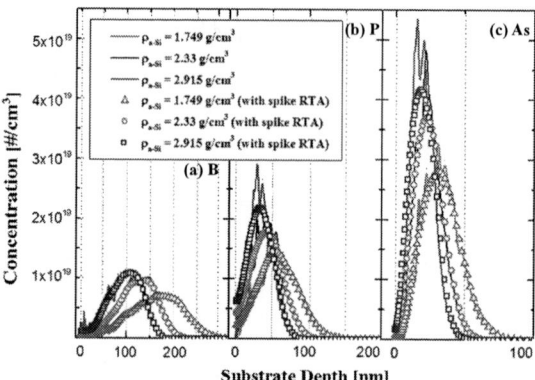

Fig. 4. The MC implant simulation results for (a) boron (b) phosphorous (c) arsenic according to the a-Si mess density (ρ_{a-Si}) variation. (energy= 10 keV, dose= 5x10^{13} cm^{-2}, spike RTA: 1050°C/1s)

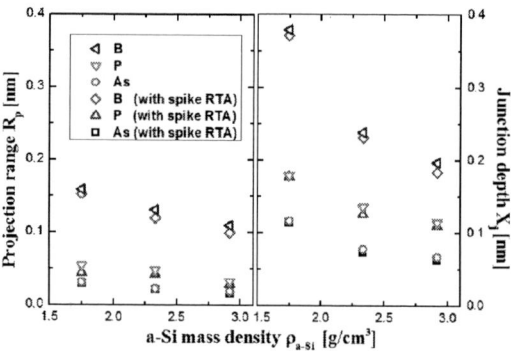

Fig. 5. Results of extracted profile mertics such as projection range (R_p) and junction depth (X_j) accoding to ρ_{a-Si} variation. (energy= 10 keV, dose= 5x10^{13} cm^{-2}, spike RTA: 1050°C/1s)

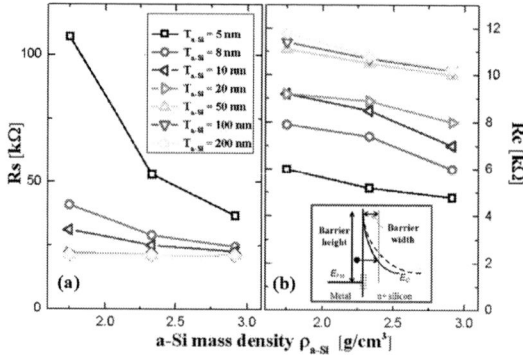

Fig. 6. Electrical characteristics of P-doped n-type nanoscale a-Si thin-film structures as a function of ρ_{a-Si}: (a) series resistance (R_s) and (b) contact resistance (R_c) for various values of the film thickness T_{a-Si} = 5 ~ 200 nm. Inset of (b) shows the schematic energy band diagram in case of Schottky contact with barrier height and width, which is determined by the doping profile near the surface. In this simulation, barrier height of Φ_B=0.2 eV is used.

Evaluation of Scattering in Asymmetric Quasi-Ballistic DG-MOSFET

Gai Liu[1], *Gang Du[1], Tiao Lu[2], Xiaoyan Liu[1], Pingwen Zhang[2] and Xing Zhang[1]

1 Institute of Microelectronics, Peking University, Beijing 100871, P. R. China

2 School of Mathematical Sciences, LMAM and CAPT, Peking University, Beijing 100871, P. R. China

* gangdu@pku.edu.cn, phone:+861062767915, fax:+861062751789

Introduction

Quasi-ballistic transport in MOSFET has been attracting much attention and scattering in quasi-ballistic transport is the major limit to the performance of MOSFET[1][2][3]. In this work, we examine important parameters that characterize scattering in quasi-ballistic transport under different scattering conditions.

Simulation Method Verification

Double-gate MOSFET with different top-gate voltage Vgt and back-gate voltage Vgb is simulated in this work, device structure and parameters are shown in Fig.1.

A time-dependent multi-subband Boltzmann transport equation solver[4][5] is used in this work, Intra-valley AP (acoustic phonon), OP (optical phonon) scattering and inter-valley OP scattering are included.

Fig.2 shows Ids-Vgt curve under different back-gate voltage Vgb. Fig.3 illustrates the potential profile along the channel. It can be seen that the back-gate voltage Vgb modulates the barrier height (BH), thus controls the current of the simulated MOSFET. According to ballistic theory [2], under high bias and ballistic condition, Ids only depends on BH, which is observed in Fig.4.

Simulated results of ballistic cases are compared with established ballistic transport theory developed by *Natori* and *Lundstrom*[1][2]: electron density at the top of barrier is shown in Fig.5 and good agreement is observed (other parameters are also compared and good agreement is achieved), which confirms the validity of our simulation.

Fig.6 shows Ids-Vds curve when scattering is present and the velocity profile along the channel is shown in Fig.7. A detailed study of average velocity at the top of barrier is presented in Fig.8&9. Average velocity deviates from ballistic theory under low Vds or when scattering is present because electrons that are back-scattered or injected from the drain reduces the average velocity at the top of barrier. Fig.10 supports this idea by showing an increasing current in –x direction when Vds decreases.

Parameter Extraction

The influence of scattering on the performance of MOSFET is characterized by the back-scattering coefficient [2]:

$$v_{scat} = \frac{1-r}{1+r} v_{ballistic} \quad (1.1), \quad r = \frac{l_{kT}}{l_{kT} + \lambda} \quad (1.2)$$

Back-scattering coefficient r, KT-layer length l_{kT} and momentum relaxation length λ are extracted from simulation using the following method:

By comparing average velocity under ballistic and non-ballistic conditions, r is extracted using (1.1). l_{kT} is extracted through self-consistent potential profile and λ is calculated by (1.2).

Result and Discussion

a. Potential Profile along the Channel Potential profile along the channel is of great importance in determining back-scattering coefficient because it directly determines kT-layer length l_{kT}. According to [6], potential profile near the top of barrier can be approximated by a power function:

$$V(x) = Kx^{\frac{1}{a}} \quad (2)$$

Where K and a are fitting parameters. Using (2), we find in our case K=-0.00164, a=0.56 best fits the simulated result (as shown in Fig.11) and (2) is a good approximation for potential profile near the top of barrier.

b. Back-Scattering Coefficient Adopting the extraction method described above, the back-scattering coefficient r is shown in Fig.12 as a function of gate voltage under high drain bias (Vds=0.5V). It is observed that r increases slightly with Vgb but this effect has little impact on the total current. On the other hand, r is 26% higher when inter-valley scattering is taken into account, and r is twice in 18nm device than in 9nm device, which illustrates the fact that the scaling of gate length plays a vital role in reducing scattering effect.

c. kT-layer Length According to [6], kT layer length can be expressed as (a and b are fitting parameters):

$$l_{kT} = L_{channel} \left(\beta \frac{k_B T/q}{V_{ds}} \right)^a$$

However, the influence of gate voltage on l_{kT} is ignored in [6]. To better describe l_{kT}, we add a term BH in the denominator to account for the influence of gate voltage, where BH is the barrier height (in volt) at the top of barrier, which is modulated by gate voltage:

$$l_{kT} = L_{channel} \left(\beta \frac{k_B T/q}{V_{ds} + BH} \right)^a \quad (3)$$

Fig.13. illustrates the simulation and theoretical results using (3), a=0.56, β=1.4 are used and good agreement is observed.

d. Momentum Relaxation Length Momentum relaxation length λ as a function of gate voltage under high Vds is shown in Fig.14. Similar to back-scattering coefficient, λ decreases slightly as gate voltage increases, but this decrease has little influence on the total current. A strong dependence of λ on scattering type is observed while different gate length results in almost the same λ.

According to [6], λ can be expressed as:

$$\lambda = \left(\frac{2\mu}{v_t} \frac{k_B T}{q} \right) \frac{\{F_0(\eta_F)\}^2}{F_{-1}(\eta_F) F_{1/2}(\eta_F)}$$

Where v_t is the thermal velocity, μ is the low field effective channel mobility (the role of μ in quasi-ballistic transport is explained in [6]), η_F is the normalized Fermi level and F is the Fermi-Dirac integral. We find μ=976cm^{2}V^{-1}s^{-1} for intra-valley scattering and μ=830cm^{2}V^{-1}s^{-1} for both intra and inter-valley scattering.

Conclusion

Quasi-ballistic asymmetric DG-MOSFET has been simulated using a multi-subband Boltzmann transport equation solver and important parameters regarding to back-scattering at the top of barrier are carefully studied in this work. It is observed that the simulated results are in good agreement with established theory and phonon scattering still plays an important role in limiting the performance of MOSFET even when gate length is scaled down to sub-10nm.

Acknowledgments

This work is supported by the National Fundamental Basic Research Program of China (Grant No 2010CBA00604)

References

[1] Natori, J. Appl.Phys., vol.76, no.8,1994.[2] Lundstrom, et al., T-ED, VOL. 49, NO. 1, 2002.[3] Lundstrom, EDL, VOL. 18, NO. 7, 1997.[4] Tiao Lu,Gang Du,et al., IWCE,2009.[5] Tiao Lu,Gang Du,et al., Commun.Comput.Phys. VOL.10, NO.2, 2011.[6] Rahman and Lundstrom, T-ED, VOL.49, NO.3, 2002.

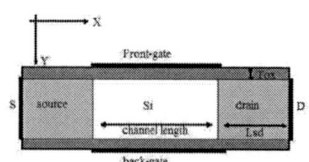

Fig.1 Simulated device structure.
EOT (effective oxide thickness)=1nm, *Lsd*=9.9nm, source/drain doping=10^{20}cm^{-3}, channel is intrinsic, gate length varies from 18nm to 9nm.

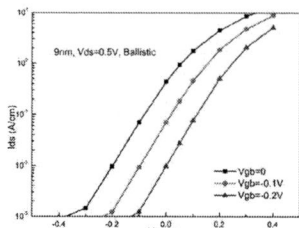

Fig.2. Simulated *Ids-Vgt* curve.

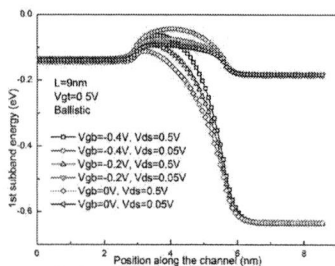

Fig.3. potential profile along the channel under different gate and drain bias

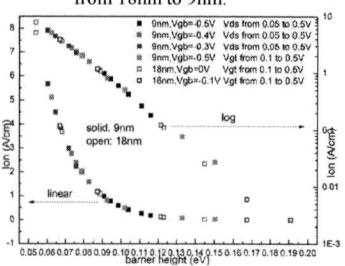

Fig.4. *Ids* dependence on barrier height for ballistic MOSFET

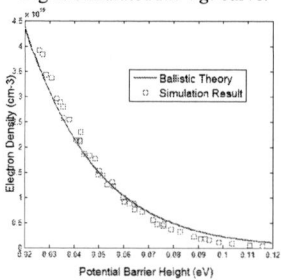

Fig.5. Electron density at the top of barrier as a function of potential barrier height

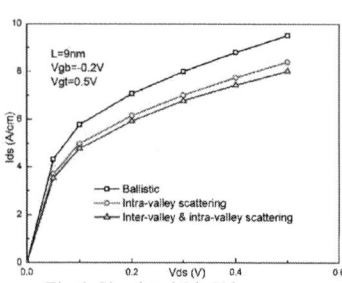

Fig.6. Simulated *Ids-Vds* curve.

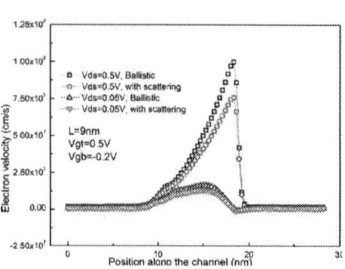

Fig.7. Electron velocity profile along the channel

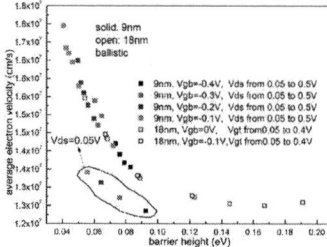

Fig.8. Average electron velocity at the top of barrier for ballistic MOSFET as a function of *BH*, deviation under *Vds*=0.05V is due to large number of electrons injected from the drain, which is rare under high drain bias.

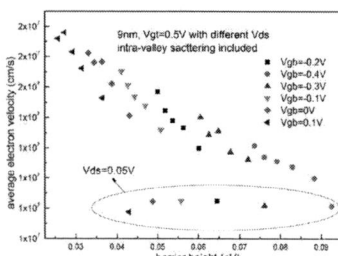

Fig.9. Average velocity as a function of barrier height (*BH*), scattering is present, deviation under low bias is due to the same reason described in Fig.8.

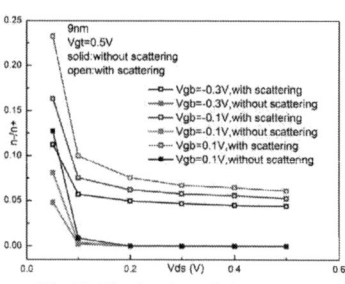

Fig.10. The fraction of electrons flowing in −x direction(n-/n+)

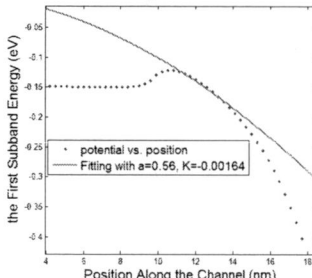

Fig.11. Fitting potential profile near the source side using (2)

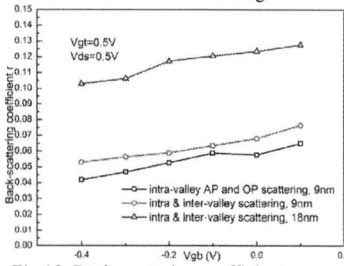

Fig.12. Back-scattering coefficient r as a function of *Vgb*, r increases when inter-valley scattering is present and *r* is twice in 18nm device than in 9nm device.

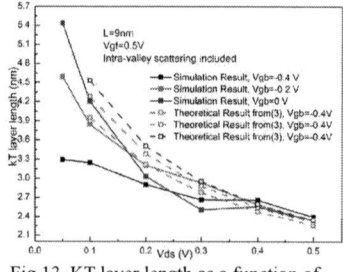

Fig.13. KT layer length as a function of Vds, dash-lines indicate theoretical calculation from (3).

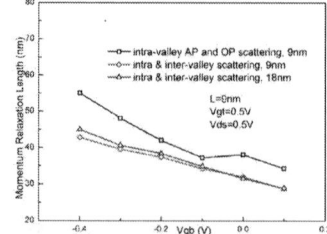

Fig.14. Momentum relaxation length λ as a function of *Vgb* under different scattering conditions, λ decreases with the increase of gate voltage, and inter-valley scattering reduces λ. λ is almost identical for devices of different gate length.

Fabrication and Characterization of a Pi-Gate Ultrathin Body Junctionless poly-Si TFTs

Jia-Jiun Wu[1], Hung-Bin Chen[1], Ming-Hung Han[1], Yung-Chun Wu[2], and Chun-Yen Chang[1]

[1]Department of Electronics Engineering and Institute of Electronics, National Chiao Tung University
[2]Department of Engineering and System Science, National Tsing Hua University
Email: boboangle2003@gmail.com

Abstract

A novel method of fabricate ultrathin body (UTB) junctionless TFTs (JLTFT) with sub-10nm poly-Si channel has been successfully demonstrated. It is no additional mask for lithography. The cost of fabrication flow can be reduced by a novel method, that demonstrate at this work. UTB JLTFT has low threshold voltage and steep subthreshold slop 160 mV/dec at W/L=0.7um/1um. An ON/OFF current ratio is about 10^6, and transconductance does not decrease rapidly at a high drain voltage.

Introduction

Recently, the junctionless(JL) MOS devices have been research for the substitute of conventional inversion mode MOSFET. Owing to the channel of JL MOS devices are high doped, the gate controllability is important. It must able to turn it off. Ultrathin body (UTB) is a way to improve gate controllability [1] [5]. Moreover, many works focus on scaling of gate length and thickness of channel [2-5] and temperature performance. But these works do not have easily fabrication flow. UTB JLTFT is promising candidates for future CMOS devices. Moreover, the UTB has better gate controllability to enhances subthreshold swing and reduces short channel effect (SCE) [6]. Formation of junctionless TFT (JLTFT) has large I_{ON}/I_{OFF} ratio [2] [6]. JL enhances the electric field and transconductance, so I_{ON} is improvement, the DIBL and SS can be reduced [1][7].

Device Fabrication

The key fabrication steps are show in fig. 1. The ultra-thin body (UTB) junction-less (JL) TFTs was fabricated on Si wafer with 400nm thermal oxide and 50nm undoped amorphous silicon (a-Si) for active area. Annealing process is for solid phase recrystallized (SPC) at 600 °C for 24 hr in N_2 ambient. Source/Drain (S/D) and Active Area (AA) defined by e-beam lithography [Fig. 1(a)]. Dip HF to etch SiO_2 for arise AA and formed the π-gate structure [Fig. 1(b)]. Dip HF twice to thinning AA to form sub-30nm active area. Dry oxidation 8nm for gate oxide and reduced active area to obtained ultra-thin channel. In situ doped n^+-poly Si 250nm for gate and define by e-beam lithography [Fig. 1(c)]. All these films were deposited by low pressure chemical vapor deposition. Fig. 2 show the TEM image of a fabricated device cross-section along gate direction. The π-gate formation and uniform thickness channel shows in Fig. 2(a). In Fig. 2(b-c), the UTB channel thickness at corner is sub-3nm, and about 10nm at middle. In Fig. 2(d) shows that the crystallization of UTB is clearly. The method of dry oxidation trimming can reduce the grain boundary to form quasi-crystal formation. Nonetheless, the method of dry oxidation trimming reduces channel width, too. Channel width from 1um reduced to 0.7um.

Result and Discussion

I-V characteristics of fabricated devices are shown in Fig. 3(a-b). In Fig. 3(a), an ON/OFF ratio is about 10^6 for various

V_D, and the transconductance (g_m) does not degrade rapidly like inversion mode TFT at high V_G. UTB JLTFT structure can improve gate controllability, and reduced electric field at surface of channel. Electric field of UTB JLTFT is more symmetrical than inversion-mode TFT, and the peak electron concentration is at the middle of channel. According from paper [1], JLTFT can reduce electron surface scattering, so the g_m degradation can be reduced, as result show in Fig. 3(b). Owing to good controllability, a low subthreshold swing (SS) of 160 V/dec is obtained. In Fig. 4(a), SS dependent on gate length and V_D. The SS proportional to supply drain voltage at different gate length. In Fig. 4(b), the short channel has worse DIBL effect at high supply drain voltage. Long channel devices have better SS and DIBL, as shown in Fig. 4(a-b).

Fig. 5 shows measured drain current as a function of gate voltage at different temperature for supply drain voltage equal 1V for (a) L=1um and (b) L=0.5um and (c) L=0.2um. Fig. 5 presents the threshold voltage decreases and subthreshold swing increases for all devices while temperature is increased. The ON current increases and OFF current increases while temperature is increased, it can be observed in Fig. 5, but the ON/OFF ratio is decreased. Comparison with Fig. 5(a-b-c), it shows long channel has less threshold degradation and ON/OFF ratio degradation. Fig. 6 shows the measured OFF current and GIDL as a function of temperature at different gate length. OFF current is defined as the current at V_G-V_{TH}=-1V and V_D=1V. GIDL current is defined as the current at V_G-V_{TH}=-1.5V and V_D=1V. GIDL current is caused by band to band tunneling, GIDL as a function of temperature and gate length.

In Table 1, comparison with important parameters, these parameters degrade as long as gate length decreases. At gate length equal to 0.2um (L=0.2um), threshold voltage has a shift, obviously. Short channel effect (SCE) of UTB JLTFT is outstanding at L=0.2um.

Conclusion

Novel fabrication flow for UTB JLTFTs with sub-10nm quasi crystal Si has been successfully demonstrated. For n-channel TFT, a low V_{TH} and good subthreshold swing are obtained. Due to UTB JL structure, the good gate controllability can be obtained and the DIBL effect can be reduced, and transconductance can keep at high level at a high drain voltage. Considering the simplification in devices fabrication, the method of dry oxidation trimming is feasible for SOP.

Reference

[1] Jean-Pierre Colinge et al., APL, 2010, pp. 073510.
[2] Horng-Chih Lin et al., EDL, 2011, pp. 53-55.
[3] Chi-Woo Lee et al., TED, 2010, pp. 620-625.
[4] C. Lee et al., SSE, 2010, pp. 97-103.
[5] van der Steen et al., TED, 2009, pp.1999-2007.
[6] B. Kim et al., IRPS, 2011, pp.126
[7] Chi-Woo Lee et al., APL, 2009, pp. 053511.

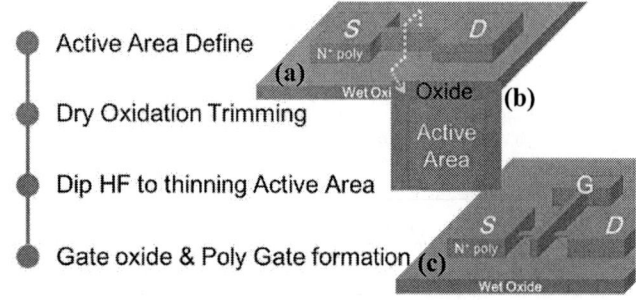

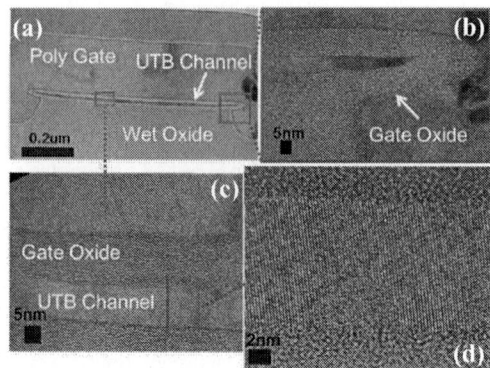

Fig.1 (a)-(c) The key fabrication flow of UTB JLTFTs device.

Fig. 2 (a) The TEM of UTB JLTFT device active area. (b)-(c) The UTB is about 3nm at corner and 10nm in middle. (d) A quasi-crystal area in UTB is observed.

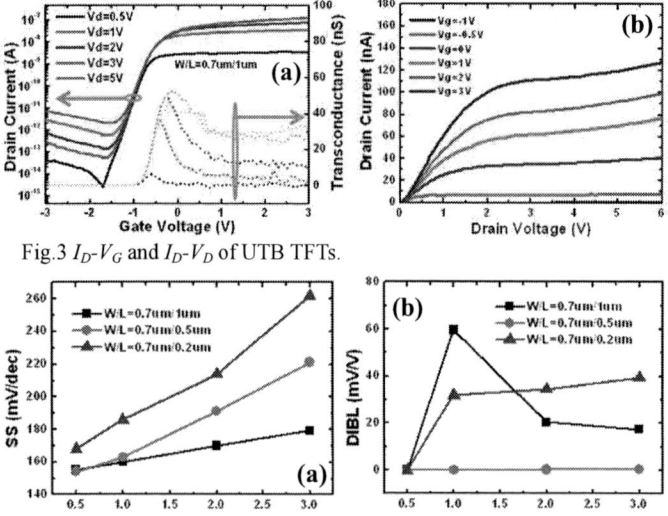

Fig.3 I_D-V_G and I_D-V_D of UTB TFTs.

	L=1.0um	L=0.5um	L=0.2um
V_{TH}(V)	-0.919	-0.987	-1.33
S.S. (mV/dec)	160	163	186
I_{ON}/I_{OFF} (V_G/V_D=3V/1V)	~10^6	~10^5	~10^5
T_{EFF} @ I_D=10^{-9} (mV/°C)	-3.3	-5.2	-6.6
T_{ION} (%/°C)	2.24	5.15	2.79

Table 1. Comparison with important parameters at difference gate length at this work. All result are at drain voltage equal to 1V. (V_{TH} @ I_{ON}=0.1nA, V_{DS}=0.5V; V_{DS}=1V, W=0.7um; $T_{ION} \equiv$ increment of I_{ON} @ V_G=3, V_D=1V per degree of temperature.)

Fig. 4 (a) The subthreshold swing degradation of UTB JLTFT devices with short channel is seriously. (b) UTB with junctionless channel reduced DIBL effect at large drain voltage.

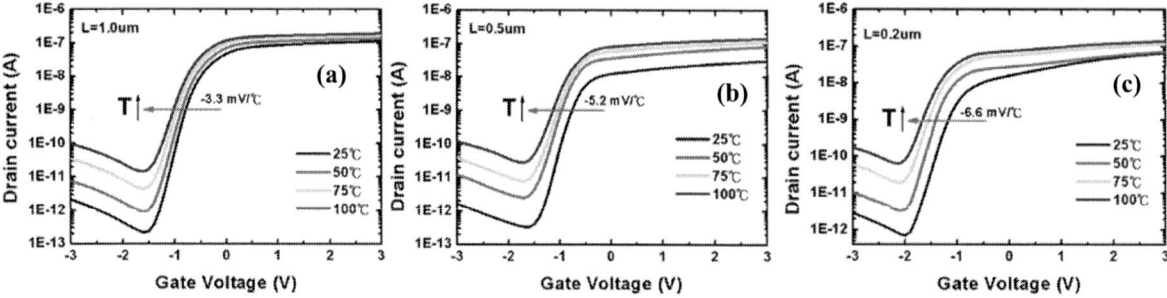

Fig. 5 Measured I_D-V_G characteristics with various temperature of UTB JLTFT with (a)L=1um and (b)L=0.5um and (c) L=0.2um.

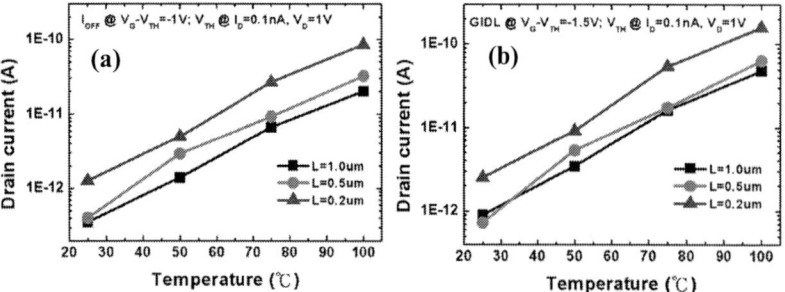

Fig. 6 I_{OFF} and GIDL of UTB JLTFT as a function of temperature and gate length.

978-1-4673-0996-7/12 $31.00 © 2012 IEEE

Quantum Drift-Diffusion and Quantum Energy Balance Simulation of Nanowire Junctionless Transistors

O. Badami[a], N. Kumar[a], D. Saha[a], S. Ganguly[a]

[a]Centre of Excellence in Nanoelectronics & Department of Electrical Engineering,
Indian Institute of Technology Bombay, India.
Email: swaroop.ganguly@gmail.com, Phone: (+91)9769597403.

Multiple gate MOSFETs (MuGFET) have gained significant attention as the scaling of the conventional MOSFET comes to an end. Of the possible architectures, the gate-all-around nanowire (NW) transistor offers the best gate control over the channel. In order to model GAA nanowire devices for channel lengths less than 10nm, while preserving a connection to the drift-diffusion framework familiar to device engineers, we have developed a quantum-corrected transport simulator that includes Quantum Drift-Diffusion (QDD) [1] and Quantum Energy Balance (QEB). This formalism is applied to the example of the NW junctionless transistor (JLT), an interesting modification to the NW-MOSFET obtained by replacing the n+-p-n+ structure by a bar of n+ region, that promises smaller variability [2].

The discretization of QDD and QEB was achieved by the Generalized Scharfetter-Gummel scheme proposed for their classical counterparts by Tang [3]. Fig. 1 shows the schematic of the cylindrical nanowire transistor, a cross-sectional cut, and the final simulation structure obtained after the exploiting its radial symmetry. A uniform doping of 5e19 cm^3 was used for the NW-JLT. For verification of the code, a NW-MOSFET structure having the same cross-sectional area as used by [4] was simulated; the results shown in Fig. 2(a) are in close agreement with what have been obtained there. Fig. 2(b) shows a TCAD simulation [5] illustrating the capability of the density-gradient method, a derivative of QDD, to model tunneling. These results also suggest that for scaled geometries, i.e. NW diameters less than 5nm, we need to consider only the 1st sub-band as the contribution from the higher energy levels is negligible. Fig. 2(c) and (d) show the probability distribution of the electrons in the 1st and 2nd sub-bands in a cross-section.

In this work, we have solved the moments of the Wigner equation, replacing the sum of the Bohm and electrostatic potential by the sub-band energy [2], resulting in the 1D transport equations

$$\frac{dJ_{sb}}{dz} = 0; \quad J_{sb} = n_{sb}\mu_{sb}\frac{dE_{sb}}{dz} + k_B\frac{d(n_{sb}T)}{dz}; \quad \frac{dS_{sb}}{dz} = \frac{J_{sb}}{q}\frac{dE_{sb}}{dz} - \frac{3}{2}n_{sb}k_B\frac{T - T_l}{\tau_\epsilon};$$

$$S_{sb} = \frac{k_B\delta T n_{sb}\mu_{sb}}{-q}\frac{dE_{sb}}{dz} + \frac{k_B^2 T^2}{q}\mu_{sb}\delta\frac{dn_{sb}}{dz} - \frac{k_B^2 T}{q}(\delta + \Delta)\frac{dT}{dz}n_{sb}\mu_{sb}$$

where E$_{sb}$ is the sub-band energy, S is the flux of energy flow, $\delta = \frac{<\tau\epsilon^2>}{<\tau\epsilon>}, \Delta = \frac{1}{(Tk_B)^2}[\frac{<\tau\epsilon^3>}{<\tau\epsilon>} - (\frac{<\tau\epsilon^2>}{<\tau\epsilon>})^2]$

and τ (ϵ) is the energy relaxation time.

The dimensionality of the transport equation is reduced to 1D by working in mode-space in the radial direction normal to transport. A coupled system of Schrodinger (for the radial direction) and Poisson (for radial and axial directions) was solved along with the 1D-transport equations comprising (a) only QDD, or, (b) QDD + QEB. A flowchart depicting the procedure for one bias point is shown in Fig. 3. Fig. 5 illustrates the potential and electron density profile obtained for Vg=0.1V and Vd=0.1V. Variation of the 1st sub-band energy along the Z direction for different values of gate and drain bias is shown in Fig. 5 for both transport options (a) and (b). Note that we have used φ$_{ms}$ = 0eV leading to a depletion-mode device in this example. Fig. 6 shows the comparison of I$_D$-V$_D$ and I$_D$-V$_G$ obtained for the NW-JLT, again for both options. We note that QEB results in significant change over QDD both in terms of the on-current that impacts digital applications, and output resistance that impacts analog. This difference is not due to the electron density which remains about the same in both models (Fig. 7) but because of the increase in electron temperature shown in Fig. 8.

A QDD + QEB transport formalism for NW devices has been developed and applied to the extremely-scaled JLT.

978-1-4673-0996-7/12 $31.00 © 2012 IEEE

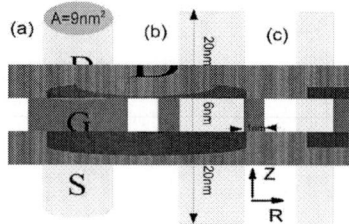

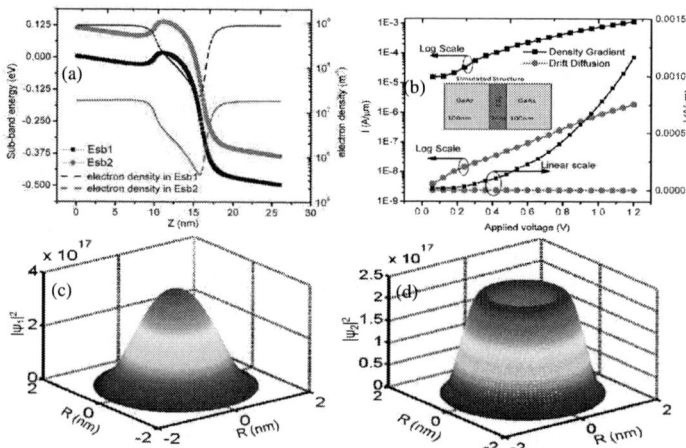

Fig. 1. (a) Schematic of the nanowire transistor. (b) Structure obtained exploiting cylindrical symmetry, and, (c) half-plane fundamental domain used for simulation.

Fig. 2. (a) Comparison of the electron density in the 1st and 2nd sub-bands. Similar results are also obtained in [1]. (b) Results from Sentaurus indicating tunneling effect through a single barrier. Probability distribution of electrons in the cross-section in (c) 1st and (d) 2nd sub-band.

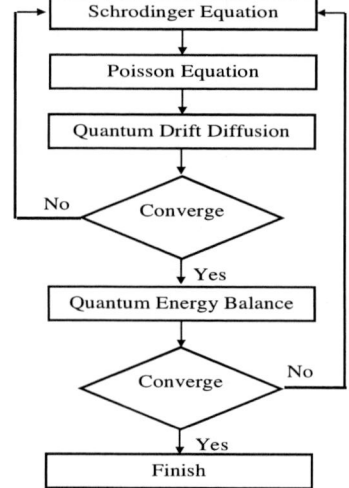

Fig. 3. Flowchart for simulation of QEB for one bias point.

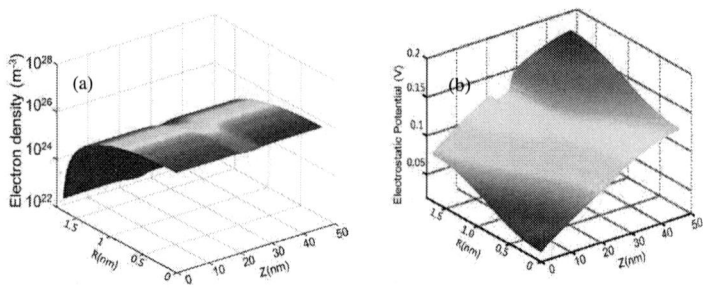

Fig. 4. (a) Electron density and (b) Potential profile in the junctionless transistor for Vd=0.1V and Vg=0.1V

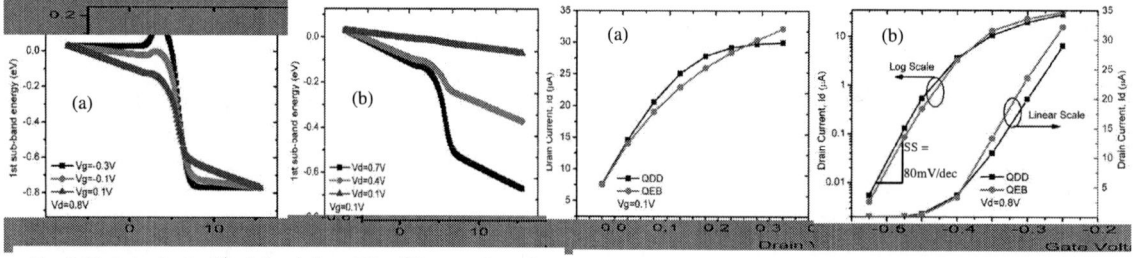

Fig. 5. Variation in the 1st sub-band along Z for different values of (a) gate and (b) drain voltage.

Fig. 6. Comparison of (a) Id-Vd and (b) Id-Vg with QEB and QDD.

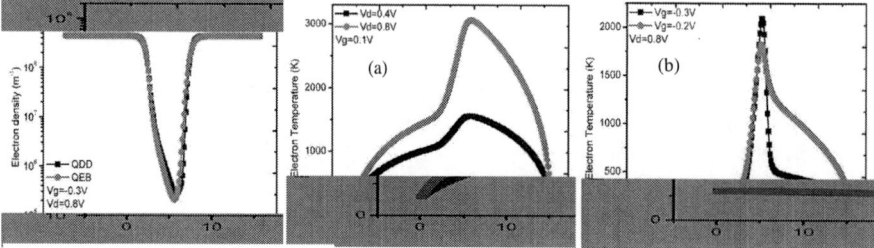

Fig. 7. 1D Electron density along Z.

Fig. 8. Electron temperature profile for different (a) drain and (b) gate bias.

References:
[1] Giorgio Baccarani et al. *Solid Stat Electronics*, vol. 52, Issue 4, pp. 526 32, APRIL 2008. [2] Akhavan, N.D. e al, SOI Conference (SOI), 2011 IEEI International. [3] Ting-Wei Tang. *IEEI Transactions on Electron Devices* VOL.ED-31, NO.12, DECEMBEI 1984. [4] Yoshihiro Yamada et al *IEEE Transactions on Electro Devices* VOL: 56, NO.7, pp. 1396 1401, JULY 2009. [5]S-Device Manua Version F 2011.09

978-1-4673-0996-7/12 $31.00 © 2012 IEEE

Characteristics and Sensitivity of p-Type Junctionless Gate-All-Around Nanowire Transistor

Ming-Hung Han[1,*], Yi-Ruei Jhan[2], Jia-Jiun Wu[1], Hung-Bin Chen[1], Yung-Chun Wu[2], and Chun-Yen Chang[1]

[1]Department of Electronics Engineering and Institute of Electronics, National Chiao Tung University
1001 Ta-Hsueh Road, Hsinchu City, Hsinchu 300, Taiwan
[2]Department of Engineering and System Science, National Tsing Hua University
101, Section 2, Kuang-Fu Road, Hsinchu City, Hsinchu 300, Taiwan
[*]Corresponding author. Tel: 886-3-571-2121 ext: 52981; Email: minghunghan@gmail.com

Abstract

In this study, we for the first time assess the characteristics and sensitivity of p-type junctionless (JL) gate-all around (GAA) nanowire transistor using 3D quantum transport device simulation for CMOS technology implementation. Since the doping concentration of p-type junctionless nanowire transistor does not as high as in n-type device due solid solubility of boron in silicon, it can be made by using midgap gate electrode material for appropriate threshold voltage. The p-type JLGAA transistor shows good on/off current ratio and better short channel characteristics compare to conventional inversion mode GAA structure. The sensitivity analyses show that the channel thickness affects the device performance such as threshold voltage (V_{th}), on current (I_{on}), and off current (I_{off}) significantly. In contrast, the channel length and oxide thickness have less impact owing to well control of short channel effect.

Introduction

Silicon nanowires with gate-all-around (GAA) are the ultimate structures, and potential candidates for next generation high-speed and low-power electronic devices due to their ideal gate controllability, low leakage, and enhanced carrier transport property [1]. Recently, the concept of junctionless (JL) MOS devices, which have their channel doping of the same type and comparable level to that of the source/drain has been proposed and explored [2-6]. GAA architecture is extremely suitable for fabricating JL devices as the gate will create depletion region from all sides to turn off the device. However, the high surface-to-volume ratio in nanowire improves their sensitivity to any changes. Moreover, many works in literature are focus on n-type device to investigate the applicable of device scaling [2-5], less of them discuss the p-type JL transistor for CMOS technology implementation. In this paper, we for the first time explore the characteristics and sensitivities of p-type JLGAA nanowire transistor using 3D quantum transport device simulation for CMOS integration.

Simulation Results and Discussions

Figure 1 shows the p-type JLGAA nanowire device structure and parameters used in this work, the gate length is 15nm, oxide thickness is 1nm, and the channel is square shape with width 5nm. Considering the solid solubility of boron in silicon, the doping concentration of channel and source/drain are $2 \times 10^{19} cm^{-3}$, which is relatively low compare with n-type device [2-4]. The gate work-function is 4.66 eV for appropriate threshold voltage, which can be easily implemented by midgap metal material such as TiN and so on. To accurately examine the numerical results with nano-scale device, the device simulation is performed by solving 3D quantum transport equations using commercial tool Synopsys DESSIS. Figure 2 shows the I_d-V_g curves of the explored p-type JLGAA device, the threshold voltage (V_{th}) is about -300mV, on current (I_{on}) and off current (I_{off}) are 238µA/µm and 0.3nA/µm without using any channel engineering or strain technology. The subthreshold slope (SS) 62mV/Dec approaches to ideal value,

and drain induced barrier lowering (DIBL) which is defined as the V_{th} difference between $V_d = -0.05V$ and $V_d = -1V$ equals to only 24.8mV. These results show that JLGAA devices have excellent short channel characteristics for device scaling, and V_{th} roll-off are better than conventional inversion mode (IM) GAA devices with n-type channel doping $10^{18} cm^{-3}$, as shown in Fig. 3. Figure 4 shows the hole density (top) and electric field (bottom) distributions in the channel at device OFF ($V_g = 0V$) and ON ($V_g = -1V$). The positions of A and A' are indicated in Fig. 1, and the electric field in oxide region is not shown. The hole density is concentrated in the middle of channel region for both device OFF and ON. In Fig. 4 (a), due to the GAA structure and channel width is thin enough; the channel region is nearly full depleted and creates relatively uniform electric field, which can reduce the bulk conduction of hole. In Fig. 4 (b), the electric field is significantly reduced in the middle of channel and is increased near the surface, which benefits the conduction of hole. Therefore, the on/off current ratio is large. Figure 5 shows the sensitivity analysis of the devices by changing the gate length (L), doping concentration (D), oxide thickness (T), and channel thickness (W) for ±10% and ±20%. The ΔV_{th}, ΔI_{on}, and ΔI_{off} are defined as the difference between maximum and minimum values divide the performance of standard device. In Fig. 5(a), the change of L shows the smallest influence to ΔV_{th}, ΔI_{on}, and ΔI_{off}, which indicate the excellent control of the short channel effect. For changing D in Fig. 5(b), a relatively high sensitivity of the V_{th} is observed; the ΔV_{th} is about 22%. In Fig. 5(c), changing the oxide thickness shows larger impact of V_{th} and less effect of device I_{on} compare the change of L. In Fig. 5(d), the ΔV_{th}, ΔI_{on}, and ΔI_{off} are 45%, 77%, and 1313%, respectively. The large deviations of characteristics vary with W show that the channel thickness is the most important parameter for JLGAA nanowire transistor due to full depletion condition of the channel.

Conclusions

In this paper, we have explored the characteristics and sensitivities of p-type JLGAA nanowire transistor for nano-CMOS technology. The simulation results show that p-type JLGAA transistors can be implemented by midgap gate material and have good on/off current ratio and excellent short channel characteristics. The lowest value of electric field and hole density positions are quite different, which benefit the current conduction. The sensitivity analyses show that the channel thickness affects the device performance significantly and channel length has less impact.

Acknowledgements

This work was supported in part by Taiwan National Science Council (NSC) under Contract NSC-100-2221-E-030-.

References

[1] F. Yang et al., in VLSI, 2004, pp. 740-742.
[2] C. Lee et al., SSE, vol. 54, 2010, pp. 97-103.
[3] J. Colinge et al., APL, vol. 96, 2010, pp. 073510.
[4] N. Akhavan et al., APL, vol. 98, 2011, pp. 103510.
[5] H. Lin et al., IEEE EDL, vol. 33, no. 1, 2012, pp. 53-55.
[6] R. Rios et al., IEEE EDL, vol. 32, no. 9, 2011, pp. 1170-1172.

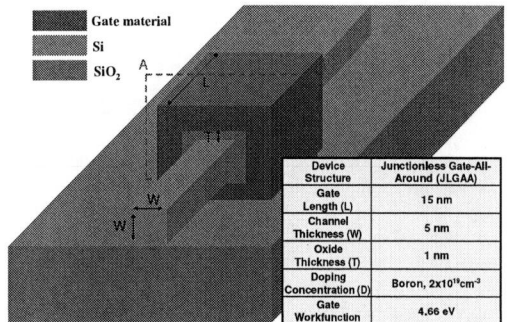

Fig. 1. The device structure and parameters of simulated junctionless gate-all-around (JLGAA) nanowire transistor.

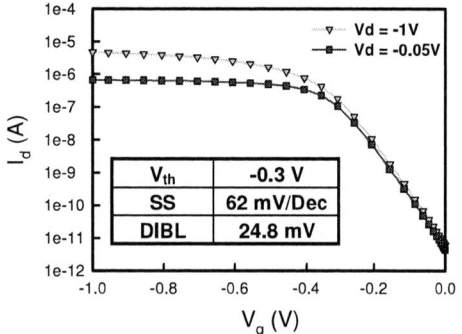

Fig. 2. The I_d-V_g curves of the p-type JLGAA nanowire transistor, the threshold voltage (V_{th}), subthreshold slope (SS), and DIBL are shown in the inset.

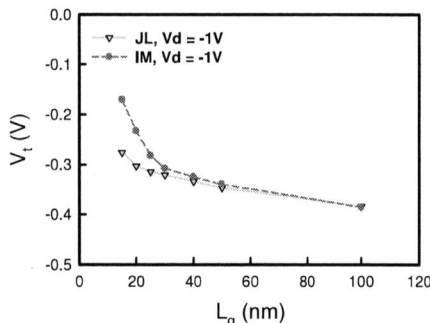

Fig. 3. The V_{th} roll-off comparison between junctionless (JL) and conventional inversion mode (IM) GAA transistor.

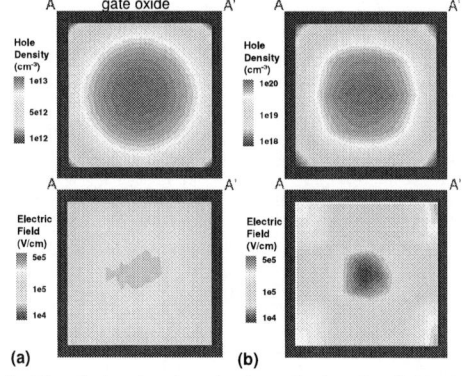

Fig. 4. The hole density (top) and electric field (bottom) distributions in the channel at device (a) OFF ($V_g = 0V$) and (b) ON ($V_g = -1V$).

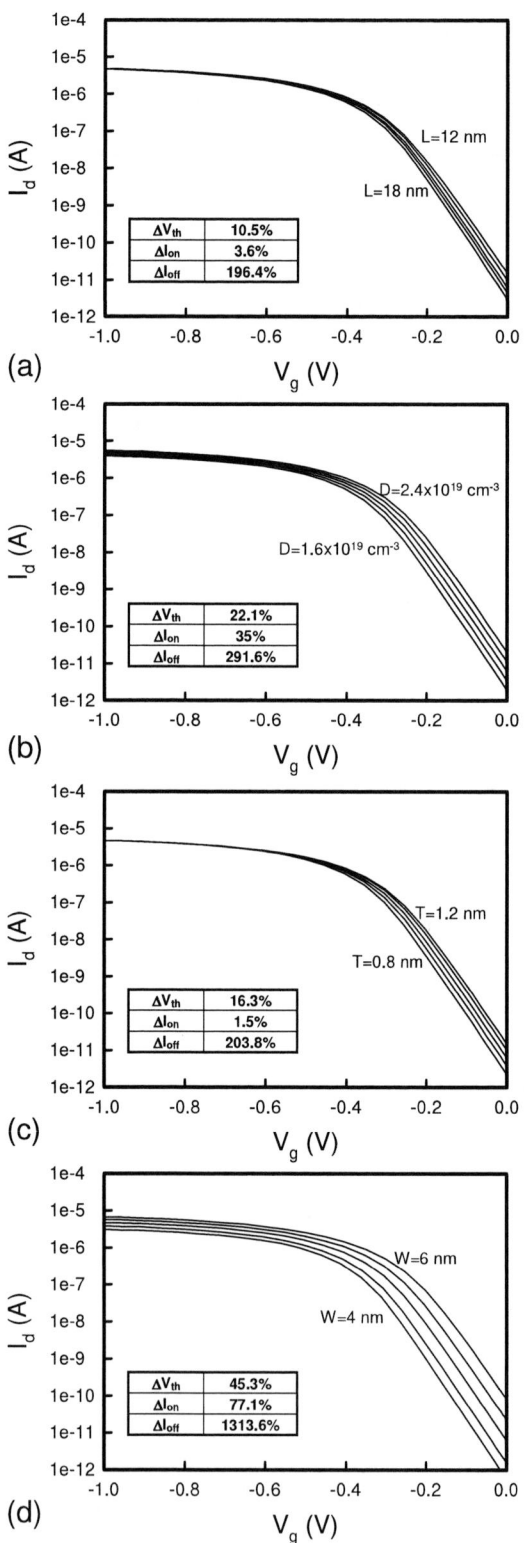

Fig. 5. The I_d-V_g curves for (a) gate length L, (b) doping concentration D, (c) oxide thickness T, and (d) channel thickness W change ±10% and ±20%; the deviations of V_{th}, I_{on}, and I_{off} are shown in the inset.

Analysis of Hysteresis Characteristics of Fabricated SiNW Biosensor in Aqueous Environment with Reference Electrode

Jung Han Lee[1]*, Jieun Lee[2], Min-Chul Sun[1,3], Won Hee Lee[2], Mihee Uhm[2], Seonwook Hwang[2], In-Young Chung[4], Dong Myong Kim[2], and Dae Hwan Kim[2] and Byung-Gook Park[1]

[1]Inter-university Semiconductor Research Center (ISRC) and School of Electrical Engineering and Computer Science, Seoul National University, Seoul 151-742, Korea.

[2]School of Electrical Engineering, Kookmin University, Seongbuk-gu, Seoul, 136-702, Korea

[3]TD (S. LSI), Semiconductor Business Group, Samsung Electronics Co., Ltd., Yongin 446-711, S. Korea

[4]Department of Electronics and Communications Engineering, Kwangwoon University, Seoul, 139-701, Korea

Tel.: +82-2-880-7279, Fax: +82-2-882-4658, E-mail address: kusa159@snu.ac.kr

Abstract

A silicon nanowire field effect transistor (SiNW FET) was fabricated through the fabrication method compatible with that of MOSFET including back-end process without lift-off process. However, when it is working in an aqueous solution, the SiNW device as well as other transducer devices has various inherent instability problems such as hysteresis characteristics. We observed the hysteresis in DI water (DW) and confirmed that it is caused by mobile ion effect in DW with various experimental results.

1. Introduction

Silicon nanowire field effect transistor has been widely researched for highly sensitive, free-labeling and real-time detection owing to good uniformity, controllable alignment, and high surface-to-volume ratio [1]. In addition, the fabrication of the SiNW FET has the compatibility with the fabrication process of complementary metal-oxide-semiconductor (CMOS), which enables the co-integration of SiNW FET sensor and CMOS described in Fig.1(a) and (b), respectively, for high performance sensor system [2]. However, when it is working in an aqueous solution, SiNW FET sensor shows hysteresis characteristics which are one of the problems of the unstable operation. The hysteresis effects caused by back gate coupling [3] and mobile ion effect [4] have been reported when operating in DW and K^+ base buffer solutions without reference electrode, respectively. Hysteresis phenomenon was also observed when measuring with a reference electrode. In this work, we experimentally investigate the mobile ion effect in the aqueous solution through DC measurement with changing integration time, stress effect, and the influence of buffered solution.

2. Experimental Details

Figure 2 and 3 describe the fabrication process of SiNW FET sensors and its SEM and optical images. After implanting channel dopants for p-region and n-region, the active regions were defined by mix-and-match process of e-beam lithography and photo lithography on a 100 nm thick (100) silicon-on-insulator (SOI) wafer as shown in Fig. 2(a) and Fig. 3(a-b). The gates of SiNW FET and MOSFET were formed through photolithography as shown in Fig. 2(b) and (c), respectively. As shown in Fig. 3(c), the

channel of SiNW FET was not covered with gate, unlike that of MOSFET, to bind chemical molecules. Then, conventional back-end process was conducted as shown in Fig. 2(d) and Fig. 3(d). Finally, the sensing areas were formed for only SiNW FET sensor. The 80 nm-width and 5 μm length SiNW FET was measured in aqueous solution with microfluidic channel and Ag/AgCl reference electrode.

3. Results and Discussion

Figure 4 plots the measured curves of the drain current (I_D) as a function of the liquid gate voltage (V_{LG}) of the SiNW sensor under the DW environments, depending on the integration time change of the sweep. After reverse sweep direction (1.2 V → -1 V), the forward sweep direction (-1 V → 1.2 V) was measured when source/drain bias voltages were applied to -1 V/0V, respectively. As the integration time (T_i) is increased, the hysteresis gap is decreased, which is different from the results of back gate sweep measurement. The drain current of reverse sweep is positively shifted and that of forward sweep is negatively shifted. It seems that the increased integration time allows current measurement after all mobile ions are stable. To clarify the effect, stress effect was measured with stress bias (V_{STR} = 1.2 V) in DW as shown in Fig. 5 which shows different results from that of integration time change. The drain current of reverse sweep is negatively shifted and that of forward sweep is negatively shifted less than that of reverse sweep since the stress effects are compensated during negative voltage sweep. Finally, the drain current was measured under buffered solution, 0.1 M potassium phosphate buffers (pH 7), depending on the integration time change of the sweep. As shown in Fig.6, no hysteresis phenomenon was observed and it shows same results as integration time is increased. The shifts of drain current are similar to stress effect since K^+ ions suppress the other mobile ion effects.

4. Conclusion

In this work, SiNW FETs for biosensor and MOSFET were fabricated simultaneously through a top-down fabrication method including conventional back-end process. However, we observed the hysteresis effects measuring fabricated SiNW FETs with reference electrode. To find out the causes, electrical measurements were conducted with changing integration time, stress effect, and in buffered solution.

978-1-4673-0996-7/12 $31.00 © 2012 IEEE

Acknowledgement

This work was supported by BK21 program and Basic Science Research Program (Grant no. 2010-0023999) through the NRF grant funded by the Korea Government (MEST).

References

[1] M. M. C. Cheng et al, Curr. Opin. Chem. Biol., vol. 10, pp. 11–19, 2006

[2] J. Lee et al, Korean Conference on Semiconductors, pp. 435-436, Feb. 2012

[3] H. Jang et al, Applied Physics Letters, Vol. 99, No. 12, pp. 252103, Dec. 2011

[4] P. G. Fernandes et al, Applied Physics Letters, Vol. 97, No. 3, pp. 034103, July, 2010

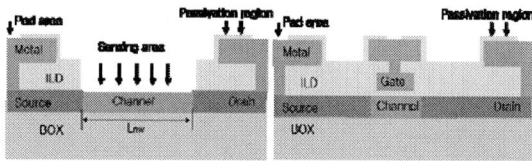

Fig. 1. Cross-sectional schematic diagrams of (a) a SiNW FET for biosensor and (b) a conventional MOSFET

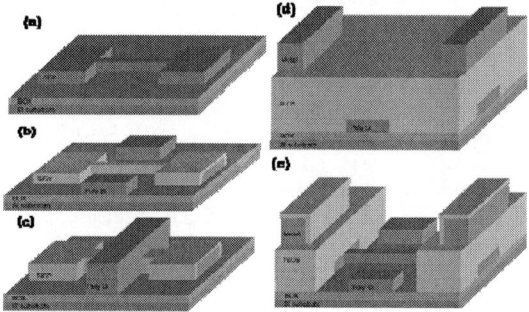

Fig. 2. Fabrication process flow of SiNW FET for biosensor and SOI MOSFET, (a) defining active region, building gates of (b) SiNW FET and (c) MOSFET, conventional back-end process (ILD and metal line), and (d) formation of sensing region for biosensor.

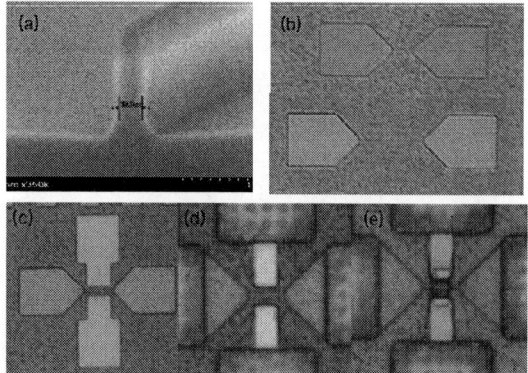

Fig. 3. (a) SEM images of active test pattern, optical images of the fabricated process, (b) defining active region, (c) gate pattering, (d) back-end process, and (e) formation of sensing region.

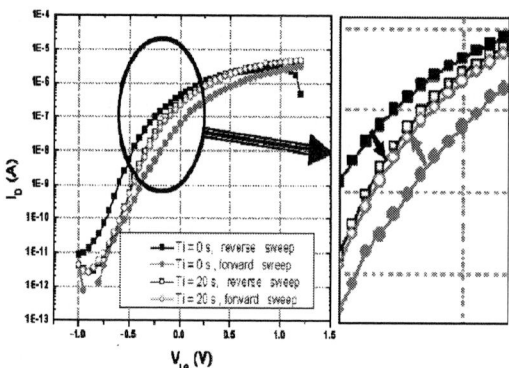

Fig. 4. Transfer characteristics in DW depending on integration time change. ($V_S = -1$ V, $V_D = 1$V)

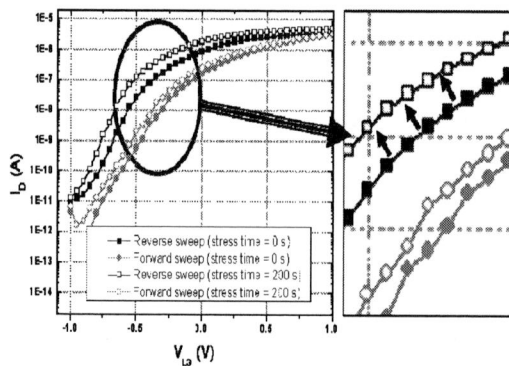

Fig. 5. Transfer characteristics in DW depending on stress effect ($V_S = -1$ V, $V_D = 1$V, $V_{STR} = 1.2$ V)

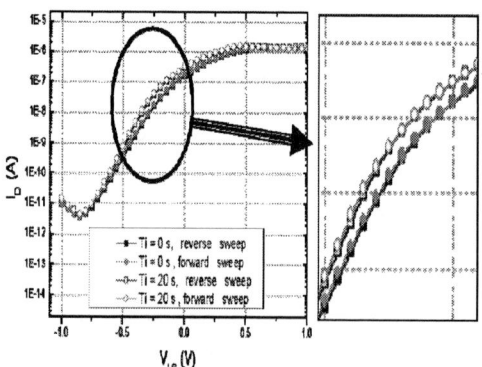

Fig. 6. Transfer characteristics in buffered solution depending on integration time change. ($V_S = -1$ V, $V_D = 1$V)

978-1-4673-0996-7/12 $31.00 © 2012 IEEE

Investigation on Hump Effects of L-shaped Tunneling Filed-Effect Transistors

Sang Wan Kim[1], Woo Young Choi[2], Hyungjin Kim[1], Min-Chul Sun[1,3], Hyun Woo Kim[1], and Byung-Gook Park[1]

[1]Inter-University Semiconductor Research Center (ISRC) and
School of Electrical Engineering and Computer Science, Seoul National University, **Seoul, Korea**
[2]Department of Electronic Engineering, Sogang University, **Seoul, Korea**
[3]TD Team (S. LSI), Device Solution Business, Samsung Electronics Co., Ltd, **Gyeonggi, Korea**
E-mail address: iskra_sw@hanmail.net

Abstract

In this paper, hump effects of L-shaped tunneling field-effect transistors (TFETs) have been investigated. It turns out that the hump effects are originated from the two different turn-on voltages ($V_{\text{turn-on}}$'s). By using device simulation, the source junction design has been optimized in order to suppress the hump effects.

Keywords: hump, tunneling field-effect transistor, TFET, L-shaped TFET and non-local tunneling

Introduction

In order to implement low-power operation while maintaining high on-current (I_{on}), both supply voltage (V_{dd}) and subthreshold swing (SS) should be reduced. However, the metal-oxide-semiconductor field-effect transistors (MOSFETs) have a fundamental limit of SS because their operation is based on thermionic carrier injection [1]. In order to overcome this limitation, several kinds of alternative devices have been proposed: impact-ionization MOS (I-MOS) devices [1]; nano-electro-mechanical FETs (NEMFETs) [2]; negative gate-capacitance FETs [3] and tunneling FETs (TFETs) [4, 5]. Among them, the TFETs are regarded as the most attractive candidate due to their compatibility with complementary MOS (CMOS) process and scalability. However, TFETs have several technical issues such as low current drivability, SS degradation at high I_{on} and low operation speed. In order to address them, we have proposed L-shaped TFETs and discussed the effects of device parameters on their electrical characteristics [6, 7]. In this paper, for the extension of our previous research, the hump effects of the L-shaped TFETs will be discussed by using device simulation [8].

Simulation

Fig. 1 shows the schematic of the simulated L-shaped TFET. It only differs from our previous device structure [6] in that the source region is extended to the bottom of silicon-on-insulator (SOI). From now on, the process margin of L-shaped TFETs and its influence on electrical characteristics will be investigated. Detailed simulation conditions are summarized in Table 1. Fig. 2 shows the transfer curve of the L-shaped TFET shown in Fig. 1. It has been observed that it shows more severe SS degradation than that in [6] and hump effects. These unique behaviors seem to be originated from the extended source regions, which means that there are two TFETs in parallel whose turn-on voltages ($V_{\text{turn-on}}$'s) are different. Thus, it is necessary to analyze the effects of the two TFETs separately as shown in the Fig. 3. For the accurate calculation of band-to-band tunneling (BTBT), non-local tunneling model

and quantum mesh have be used. In the case of the non-local tunneling model, the BTBT current is calculated only in the quantum mesh. As shown Fig. 3, by using differently defined quantum meshes, a TFET with quantum mesh in whole region (TFET$_{\text{control}}$) has been dissected into a bulk TFET (TFET$_{\text{bulk}}$) and mesa TFET (TFET$_{\text{mesa}}$).

Results and Discussion

Fig. 4 shows the transfer curves of the TFET$_{\text{bulk}}$ and TFET$_{\text{mesa}}$. It has been observed that the TFET$_{\text{bulk}}$ suffers from large SS and low current drivability. On the other hand, the TFET$_{\text{mesa}}$ shows better performance than TFET$_{\text{bulk}}$ in terms of SS and I_{on}. It is because the tunneling barrier width (W_t) and cross-sectional area of tunneling junction are defined by L_i and H_i, respectively, as explained in [6]. It should be noted that the different types of TFETs have different $V_{\text{turn-on}}$'s.

Fig. 5 presents the three transfer curves of the L-shaped TFETs with quantum mesh in the whole, bulk SOI and mesa region. As expected, the transfer curve of the TFET$_{\text{control}}$ is equal to the sum of the transfer curve of the TFET$_{\text{bulk}}$ and TFET$_{\text{mesa}}$. In other words, the TFET$_{\text{bulk}}$ and TFET$_{\text{mesa}}$ connected in parallel have different $V_{\text{turn-on}}$'s and they result in hump effects. In order to suppress the hump effects, the source region needs to be confined in the mesa region.

Summary

The hump effects of L-shaped TFETs have been discussed. Based on the TCAD simulation results, it has been found that the hump effects are originated from the different $V_{\text{turn-on}}$'s of parallel connected TFETs. For the further improvement L-shaped TFETs, the source junction should be designed carefully during the fabrication process.

Acknowledgments

This work was supported in part by the Smart IT Convergence System Research Center (Global Frontier Project), in part by the National Research Foundation (NRF) of Korea funded by the Ministry of Education, Science and Technology (MEST) under Grants 2011-0019107 (Development of Future-Oriented Technology) and 2011-0027471 (Mid-Career Researcher Program) and in part by the Ministry of Knowledge Economy (MKE) of Korea under Grant NIPA-2012-H0301-12-1007 (University ITRC support program supervised by the National IT Industry Promotion Agency).

References

[1] K. Gopalakrishnan, P. B. Griffin, and J. D. Plummer, "I-MOS: a novel semiconductor device with a subthreshold slope lower than

kT/q," in *IEDM Tech. Dig.*, 2002, pp. 289-292.

[2] H. Kam, D. T. Lee, R. T. Howe, and T.-J. King, "A new nano-electro-mechanical field effect transistor (NEMFET) design for low-power electronics," in *IEDM Tech. Dig.*, 2005, pp. 463-466.

[3] S. Salahuddin and S. Datta, "Use of negative capacitance to provide voltage amplification for low power nanoscale devices," *Nano Lett.*, vol. 8, no. 2, pp. 405-410, Feb. 2008.

[4] P.-F. Wang, K. Hilsenbeck, Th. Nirschl, M. Oswald, C. Stepper, M. Weiss, D. Schmitt-Landsiedel, and W. Hansch, "Complimentary tunneling transistor for low power application", *Solid-State Electronics*, vol. 48, no. 12, pp. 2181-2186, Dec. 2004.

[5] W. Y. Choi, B.-G. Park, J. D. Lee, and T.-J. K. Liu, "Tunneling field-effect transistors (TFETs) with subthreshold swing (SS) less than 60 mV/dec", vol. 28, no. 8, *IEEE Electron Device Lett.*, pp. 743-745, August 2007.

[6] S. W. Kim, W. Y. Choi, M.-C. Sun, H. W. Kim, and B.-G. Park, "L-shaped tunneling field-effect transistors (TFETs) for low subthreshold swing and high current drivability," *proc. Int. Microprocesses and Nanotechnology Conf.*, p. 26C-4-5L, Oct. 2011.

[7] S. W. Kim, W. Y. Choi, M.-C. Sun, H. W. Kim, and B.-G. Park, "Design guidelines of Si-based L-shaped tunneling field-effect transistors," in press.

[8] ATLAS user's manual (SILVACO International, Santa Clara, CA, 2009)

Table 1. Simulated device parameters

Parameters	Value	Parameters	Value
L_g	50 nm	N_{gate}	n-type, 10^{21} cm^{-3}
L_i	4 nm	N_{source}	p-type, 10^{20} cm^{-3}
H_i	50 nm	N_{drain}	n-type, 10^{20} cm^{-3}
V_{dd}	0.7 V	N_{body}	p-type, 10^{15} cm^{-3}
T_{ox}	2 nm (SiO$_2$)	T_{SOI}	20 nm

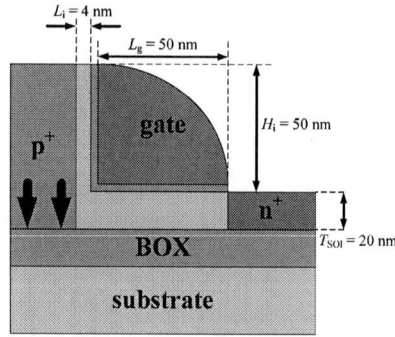

Fig. 1. Schematic diagram of simulated L-shaped TFET.

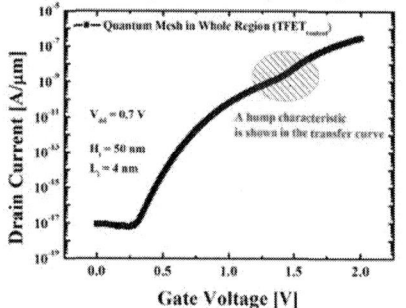

Fig. 2. Transfer curve of the L-shaped TFET in Fig. 2.

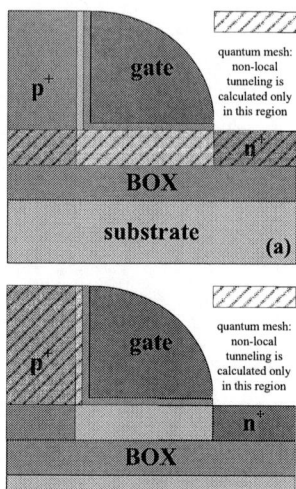

Fig. 3. Simulation strategy for analyzing the hump characteristic. TFET$_{control}$ has been dissected into a (a) TFET$_{bulk}$ and (b) TFET$_{mesa}$

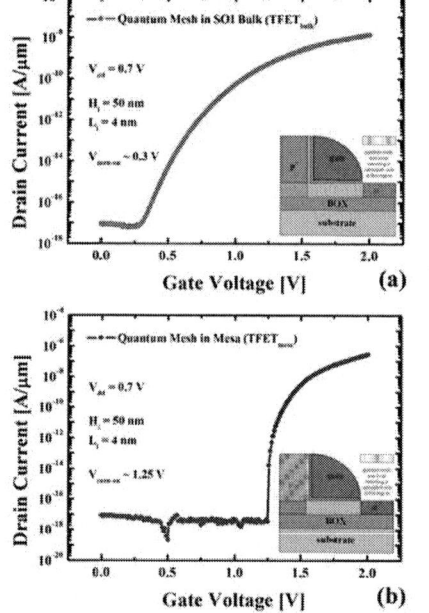

Fig. 4. Simulation results with different quantum meshes. (a) Quantum mesh in bulk region. (b) Quantum mesh in mesa region.

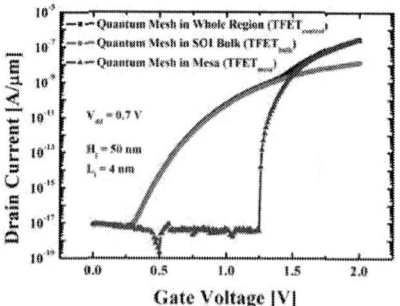

Fig. 5. Comparison of transfer curves of the three different TFETs.

Device Structure for the Characterization of Nanowire Thermocouples

Gergo P. Szakmany, Peter M. Krenz*, Alexei O. Orlov, Gary H. Bernstein, Wolfgang Porod

Department of Electrical Engineering, University of Notre Dame, Notre Dame, IN 46556 USA
email: pkrenz@nd.edu phone: (574)631-2420 fax: (574)631-4393

Nanowire thermocouples are attractive candidates for various applications including temperature measurements on the sub-micrometer scale [1], energy harvesting [2], and infrared detection [3]. Thermocouples, which consist of two wires made of dissimilar materials, produce an open-circuit voltage when a temperature difference ΔT is generated across its hot and cold junctions, expressed by

$$V_{OC} = (S_1 - S_2)\Delta T. \tag{1}$$

When reducing the dimensions of thermocouple wires, the Seebeck coefficients, S, are reduced when compared to those of bulk materials [4]. Therefore, an experimental method is required to accurately characterize the Seebeck coefficients of nanowires. For this characterization, precise knowledge of the temperature difference between the hot and cold junctions and the generated open-circuit voltage is necessary. We have developed a device structure that includes the nanowire thermocouple, a resistive heating element, and thermometers located at the hot and cold junctions. The structure of this device is shown in the scanning electron micrograph of Figure 1.

Passing current through the heater locally increases the temperature of the thermometer and hot junction of the thermocouple. Due to the symmetric device layout, the thermometer and thermocouple are heated equally, as shown by numerical simulations using COMSOL Multiphysics. Figure 2a shows the temperature increase on the surface of the device as a result of 150 µA flowing through the heater. Figure 2b shows the temperature distribution along the thermometer and the thermocouple. The temperature distribution along the thermocouple is uneven, since it consists of two different metals. During the measurement, the average temperature of the calibrated metal resistance thermometer is recorded along with a temperature difference between the hot nanojunction formed at the overlap area and the remote cold junction located more than 20 µm away from the heater. The simulation shows that for the chosen geometry the average temperature of the thermometer is very close to the temperature of the hot junction.

In order to demonstrate the feasibility of this approach, we report the measurements of the Seebeck coefficients of a palladium-gold and a palladium-chrome nanowire (70 nm wide and 50 nm thick) thermocouple. The thermocouple, heater, and thermometers were fabricated on top of 640 nm of thermally grown SiO_2 on a silicon wafer using electron beam lithography and electron beam evaporation.

The measurements start by calibrating the resistive thermometer. A reference thermometer is placed in close proximity to the resistive thermometer and the device is lowered into a N_2 cryostat. The resistance at different temperatures of the four-terminal thermometer is measured using a Wheatstone bridge.

The heater is calibrated next. A small AC current, i, at a frequency of $f = 4$ Hz is passed through the heater. Since the power dissipated in the heater due to Joule heating is proportional to i^2, the temperature varies at twice the frequency of the current, $2f = 8$ Hz. Using the resistive thermometer, the temperature is recorded for different magnitudes of currents. A second resistive thermometer was located near the cold junction of the thermocouple. Monitoring its temperature showed that the heater does not significantly alter the temperature of the cold junction. This is shown in Figure 3a.

The relative Seebeck coefficient of the nanowire thermocouple is then measured by passing a current at frequency f through the heater, while recording the open-circuit voltage across the thermocouple at $2f$. Figure 3b shows the measured open-circuit voltage for several temperatures. The relative Seebeck coefficient, $S_{Pd-Au} = 3.0 \pm 0.1$ µV/K (25 % of bulk value), is calculated using Equation 1. The experiment was repeated for a palladium-chrome thermocouple, and a value of $S_{Pd-Cr} = 15.2 \pm 0.1$ µV/K (40 % of bulk value) was obtained.

References:

[1] E Shapira, D Marchak, A Tsukernik, and Y Selzer, "Segmented metal nanowires as nanoscale thermocouples," *Nanotechnol.*, vol. 19, no. 12, pp. 125501, Feb. 2008.

[2] R.J.M. Vullers, R. van Schaijk, I. Doms, C. Van Hoof, and R. Mertens, "Micropower energyharvesting," *Solid State Electron.*, vol. 53, no. 7, pp. 684–693, Jul. 2009.

[3] P. M. Krenz, B. Tiwari, G. Szakmany, A. O. Orlov, F. J. Gonzalez, G. D. Boreman, W. Porod, "Response increase of IR antenna-coupled thermocouple using impedance matching," *IEEE J. Quantum Electron.*, accepted for publication.

[4] M. C. Salvadori, A. R. Vaz, F. S. Teixeira, I. G. Brown, and M. Cattani, "Thermoelectric effect in very thin film Pt/Au thermocouples," *App. Phys. Lett.*, vol. 88, no. 13, pp. 113106, Mar. 2006.

Figures:

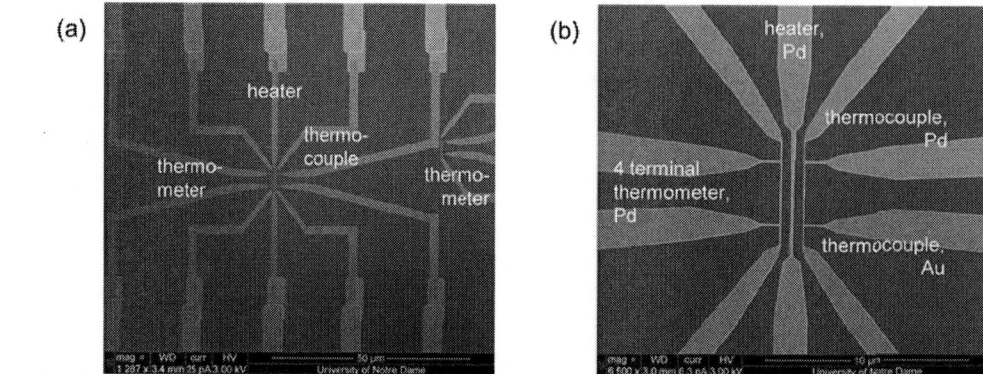

Figure 1: Scanning electron micrograph of device used to measure the relative Seebeck coefficient of two metal nanowires. (a) A thermometer is shown near the cold junction of the thermocouple. (b) Close up view showing the thermometer, heater, and thermocouple.

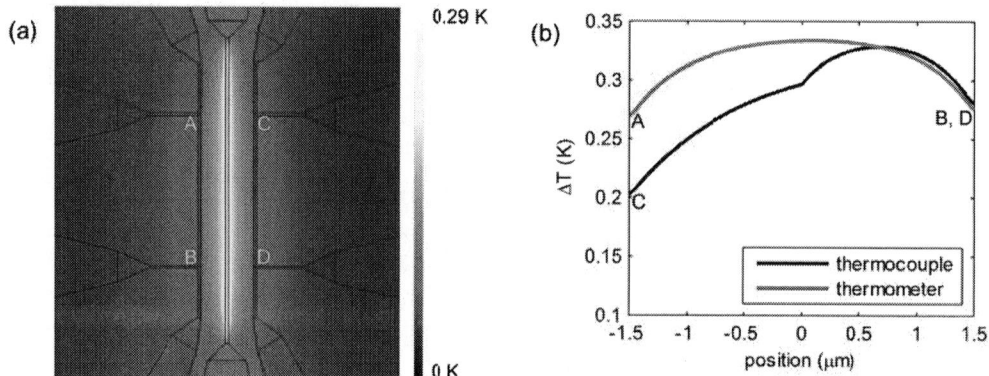

Figure 2: Simulated temperature increase caused by 150 μA passing through heater on (a) device surface and (b) along thermometer between points A, B and thermocouple between points C, D.

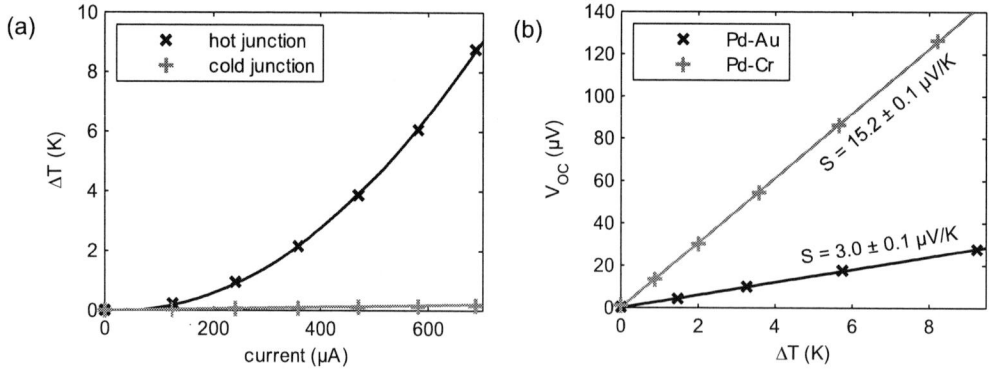

Figure 3: (a) Current passed through heater vs. measured temperature at the hot and cold junction of the thermocouple. (b) Measured open-circuit voltage for several temperature increases at the hot junction of the thermocouple.

CURRAN ASSOCIATES INC.
proceedings
.com

9781467309967